A Course in
LINEAR ALGEBRA
with Applications

2nd Edition

A Course in
LINEAR ALGEBRA
with Applications

2nd Edition

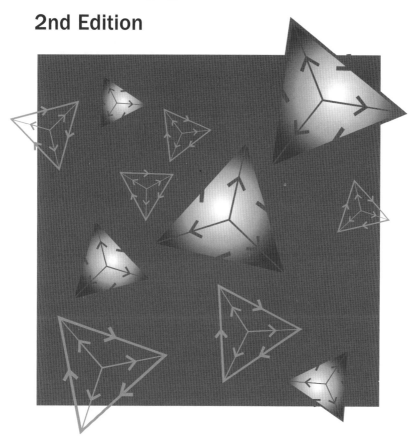

Derek J. S. Robinson

University of Illinois in Urbana-Champaign, USA

 **World Scientific**

NEW JERSEY · LONDON · SINGAPORE · BEIJING · SHANGHAI · HONG KONG · TAIPEI · CHENNAI

Published by

World Scientific Publishing Co. Pte. Ltd.

5 Toh Tuck Link, Singapore 596224

USA office: 27 Warren Street, Suite 401-402, Hackensack, NJ 07601

UK office: 57 Shelton Street, Covent Garden, London WC2H 9HE

British Library Cataloguing-in-Publication Data
A catalogue record for this book is available from the British Library.

A COURSE IN LINEAR ALGEBRA WITH APPLICATIONS (2nd Edition)

ISBN 981-270-023-4
ISBN 981-270-024-2 (pbk)

Printed in Singapore by B & JO Enterprise

For

JUDITH, EWAN and GAVIN

PREFACE TO THE SECOND EDITION

The principal change from the first edition is the addition of a new chapter on linear programming. While linear programming is one of the most widely used and successful applications of linear algebra, it rarely appears in a text such as this. In the new Chapter Ten the theoretical basis of the simplex algorithm is carefully explained and its geometrical interpretation is stressed.

Some further applications of linear algebra have been added, for example the use of Jordan normal form to solve systems of linear differential equations and a discussion of extremal values of quadratic forms.

On the theoretical side, the concepts of coset and quotient space are thoroughly explained in Chapter 5. Cosets have useful interpretations as solutions sets of systems of linear equations. In addition the Isomorphisms Theorems for vector spaces are developed in Chapter Six: these shed light on the relationship between subspaces and quotient spaces.

The opportunity has also been taken to add further exercises, revise the exposition in several places and correct a few errors. Hopefully these improvements will increase the usefulness of the book to anyone who needs to have a thorough knowledge of linear algebra and its applications.

I am grateful to Ms. Tan Rok Ting of World Scientific for assistance with the production of this new edition and for patience in the face of missed deadlines. I thank my family for their support during the preparation of the manuscript.

Derek Robinson
Urbana, Illinois
May 2006

PREFACE TO THE FIRST EDITION

A rough and ready definition of linear algebra might be: that part of algebra which is concerned with quantities of the first degree. Thus, at the very simplest level, it involves the solution of systems of linear equations, and in a real sense this elementary problem underlies the whole subject. Of all the branches of algebra, linear algebra is the one which has found the widest range of applications. Indeed there are few areas of the mathematical, physical and social sciences which have not benefitted from its power and precision. For anyone working in these fields a thorough knowledge of linear algebra has become an indispensable tool. A recent feature is the greater mathematical sophistication of users of the subject, due in part to the increasing use of algebra in the information sciences. At any rate it is no longer enough simply to be able to perform Gaussian elimination and deal with real vector spaces of dimensions two and three.

The aim of this book is to give a comprehensive introduction to the core areas of linear algebra, while at the same time providing a selection of applications. We have taken the point of view that it is better to consider a few quality applications in depth, rather than attempt the almost impossible task of covering all conceivable applications that potential readers might have in mind.

The reader is not assumed to have any previous knowledge of linear algebra - though in practice many will - but is expected to have at least the mathematical maturity of a student who has completed the calculus sequence. In North America such a student will probably be in the second or third year of study.

The book begins with a thorough discussion of matrix operations. It is perhaps unfashionable to precede systems of linear equations by matrices, but I feel that the central position of matrices in the entire theory makes this a logical and reasonable course. However the motivation for the introduction of matrices, by means of linear equations, is still provided informally. The second chapter forms a basis for the whole subject with a full account of the theory of linear equations. This is followed by a chapter on determinants, a topic that has been unfairly neglected recently. In practice it is hard to give a satisfactory definition of the general $n \times n$ determinant without using permutations, so a brief account of these is given.

Chapters Five and Six introduce the student to vector spaces. The concept of an abstract vector space is probably the most challenging one in the entire subject for the non-mathematician, but it is a concept which is well worth the effort of mastering. Our approach proceeds in gentle stages, through a series of examples that exhibit the essential features of a vector space; only then are the details of the definition written down. However I feel that nothing is gained by ducking the issue and omitting the definition entirely, as is sometimes done.

Linear tranformations are the subject of Chapter Six. After a brief introduction to functional notation, and numerous examples of linear transformations, a thorough account of the relation between linear transformations and matrices is given. In addition both kernel and image are introduced and are related to the null and column spaces of a matrix.

Orthogonality, perhaps the heart of the subject, receives an extended treatment in Chapter Seven. After a gentle introduction by way of scalar products in three dimensions — which will be familiar to the student from calculus — inner product spaces are defined and the Gram-Schmidt procedure is described. The chapter concludes with a detailed account of The Method of Least Squares, including the problem of finding optimal solutions, which texts at this level often fail to cover.

Chapter Eight introduces the reader to the theory of eigenvectors and eigenvalues, still one of the most powerful tools in linear algebra. Included is a detailed account of applications to systems of linear differential equations and linear recurrences, and also to Markov processes. Here we have not shied away from the more difficult case where the eigenvalues of the coefficient matrix are not all different.

The final chapter contains a selection of more advanced topics in linear algebra, including the crucial Spectral Theorem on the diagonalizability of real symmetric matrices. The usual applications of this result to quadratic forms, conics and quadrics, and maxima and minima of functions of several variables follow.

Also included in Chapter Nine are treatments of bilinear forms and Jordan Normal Form, topics that are often not considered in texts at this level, but which should be more widely known. In particular, canonical forms for both symmetric and skew-symmetric bilinear forms are obtained. Finally, Jordan Normal Form is presented by an accessible approach that requires only an elementary knowledge of vector spaces.

Chapters One to Eight, together with Sections 9.1 and 9.2, correspond approximately to a one semester course taught by the author over a period of many years. As time allows, other topics from Chapter Nine may be included. In practice some of the contents of Chapters One and Two will already be familiar to many readers and can be treated as review. Full proofs are almost always included: no doubt some instructors may not wish to cover all of them, but it is stressed that for maximum understanding of the material as many proofs as possible should be read. A good supply of problems appears at the end of each section. As always in mathematics, it is an indispensible part of learning the subject to attempt as many problems as possible.

This book was originally begun at the suggestion of Harriet McQuarrie. I thank Ms. Ho Hwei Moon of World Scientific Publishing Company for her advice and for help with editorial work. I am grateful to my family for their patience, and to my wife Judith for her encouragement, and for assistance with the proof-reading.

Derek Robinson\
Singapore\
March 1991

CONTENTS

Chapter Four   Introduction to Vector Spaces

Chapter Five   Basis and Dimension

Chapter Six   Linear Transformations

Chapter Seven   Orthogonality in Vector Spaces

Chapter Eight   Eigenvectors and Eigenvalues

Chapter Nine   More Advanced Topics

Chapter Ten   Linear Programming

Appendix   Mathematical Induction

Answers to the Exercises

Bibliography

Index

Chapter One

MATRIX ALGEBRA

In this first chapter we shall introduce one of the principal objects of study in linear algebra, a *matrix* or rectangular array of numbers, together with the standard matrix operations. Matrices are encountered frequently in many areas of mathematics, engineering, and the physical and social sciences, typically when data is given in tabular form. But perhaps the most familiar situation in which matrices arise is in the solution of systems of linear equations.

1.1 Matrices

An $m \times n$ *matrix* A is a rectangular array of numbers, real or complex, with m rows and n columns. We shall write a_{ij} for the number that appears in the ith row and the jth column of A; this is called the (i, j) *entry* of A. We can either write A in the extended form

$$\begin{pmatrix} a_{11} & a_{12} & \cdots & a_{1n} \\ a_{21} & a_{22} & \cdots & a_{2n} \\ \vdots & \vdots & \ddots & \vdots \\ a_{m1} & a_{m2} & \cdots & a_{mn} \end{pmatrix}$$

or in the more compact form

$$(a_{ij})_{m,n}.$$

Thus in the compact form a formula for the (i, j) entry of A is given inside the round brackets, while the subscripts m and n tell us the respective numbers of rows and columns of A.

1

Explicit examples of matrices are

$$\begin{pmatrix} 4 & 3 \\ 1 & 2 \end{pmatrix} \text{ and } \begin{pmatrix} 0 & 2.4 & 6 \\ \sqrt{-2} & 3/5 & -1 \end{pmatrix}.$$

Example 1.1.1

Write down the extended form of the matrix $((-1)^i j + i)_{3,2}$.

The (i, j) entry of the matrix is $(-1)^i j + i$ where $i = 1$, 2, 3, and $j = 1$, 2. So the matrix is

$$\begin{pmatrix} 0 & -1 \\ 3 & 4 \\ 2 & 1 \end{pmatrix}.$$

It is necessary to decide when two matrices A and B are to be regarded as *equal*; in symbols $A = B$. Let us agree this will mean that the matrices A and B have the same numbers of rows and columns, and that, for all i and j, the (i, j) entry of A equals the (i, j) entry of B. In short, two matrices are equal if they look exactly alike.

As has already been mentioned, matrices arise when one has to deal with linear equations. We shall now explain how this comes about. Suppose we have a set of m linear equations in n unknowns x_1, x_2, ..., x_n. These may be written in the form

$$\begin{cases} a_{11}x_1 & + & a_{12}x_2 & + & \cdots & + & a_{1n}x_n & = & b_1 \\ a_{21}x_1 & + & a_{22}x_2 & + & \cdots & + & a_{2n}x_n & = & b_2 \\ & \cdot & & \cdot & \cdots & & \cdot & & \cdot \\ a_{m1}x_1 & + & a_{m2}x_2 & + & \cdots & + & a_{mn}x_n & = & b_m \end{cases}$$

Here the a_{ij} and b_i are to be regarded as given numbers. The problem is to *solve* the system, that is, to find all n-tuples of numbers x_1, x_2, ..., x_n that satisfy every equation of the system, or to show that no such numbers exist. Solving a set of linear equations is in many ways the most basic problem of linear algebra.

The reader will probably have noticed that there is a matrix involved in the above linear system, namely the *coefficient matrix*

$$A = (a_{ij})_{m,n}.$$

In fact there is a second matrix present; it is obtained by using the numbers b_1, b_2, ..., b_m to add a new column, the $(n+1)$th, to the coefficient matrix A. This results in an $m \times (n+1)$ matrix called the *augmented matrix* of the linear system. The problem of solving linear systems will be taken up in earnest in Chapter Two, where it will emerge that the coefficient and augmented matrices play a critical role. At this point we merely wish to point out that here is a natural problem in which matrices are involved in an essential way.

Example 1.1.2

The coefficient and augmented matrices of the pair of linear equations

$$\begin{cases} 2x_1 & -3x_2 & +5x_3 & = 1 \\ -x_1 & + x_2 & - x_3 & = 4 \end{cases}$$

are respectively

$$\begin{pmatrix} 2 & -3 & 5 \\ -1 & 1 & -1 \end{pmatrix} \text{ and } \begin{pmatrix} 2 & -3 & 5 & 1 \\ -1 & 1 & -1 & 4 \end{pmatrix}.$$

Some special matrices

Certain special types of matrices that occur frequently will now be recorded.

(i) A $1 \times n$ matrix, or $n - row\ vector$, A has a single row

$$A = (a_{11}\ a_{12}\ ...\ a_{1n}).$$

(ii) An $m \times 1$ matrix, or *m-column vector*, B has just one column

$$B = \begin{pmatrix} b_{11} \\ b_{21} \\ \vdots \\ b_{m1} \end{pmatrix}.$$

(iii) A matrix with the same number of rows and columns is said to be *square*.

(iv) A *zero matrix* is a matrix all of whose entries are zero. The zero $m \times n$ matrix is denoted by

$$0_{mn} \quad \text{or} \quad \text{simply } 0.$$

Sometimes 0_{nn} is written 0_n. For example, 0_{23} is the matrix

$$\begin{pmatrix} 0 & 0 & 0 \\ 0 & 0 & 0 \end{pmatrix}.$$

(v) *The identity $n \times n$ matrix* has 1's on *the principal diagonal*, that is, from top left to bottom right, and zeros elsewhere; thus it has the form

$$\begin{pmatrix} 1 & 0 & \cdots & 1 \\ 0 & 1 & \cdots & 0 \\ \cdot & \cdot & \cdots & \cdot \\ 0 & 0 & \cdots & 1 \end{pmatrix}.$$

This matrix is written

$$I_n \quad \text{or} \quad \text{simply } I.$$

The identity matrix plays the role of the number 1 in matrix multiplication.

(vi) A square matrix is called *upper triangular* if it has only zero entries below the principal diagonal. Similarly a matrix is *lower triangular* if all entries above the principal diagonal are zero. For example, the matrices

$$\begin{pmatrix} 2 & 1 \\ 0 & 3 \end{pmatrix} \quad \text{and} \quad \begin{pmatrix} a & 0 & 0 \\ b & d & 0 \\ c & e & f \end{pmatrix}$$

are upper triangular and lower triangular respectively.

(vii) A square matrix in which all the non-zero elements lie on the principal diagonal is called a *diagonal matrix*. A *scalar matrix* is a diagonal matrix in which the elements on the principal diagonal are all equal. For example, the matrices

$$\begin{pmatrix} a & 0 & 0 \\ 0 & b & 0 \\ 0 & 0 & c \end{pmatrix} \quad \text{and} \quad \begin{pmatrix} a & 0 & 0 \\ 0 & a & 0 \\ 0 & 0 & a \end{pmatrix}$$

are respectively diagonal and scalar. Diagonal matrices have much simpler algebraic properties than general square matrices.

Exercises 1.1

1. Write out in extended form the matrix $((-1)^{i-j}(i+j))_{2,4}$.

2. Find a formula for the (i, j) entry of each of the following matrices:

$$\text{(a)} \begin{pmatrix} -1 & 1 & -1 \\ 1 & -1 & 1 \\ -1 & 1 & -1 \end{pmatrix}, \quad \text{(b)} \begin{pmatrix} 1 & 2 & 3 & 4 \\ 5 & 6 & 7 & 8 \\ 9 & 10 & 11 & 12 \\ 13 & 14 & 15 & 16 \end{pmatrix}.$$

3. Using the fact that matrices have a rectangular shape, say how many different zero matrices can be formed using a total of 12 zeros.

4. For every integer $n > 1$ there are always at least two zero matrices that can be formed using a total of n zeros. For which n are there *exactly* two such zero matrices?

5. Which matrices are both upper and lower triangular?

1.2 Operations with Matrices

We shall now introduce a number of standard operations that can be performed on matrices, among them addition, scalar multiplication and multiplication. We shall then describe the principal properties of these operations. Our object in so doing is to develop a systematic means of performing calculations with matrices.

(i) *Addition and subtraction*

Let A and B be two $m \times n$ matrices; as usual write a_{ij} and b_{ij} for their respective (i, j) entries. Define the *sum* $A + B$ to be the $m \times n$ matrix whose (i, j) entry is $a_{ij} + b_{ij}$; thus to form the matrix $A + B$ we simply add corresponding entries of A and B. Similarly, the *difference* $A - B$ is the $m \times n$ matrix whose (i, j) entry is $a_{ij} - b_{ij}$. However $A + B$ and $A - B$ are not defined if A and B do not have the same numbers of rows and columns.

(ii) *Scalar multiplication*

By a *scalar* we shall mean a number, as opposed to a matrix or array of numbers. Let c be a scalar and A an $m \times n$ matrix. The *scalar multiple* cA is the $m \times n$ matrix whose (i, j) entry is ca_{ij}. Thus to form cA we multiply every entry of A by the scalar c. The matrix $(-1)A$ is usually written $-A$; it is called the *negative* of A since it has the property that $A + (-A) = 0$.

Example 1.2.1

If

$$A = \begin{pmatrix} 1 & 2 & 0 \\ -1 & 0 & 1 \end{pmatrix} \text{ and } B = \begin{pmatrix} 1 & 1 & 1 \\ 0 & -3 & 1 \end{pmatrix},$$

then

$$2A + 3B = \begin{pmatrix} 5 & 7 & 3 \\ -2 & -9 & 5 \end{pmatrix} \text{ and } 2A - 3B = \begin{pmatrix} -1 & 1 & -3 \\ -2 & 9 & -1 \end{pmatrix}.$$

(iii) *Matrix multiplication*

It is less obvious what the "natural" definition of the product of two matrices should be. Let us start with the simplest interesting case, and consider a pair of 2×2 matrices

$$A = \begin{pmatrix} a_{11} & a_{12} \\ a_{21} & a_{22} \end{pmatrix} \text{ and } B = \begin{pmatrix} b_{11} & b_{12} \\ b_{21} & b_{22} \end{pmatrix}.$$

In order to motivate the definition of the matrix product AB we consider two sets of linear equations

$$\begin{cases} a_{11}y_1 + a_{12}y_2 = x_1 \\ a_{21}y_1 + a_{22}y_2 = x_2 \end{cases} \text{ and } \begin{cases} b_{11}z_1 + b_{12}z_2 = y_1 \\ b_{21}z_1 + b_{22}z_2 = y_2 \end{cases}$$

Observe that the coefficient matrices of these linear systems are A and B respectively. We shall think of these equations as representing changes of variables from y_1, y_2 to x_1, x_2, and from z_1, z_2 to y_1, y_2 respectively.

Suppose that we replace y_1 and y_2 in the first set of equations by the values specified in the second set. After simplification we obtain a new set of equations

$$\begin{cases} (a_{11}b_{11} + a_{12}b_{21})z_1 + (a_{11}b_{12} + a_{12}b_{22})z_2 = x_1 \\ (a_{21}b_{11} + a_{22}b_{21})z_1 + (a_{21}b_{12} + a_{22}b_{22})z_2 = x_2 \end{cases}$$

This has coefficient matrix

$$\begin{pmatrix} a_{11}b_{11} + a_{12}b_{21} & a_{11}b_{12} + a_{12}b_{22} \\ a_{21}b_{11} + a_{22}b_{21} & a_{21}b_{12} + a_{22}b_{22} \end{pmatrix}$$

and represents a change of variables from z_1, z_2 to x_1, x_2 which may be thought of as the composite of the original changes of variables.

At first sight this new matrix looks formidable. However it is in fact obtained from A and B in quite a simple fashion, namely by the row-times-column rule. For example, the $(1, 2)$ entry arises from multiplying corresponding entries of row 1 of A and column 2 of B, and then adding the resulting numbers; thus

$$(a_{11} \quad a_{12}) \begin{pmatrix} b_{12} \\ b_{22} \end{pmatrix} \rightarrow a_{11}b_{12} + a_{12}b_{22}.$$

Other entries arise in a similar fashion from a row of A and a column of B.

Having made this observation, we are now ready to define the product AB where A is an $m \times n$ matrix and B is an $n \times p$ matrix. The rule is that the (i, j) entry of AB is obtained by multiplying corresponding entries of row i of A and column j of B, and then adding up the resulting products. This is the *row-times-column rule*. Now row i of A and column j of B are

$$(a_{i1} \quad a_{i2} \quad \ldots \quad a_{in}) \text{ and } \begin{pmatrix} b_{1j} \\ b_{2j} \\ \vdots \\ b_{nj} \end{pmatrix}.$$

Hence the (i, j) entry of AB is

$$a_{i1}b_{1j} + a_{i2}b_{2j} + \cdots + a_{in}b_{nj},$$

which can be written more concisely using the summation notation as

$$\sum_{k=1}^{n} a_{ik}b_{kj}.$$

Notice that the rule only makes sense if *the number of columns of A equals the number of rows of B.* Also the product of an $m \times n$ matrix and an $n \times p$ matrix is an $m \times p$ matrix.

Example 1.2.2

Let

$$A = \begin{pmatrix} 2 & 1 & -1 \\ 3 & 0 & 2 \end{pmatrix} \text{ and } B = \begin{pmatrix} 0 & 2 & 0 \\ 1 & 1 & 1 \\ 1 & 5 & -1 \end{pmatrix}.$$

Since A is 2×3 and B is 3×3, we see that AB is defined and is a 2×3 matrix. However BA is not defined. Using the row-times-column rule, we quickly find that

$$AB = \begin{pmatrix} 0 & 0 & 2 \\ 2 & 16 & -2 \end{pmatrix}.$$

Example 1.2.3

Let

$$A = \begin{pmatrix} 0 & 1 \\ 0 & 0 \end{pmatrix} \text{ and } B = \begin{pmatrix} 1 & 1 \\ 0 & 0 \end{pmatrix}.$$

In this case both AB and BA are defined, but these matrices are different:

$$AB = \begin{pmatrix} 0 & 0 \\ 0 & 0 \end{pmatrix} = 0_{22} \text{ and } BA = \begin{pmatrix} 0 & 1 \\ 0 & 0 \end{pmatrix}.$$

Thus already we recognise some interesting features of matrix multiplication. The matrix product is not *commutative*, that is, AB and BA may be different when both are defined; also the product of two non-zero matrices can be zero, a phenomenon which indicates that any theory of division by matrices will face considerable difficulties.

Next we show how matrix mutiplication provides a way of representing a set of linear equations by a single matrix equation. Let $A = (a_{ij})_{m,n}$ and let X and B be the column vectors with entries $x_1, x_2, ..., x_n$ and $b_1, b_2, ..., b_m$ respectively. Then the matrix equation

$$AX = B$$

is equivalent to the linear system

$$\begin{cases} a_{11}x_1 & + & a_{12}x_2 & + & \cdots & + & a_{1n}x_n & = & b_1 \\ a_{21}x_1 & + & a_{22}x_2 & + & \cdots & + & a_{2n}x_n & = & b_2 \\ & & & & \cdots & & & & \\ a_{m1}x_1 & + & a_{m2}x_2 & + & \cdots & + & a_{mn}x_n & = & b_m \end{cases}$$

For if we form the product AX and equate its entries to the corresponding entries of B, we recover the equations of the linear system. Here is further evidence that we have got the definition of the matrix product right.

Example 1.2.4

The matrix form of the pair of linear equations

$$\begin{cases} 2x_1 & - & 3x_2 & + & 5x_3 & = 1 \\ -x_1 & + & x_2 & - & x_3 & = 4 \end{cases}$$

is

$$\begin{pmatrix} 2 & -3 & 5 \\ -1 & 1 & -1 \end{pmatrix} \begin{pmatrix} x_1 \\ x_2 \\ x_3 \end{pmatrix} = \begin{pmatrix} 1 \\ 4 \end{pmatrix}.$$

(iv) *Powers of a matrix*

Once matrix products have been defined, it is clear how to define a non-negative power of a square matrix. Let A be an $n \times n$ matrix; then the mth *power* of A, where m is a non-negative integer, is defined by the equations

$$A^0 = I_n \quad \text{and} \quad A^{m+1} = A^m A.$$

This is an example of a *recursive definition*: the first equation specifies A^0, while the second shows how to define A^{m+1}, under the assumption that A^m has already been defined. Thus $A^1 = A$, $A^2 = AA$, $A^3 = A^2 A$ etc. We do not attempt to define negative powers at this juncture.

Example 1.2.5

Let

$$A = \begin{pmatrix} 0 & 1 \\ -1 & 0 \end{pmatrix}.$$

Then

$$A^2 = \begin{pmatrix} -1 & 0 \\ 0 & -1 \end{pmatrix}, \quad A^3 = \begin{pmatrix} 0 & -1 \\ 1 & 0 \end{pmatrix} \quad \text{and} \quad A^4 = \begin{pmatrix} 1 & 0 \\ 0 & 1 \end{pmatrix}.$$

The reader can verify that higher powers of A do not lead to new matrices in this example. Therefore A has just four distinct powers, $A^0 = I_2$, $A^1 = A$, A^2 and A^3.

(v) *The transpose of a matrix*

If A is an $m \times n$ matrix, the *transpose* of A,

$$A^T,$$

is the $n \times m$ matrix whose (i, j) entry equals the (j, i) entry of A. Thus the columns of A become the rows of A^T. For example, if

$$A = \begin{pmatrix} a & b \\ c & d \\ e & f \end{pmatrix},$$

then the transpose of A is

$$A^T = \begin{pmatrix} a & c & e \\ b & d & f \end{pmatrix}.$$

A matrix which equals its transpose is called *symmetric*. On the other hand, if A^T equals $-A$, then A is said to be *skew-symmetric*. For example, the matrices

$$\begin{pmatrix} a & b \\ b & c \end{pmatrix} \text{ and } \begin{pmatrix} 0 & -a \\ a & 0 \end{pmatrix}.$$

are symmetric and skew-symmetric respectively. Clearly symmetric matrices and skew-symmetric matrices must be square. We shall see in Chapter Nine that symmetric matrices can in a real sense be reduced to diagonal matrices.

The laws of matrix algebra

We shall now list a number of properties which are satisfied by the various matrix operations defined above. These properties will allow us to manipulate matrices in a systematic manner. Most of them are familiar from arithmetic; note however the absence of the commutative law for multiplication.

In the following theorem A, B, C are matrices and c, d are scalars; it is understood that the numbers of rows and columns of the matrices are such that the various matrix products and sums mentioned make sense.

Theorem 1.2.1

(a) $A + B = B + A$, (*commutative law of addition*);
(b) $(A + B) + C = A + (B + C)$, (*associative law of addition*);
(c) $A + 0 = A$;
(d) $(AB)C = A(BC)$, (*associative law of multiplication*);
(e) $AI = A = IA$;

(f) $A(B + C) = AB + AC$, (*distributive law*);
(g) $(A + B)C = AC + BC$, (*distributive law*);
(h) $A - B = A + (-1)B$;
(i) $(cd)A = c(dA)$;
(j) $c(AB) = (cA)B = A(cB)$;
(k) $c(A + B) = cA + cB$;
(l) $(c + d)A = cA + dA$;
(m) $(A + B)^T = A^T + B^T$;
(n) $(AB)^T = B^T A^T$.

Each of these laws is a logical consequence of the definitions of the various matrix operations. To give formal proofs of them all is a lengthy, but routine, task; an example of such a proof will be given shortly. It must be stressed that familiarity with these laws is essential if matrices are to be manipulated correctly.

We remark that it is unambiguous to use the expression $A + B + C$ for both $(A + B) + C$ and $A + (B + C)$. For by the associative law of addition these matrices are equal. The same comment applies to sums like $A + B + C + D$, and also to matrix products such as $(AB)C$ and $A(BC)$, both of which are written as ABC.

In order to illustrate the use of matrix operations, we shall now work out three problems.

Example 1.2.6

Prove the associative law for matrix multiplication, $(AB)C = A(BC)$ where A, B, C are $m \times n$, $n \times p$, $p \times q$ matrices respectively.

In the first place observe that all the products mentioned exist, and that both $(AB)C$ and $A(BC)$ are $m \times q$ matrices. To show that they are equal, we need to verify that their (i, j) entries are the same for all i and j.

Let d_{ik} be the (i, k) entry of AB; then $d_{ik} = \sum_{l=1}^{n} a_{il}b_{lk}$. Thus the (i, j) entry of $(AB)C$ is $\sum_{k=1}^{p} d_{ik}c_{kj}$, that is

$$\sum_{k=1}^{p}(\sum_{l=1}^{n} a_{il}b_{lk})c_{kj}.$$

After a change in the order of summation, this becomes

$$\sum_{l=1}^{n} a_{il}(\sum_{k=1}^{p} b_{lk}c_{kj}).$$

Here it is permissible to change the order of the two summations since this just corresponds to adding up the numbers $a_{il}b_{lk}c_{kj}$ in a different order. Finally, by the same procedure we recognise the last sum as the (i, j) entry of the matrix $A(BC)$.

The next two examples illustrate the use of matrices in real-life situations.

Example 1.2.7

A certain company manufactures three products P, Q, R in four different plants W, X, Y, Z. The various costs (in whole dollars) involved in producing a single item of a product are given in the table

	P	Q	R
material	1	2	1
labor	3	2	2
overheads	2	1	2

The numbers of items produced in one month at the four locations are as follows:

	W	X	Y	Z
P	2000	3000	1500	4000
Q	1000	500	500	1000
	2000	2000	2500	2500

The problem is to find the total monthly costs of material, labor and overheads at each factory.

Let C be the "cost" matrix formed by the first set of data and let N be the matrix formed by the second set of data. Thus

$$C = \begin{pmatrix} 1 & 2 & 1 \\ 3 & 2 & 2 \\ 2 & 1 & 2 \end{pmatrix} \text{ and } N = \begin{pmatrix} 2000 & 3000 & 1500 & 4000 \\ 1000 & 500 & 500 & 1000 \\ 2000 & 2000 & 2500 & 2500 \end{pmatrix}.$$

The total costs per month at factory W are clearly

$$\text{material} : 1 \times 2000 + 2 \times 1000 + 1 \times 2000 = 6000$$
$$\text{labor} : 3 \times 2000 + 2 \times 1000 + 2 \times 2000 = 12000$$
$$\text{overheads} : 2 \times 2000 + 1 \times 1000 + 2 \times 2000 = 9000$$

Now these amounts arise by multiplying rows 1, 2 and 3 of matrix C times column 1 of matrix N, that is, as the $(1, 1)$, $(2, 1)$, and $(3, 1)$ entries of matrix product CN. Similarly the costs at the other locations are given by entries in the other columns of the matrix CN. Thus the complete answer can be read off from the matrix product

$$CN = \begin{pmatrix} 6000 & 6000 & 5000 & 8500 \\ 12000 & 14000 & 10500 & 19000 \\ 9000 & 10500 & 8500 & 14000 \end{pmatrix}.$$

Here of course the rows of CN correspond to material, labor and overheads, while the columns correspond to the four plants W, X, Y, Z.

Example 1.2.8

In a certain city there are 10,000 people of employable age. At present 7000 are employed and the rest are out of work. Each year 10% of those employed become unemployed, while 60% of the unemployed find work. Assuming that the total pool of people remains the same, what will the employment picture be in three years time?

Let e_n and u_n denote the numbers of employed and unemployed persons respectively after n years. The information given translates into the equations

$$\begin{cases} e_{n+1} = .9e_n + .6u_n \\ u_{n+1} = .1e_n + .4u_n \end{cases}$$

These linear equations are converted into a single matrix equation by introducing matrices

$$X_n = \begin{pmatrix} e_n \\ u_n \end{pmatrix} \text{ and } A = \begin{pmatrix} .9 & .6 \\ .1 & .4 \end{pmatrix}.$$

The equivalent matrix equation is

$$X_{n+1} = AX_n.$$

Taking n to be 0, 1, 2 successively, we see that $X_1 = AX_0$, $X_2 = AX_1 = A^2 X_0$, $X_3 = AX_2 = A^3 X_0$. In general

$$X_n = A^n X_0.$$

Now we were told that $e_0 = 7000$ and $u_0 = 3000$, so

$$X_0 = \begin{pmatrix} 7000 \\ 3000 \end{pmatrix}.$$

Thus to find X_3 all that we need to do is to compute the power A^3. This turns out to be

$$\begin{pmatrix} .861 & .834 \\ .139 & .166 \end{pmatrix}.$$

Hence

$$X^3 = A^3 X_0 = \begin{pmatrix} 8529 \\ 1471 \end{pmatrix},$$

so that 8529 of the 10,000 will be in work after three years.

At this point an interesting question arises: what will the numbers of employed and unemployed be in the long run? This problem is an example of a Markov process; these processes will be studied in Chapter Eight as an application of the theory of eigenvalues.

The inverse of a square matrix

An $n \times n$ matrix A is said to be *invertible* if there is an $n \times n$ matrix B such that

$$AB = I_n = BA.$$

Then B is called an *inverse* of A. A matrix which is not invertible is sometimes called *singular*, while an invertible matrix is said to be *non-singular*.

Example 1.2.9

Show that the matrix

$$\begin{pmatrix} 1 & 3 \\ 3 & 9 \end{pmatrix}$$

is not invertible.

If $\begin{pmatrix} a & b \\ c & d \end{pmatrix}$ were an inverse of the matrix, then we should have

$$\begin{pmatrix} 1 & 3 \\ 3 & 9 \end{pmatrix} \cdot \begin{pmatrix} a & b \\ c & d \end{pmatrix} = \begin{pmatrix} 1 & 0 \\ 0 & 1 \end{pmatrix},$$

which leads to a set of linear equations with no solutions,

$$\begin{cases} a + 3c & = 1 \\ b + 3d & = 0 \\ 3a + 9c & = 0 \\ 3b + 9d & = 1 \end{cases}$$

Indeed the first and third equations clearly contradict each other. Hence the matrix is not invertible.

Example 1.2.10

Show that the matrix

$$A = \begin{pmatrix} 1 & -2 \\ 0 & 1 \end{pmatrix}$$

is invertible and find an inverse for it.

Suppose that $B = \begin{pmatrix} a & b \\ c & d \end{pmatrix}$ is an inverse of A. Write out the product AB and set it equal to I_2, just as in the previous example. This time we get a set of linear equations that has a solution,

$$\begin{cases} a - 2c & = 1 \\ b - 2d & = 0 \\ c & = 0 \\ d & = 1 \end{cases}$$

Indeed there is a unique solution $a = 1$, $b = 2$, $c = 0$, $d = 1$. Thus the matrix

$$B = \begin{pmatrix} 1 & 2 \\ 0 & 1 \end{pmatrix}$$

is a candidate. To be sure that B is really an inverse of A, we need to verify that BA is also equal to I_2; this is in fact true, as the reader should check.

At this point the natural question is: how can we tell if a square matrix is invertible, and if it is, how can we find an inverse? From the examples we have seen enough to realise that the question is intimately connected with the problem of solving systems of linear systems, so it is not surprising that we must defer the answer until Chapter Two.

We now present some important facts about inverses of matrices.

Theorem 1.2.2

A square matrix has at most one inverse.

Proof
Suppose that a square matrix A has two inverses B_1 and B_2. Then

$$AB_1 = AB_2 = I = B_1A = B_2A.$$

The idea of the proof is to consider the product $(B_1A)B_2$; since $B_1A = I$, this equals $IB_2 = B_2$. On the other hand, by the associative law it also equals $B_1(AB_2)$, which equals $B_1I = B_1$. Therefore $B_1 = B_2$.

From now on we shall write

$$A^{-1}$$

for the unique inverse of an invertible matrix A.

Theorem 1.2.3

(a) *If A is an invertible matrix, then A^{-1} is invertible and $(A^{-1})^{-1} = A$.*

(b) *If A and B are invertible matrices of the same size, then AB is invertible and $(AB)^{-1} = B^{-1}A^{-1}$.*

Proof

(a) Certainly we have $AA^{-1} = I = A^{-1}A$, equations which can be viewed as saying that A is an inverse of A^{-1}. Therefore, since A^{-1} cannot have more than one inverse, its inverse must be A.

(b) To prove the assertions we have only to check that $B^{-1}A^{-1}$ is an inverse of AB. This is easily done: $(AB)(B^{-1}A^{-1}) = A(BB^{-1})A^{-1}$, by two applications of the associative law; the latter matrix equals $AIA^{-1} = AA^{-1} = I$. Similarily $(B^{-1}A^{-1})(AB) = I$. Since inverses are unique, $(AB)^{-1} = B^{-1}A^{-1}$.

Partitioned matrices

A matrix is said to be *partitioned* if it is subdivided into a rectangular array of submatrices by a series of horizontal or vertical lines. For example, if A is the matrix $(a_{ij})_{3,3}$, then

$$
\begin{pmatrix}
a_{11} & a_{12} & | & a_{13} \\
a_{21} & a_{22} & | & a_{23} \\
-- & -- & | & -- \\
a_{31} & a_{32} & | & a_{33}
\end{pmatrix}
$$

is a partitioning of A. Another example of a partitioned matrix is the augmented matrix of the linear system whose matrix form is $AX = B$; here the partitioning is $[A|B]$.

There are occasions when it is helpful to think of a matrix as being partitioned in some particular manner. A common one is when an $m \times n$ matrix A is partitioned into its columns $A_1, A_2, \ldots, A_n,$

$$A = (A_1|A_2| \ ... \ |A_n).$$

Because of this it is important to observe the following fact.

Theorem 1.2.4

Partitioned matrices can be added and multiplied according to the usual rules of matrix algebra.

Thus to add two partitioned matrices, we add corresponding entries, although these are now matrices rather than scalars. To multiply two partitioned matrices use the row-times-column rule. Notice however that the partitions of the matrices must be compatible if these operations are to make sense.

Example 1.2.11

Let $A = (a_{ij})_{4,4}$ be partitioned into four 2×2 matrices

$$A = \begin{pmatrix} A_{11} & A_{12} \\ A_{21} & A_{22} \end{pmatrix}$$

where

$$A_{11} = \begin{pmatrix} a_{11} & a_{12} \\ a_{21} & a_{22} \end{pmatrix}, \ A_{12} = \begin{pmatrix} a_{13} & a_{14} \\ a_{23} & a_{24} \end{pmatrix},$$

$$A_{21} = \begin{pmatrix} a_{31} & a_{32} \\ a_{41} & a_{42} \end{pmatrix}, \ A_{22} = \begin{pmatrix} a_{33} & a_{34} \\ a_{43} & a_{44} \end{pmatrix}.$$

Let $B = (b_{ij})_{4,4}$ be similarly partitioned into submatrices B_{11}, B_{12}, B_{21}, B_{22}

$$B = \begin{pmatrix} B_{11} & B_{12} \\ B_{21} & B_{22} \end{pmatrix}.$$

Then

$$A + B = \begin{pmatrix} A_{11} + B_{11} & A_{12} + B_{12} \\ A_{21} + B_{21} & A_{22} + B_{22} \end{pmatrix}$$

by the rule of addition for matrices.

Example 1.2.12

Let A be an $m \times n$ matrix and B an $n \times p$ matrix; write B_1, B_2, ..., B_p for the columns of B. Then, using the partition of B into columns $B = [B_1|B_2| \ ... \ |B_p]$, we have

$$AB = (AB_1|AB_2| \ ... \ |AB_p).$$

This follows at once from the row-times-column rule of matrix multiplication.

Exercises 1.2

1. Define matrices

$$A = \begin{pmatrix} 1 & 2 & 3 \\ 0 & 1 & -1 \\ 2 & 1 & 0 \end{pmatrix}, \quad B = \begin{pmatrix} 2 & 1 \\ 1 & 2 \\ 1 & 1 \end{pmatrix}, \quad C = \begin{pmatrix} 3 & 0 & 4 \\ 0 & 1 & 0 \\ 2 & -1 & 3 \end{pmatrix}.$$

    (a) Compute $3A - 2C$.
    (b) Verify that $(A + C)B = AB + CB$.
    (c) Compute A^2 and A^3 .   (d) Verify that $(AB)^T = B^T A^T$.

2. Establish the *laws of exponents*: $A^m A^n = A^{m+n}$ and $(A^m)^n = A^{mn}$ where A is any square matrix and m and n are non-negative integers. [Use induction on n : see Appendix.]

3. If the matrix products AB and BA both exist, what can you conclude about the sizes of A and B?

4. If $A = \begin{pmatrix} 0 & -1 \\ 1 & 1 \end{pmatrix}$, what is the first positive power of A that equals I_2?

5. Show that *no* positive power of the matrix $\begin{pmatrix} 1 & 1 \\ 0 & 1 \end{pmatrix}$ equals I_2 .

6. Prove the distributive law $A(B+C) = AB + AC$ where A is $m \times n$, and B and C are $n \times p$.

7. Prove that $(AB)^T = B^T A^T$ where A is $m \times n$ and B is $n \times p$.

8. Establish the rules $c(AB) = (cA)B = A(cB)$ and $(cA)^T = cA^T$.

9. If A is an $n \times n$ matrix some power of which equals I_n, then A is invertible. Prove or disprove.

10. Show that any two $n \times n$ diagonal matrices commute.

11. Prove that a scalar matrix commutes with every square matrix of the same size.

12. A certain library owns 10,000 books. Each month 20% of the books in the library are lent out and 80% of the books lent out are returned, while 10% remain lent out and 10% are reported lost. Finally, 25% of the books listed as lost the previous month are found and returned to the library. At present 9000 books are in the library, 1000 are lent out, and none are lost. How many books will be in the library, lent out, and lost after two months ?

13. Let A be any square matrix. Prove that $\frac{1}{2}(A + A^T)$ is symmetric, while the matrix $\frac{1}{2}(A - A^T)$ is skew-symmetric.

14. Use the last exercise to show that every square matrix can be written as the sum of a symmetric matrix and a skew-symmetric matrix. Illustrate this fact by writing the matrix

$$\begin{pmatrix} 1 & 3 & 4 \\ 2 & 5 & -1 \\ 7 & -6 & 5 \end{pmatrix}$$

as the sum of a symmetric and a skew-symmetric matrix.

15. Prove that the sum referred to in Exercise 14 is always unique.

16. Show that an $n \times n$ matrix A which commutes with every other $n \times n$ matrix must be scalar. [Hint: A commutes with the matrix whose (i, j) entry is 1 and whose other entries are all 0.]

17. (*Negative powers of matrices*) Let A be an invertible matrix. If $n > 0$, define the power A^{-n} to be $(A^{-1})^n$. Prove that $A^{-n} = (A^n)^{-1}$.

18. For each of the following matrices find the inverse or show that the matrix is not invertible:

$$\text{(a)} \begin{pmatrix} 1 & 2 \\ 3 & 4 \end{pmatrix} : \quad \text{(b)} \begin{pmatrix} 1 & -3 \\ -2 & 6 \end{pmatrix}.$$

19. Generalize the laws of exponents to negative powers of an invertible matrix [see Exercise 2.]

20. Let A be an invertible matrix. Prove that A^T is invertible and $(A^T)^{-1} = (A^{-1})^T$.

21. Give an example of a 3×3 matrix A such that $A^3 = 0$, but $A^2 \neq 0$.

1.3 Matrices over Rings and Fields

Up to this point we have assumed that all our matrices have as their entries real or complex numbers. Now there are circumstances under which this assumption is too restrictive; for example, one might wish to deal only with matrices whose entries are integers. So it is desirable to develop a theory of matrices whose entries belong to certain abstract algebraic systems. If we review all the definitions given so far, it becomes clear that what we really require of the entries of a matrix is that they belong to a "system" in which we can add and multiply, subject of course to reasonable rules. By this we mean rules of such a nature that the laws of matrix algebra listed in Theorem 1.2.1 will hold true.

The type of abstract algebraic system for which this can be done is called a *ring with identity*. By this is meant a set R, with a *rule of addition* and a *rule of multiplication*; thus if r_1 and r_2 are elements of the set R, then there is a unique *sum* $r_1 + r_2$ and a unique *product* $r_1 r_2$ in R. In addition the following laws are required to hold:

(a) $r_1 + r_2 = r_2 + r_1$, (*commutative law of addition*):

(b) $(r_1 + r_2) + r_3 = r_1 + (r_2 + r_3)$, (*associative law of addition*):

(c) R contains a *zero element* 0_R with the property $r + 0_R = r$:

(d) each element r of R has a *negative*, that is, an element $-r$ of R with the property $r + (-r) = 0_R$:

(e) $(r_1 r_2) r_3 = r_1 (r_2 r_3)$, (*associative law of multiplication*):

(f) R contains an *identity element* 1_R, different from 0_R, such that $r 1_R = r = 1_R r$:

(g) $(r_1 + r_2) r_3 = r_1 r_3 + r_2 r_3$, (*distributive law*):

(h) $r_1 (r_2 + r_3) = r_1 r_2 + r_1 r_3$, (*distributive law*).

These laws are to hold for all elements r_1, r_2, r_3, r of the ring R. The list of rules ought to seem reasonable since all of them are familiar laws of arithmetic.

If two further rules hold, then the ring is called a *field*:

(i) $r_1 r_2 = r_2 r_1$, (*commutative law of multiplication*):

(j) each element r in R other than the zero element 0_R has an *inverse*, that is, an element r^{-1} in R such that $r r^{-1} = 1_R = r^{-1} r$.

So the additional rules require that multiplication be a commutative operation, and that each non-zero element of R have an inverse. Thus *a field is essentially an abstract system in which one can add, multiply and divide, subject to the usual laws of arithmetic.*

Of course the most familiar examples of fields are

C and R,

the fields of complex numbers and real numbers respectively, where the addition and multiplication used are those of arithmetic. These are the examples that motivated the definition of a field in the first place. Another example is the *field of rational numbers*

Q.

(Recall that a *rational number* is a number of the form a/b where a and b are integers). On the other hand, the set of all integers **Z**, (with the usual sum and product), is a ring with identity, but it is not a field since 2 has no inverse in this ring.

All the examples given so far are infinite fields. But there are also finite fields, the most familiar being the field of two elements. This field has the two elements 0 and 1, sums and products being calculated according to the tables

+	0	1
0	0	1
1	1	0

and

×	0	1
0	0	0
1	0	1

respectively. For example, we read off from the tables that $1 + 1 = 0$ and $1 \times 1 = 1$. In recent years finite fields have become of importance in computer science and in coding theory. Thus the significance of fields extends beyond the domain of pure mathematics.

Suppose now that R is an arbitrary ring with identity. An $m \times n$ *matrix over* R is a rectangular $m \times n$ array of elements belonging to the ring R. It is possible to form sums and products of matrices over R, and the scalar multiple of a matrix over R by an element of R, by using exactly the same definitions as in the case of matrices with numerical entries. That the laws of matrix algebra listed in Theorem 1.2.1 are still valid is guaranteed by the ring axioms. Thus in the general theory the only change is that the scalars which appear as entries of a matrix are allowed to be elements of an arbitrary ring with identity.

Some readers may feel uncomfortable with the notion of a matrix over an abstract ring. However, if they wish, they may safely assume in the sequel that the field of scalars is either $\mathbf{R}$ or $\mathbf{C}$. Indeed there are places where we will definitely want to assume this. Nevertheless we wish to make the point that much of linear algebra can be done in far greater generality than over $\mathbf{R}$ and $\mathbf{C}$.

Example 1.3.1

Let $A = \begin{pmatrix} 1 & 0 \\ 1 & 1 \end{pmatrix}$ and $B = \begin{pmatrix} 0 & 1 \\ 0 & 1 \end{pmatrix}$ be matrices *over the field of two elements*. Using the tables above and the rules of matrix addition and multiplication, we find that

$$A + B = \begin{pmatrix} 1 & 1 \\ 1 & 0 \end{pmatrix} \text{ and } AB = \begin{pmatrix} 0 & 1 \\ 0 & 0 \end{pmatrix}.$$

Algebraic structures in linear algebra

There is another reason for introducing the concept of a ring at this stage. For rings, one of the fundamental structures of algebra, occur naturally at various points in linear algebra. To illustrate this, let us write

$$M_n(R)$$

for the set of all $n \times n$ matrices over a fixed ring with identity R. If the standard matrix operations of addition and multiplication are used, this set becomes a ring, *the ring of all $n \times n$*

matrices over R. The validity of the ring axioms follows from Theorem 1.2.1. An obviously important example of a ring is $M_n(\mathbf{R})$. Later we shall discover other places in linear algebra where rings occur naturally.

Finally, we mention another important algebraic structure that appears naturally in linear algebra, a group. Consider the set of all *invertible* $n \times n$ matrices over a ring with identity R; denote this by

$$GL_n(R).$$

This is a set equipped with a rule of multiplication; for if A and B are two invertible $n \times n$ matrices over R, then AB is also invertible and so belongs to $GL_n(R)$, as the proof of Theorem 1.2.3 shows. In addition, each element of this set has an inverse which is also in the set. Of course the identity $n \times n$ matrix belongs to $GL_n(R)$, and multiplication obeys the associative law.

All of this means that $GL_n(R)$ is a group. The formal definition is as follows. A *group* is a set G with a rule of multiplication; thus if g_1 and g_2 are elements of G, there is a unique *product* $g_1 g_2$ in G. The following axioms must be satisfied:

(a) $(g_1 g_2)g_3 = (g_1 g_2)g_3$, (*associative law*):
(b) there is an *identity element* 1_G with the property $1_G g = g = g 1_G$:
(c) each element g of G has an *inverse element* g^{-1} in G such that $g g^{-1} = 1_G = g^{-1} g$.

These statements must hold for all elements g, g_1, g_2, g_3 of G.

Thus the set $GL_n(R)$ of all invertible matrices over R, a ring with identity, is a group; this important group is known as the *general linear group of degree n over R*. Groups occur in many areas of science, particularly in situations where symmetry is important.

Exercises 1.3

1. Show that the following sets of numbers are fields if the usual addition and multiplication of arithmetic are used:
    (a) the set of all rational numbers;
    (b) the set of all numbers of the form $a + b\sqrt{2}$ where a and b are rational numbers;
    (c) the set of all numbers of the form $a + b\sqrt{-1}$ where where a and b are rational numbers.

2. Explain why the ring $M_n(\mathbf{C})$ is *not* a field if $n > 1$.

3. How many $n \times n$ matrices are there over the field of two elements? How many of these are symmetric ? [You will need the formula $1 + 2 + 3 + \cdots + n = n(n+1)/2$; for this see Example A.1 in the Appendix].

4. Let

$$A = \begin{pmatrix} 1 & 1 & 1 \\ 0 & 1 & 1 \\ 0 & 1 & 0 \end{pmatrix} \text{ and } B = \begin{pmatrix} 0 & 1 & 1 \\ 1 & 1 & 1 \\ 1 & 1 & 0 \end{pmatrix}$$

be matrices over the field of two elements. Compute $A + B$, A^2 and AB.

5. Show that the set of all $n \times n$ scalar matrices over $\mathbf{R}$ with the usual matrix operations is a field.

6. Show that the set of all non-zero $n \times n$ scalar matrices over $\mathbf{R}$ is a group with respect to matrix multiplication.

7. Explain why the set of all non-zero integers with the usual multiplication is not a group.

Chapter Two

SYSTEMS OF LINEAR EQUATIONS

In this chapter we address what has already been described as one of the fundamental problems of linear algebra: to determine if a system of linear equations - or *linear system* - has a solution, and, if so, to find all its solutions. Almost all the ensuing chapters depend, directly or indirectly, on the results that are described here.

2.1 Gaussian Elimination

We begin by considering in detail three examples of linear systems which will serve to show what kind of phenomena are to be expected; they will also give some idea of the techniques that are available for solving linear systems.

Example 2.1.1

$$\begin{cases} x_1 & - & x_2 & + \ x_3 & + & x_4 & = 2 \\ x_1 & + & x_2 & + \ x_3 & - & x_4 & = 3 \\ x_1 & + & 3x_2 & + \ x_3 & - & 3x_4 & = 1 \end{cases}$$

To determine if the system has a solution, we apply certain operations to the equations of the system which are designed to eliminate unknowns from as many equations as possible. The important point about these operations is that, although they change the linear system, they do not change its solutions.

We begin by subtracting equation 1 from equations 2 and 3 in order to eliminate x_1 from the last two equations. These operations can be conveniently denoted by $(2) - (1)$ and $(3) - (1)$ respectively. The effect is to produce a new linear system

$$\begin{cases} x_1 & -\ x_2 & +x_3 & +\ x_4 & = & 2 \\ & 2x_2 & & -\ 2x_4 & = & 1 \\ & 4x_2 & & -\ 4x_4 & = & -1 \end{cases}$$

Next multiply equation 2 of this new system by $\frac{1}{2}$, an operation which is denoted by $\frac{1}{2}(2)$, to get

$$\begin{cases} x_1 & -\ x_2 & +x_3 & +\ x_4 & = & 2 \\ & x_2 & & -\ x_4 & = & \frac{1}{2} \\ & 4x_2 & & -\ 4x_4 & = & -1 \end{cases}$$

Finally, eliminate x_2 from equation 3 by performing the operation $(3) - 4(2)$, that is, subtract 4 times equation 2 from equation 3; this yields the linear system

$$\begin{cases} x_1 & -\ x_2 & +\ x_3 & +\ x_4 & = & 2 \\ & x_2 & & -\ x_4 & = & \frac{1}{2} \\ & & & 0 & = & -3 \end{cases}$$

Of course the third equation is false, so the original linear system has no solutions, that is, it is *inconsistent*.

Example 2.1.2

$$\begin{cases} x_1 & +\ 4x_2 & +\ 2x_3 & = & -2 \\ -2x_1 & -\ 8x_2 & +\ 3x_3 & = & 32 \\ & x_2 & +\ x_3 & = & 1 \end{cases}$$

Add two times equation 1 to equation 2, that is, perform the operation $(2) + 2(1)$, to get

$$\begin{cases} x_1 & +\ 4x_2 & +\ 2x_3 & = & -2 \\ & & 7x_3 & = & 28 \\ & x_2 & +\ x_3 & = & 1 \end{cases}$$

At this point we should have liked x_2 to appear in the second equation: however this is not the case. To remedy the situation we interchange equations 2 and 3, in symbols $(2)\leftrightarrow(3)$. The linear system now takes the form

$$\begin{cases} x_1 & + 4x_2 & + 2x_3 & = -2 \\ & x_2 & + x_3 & = 1 \\ & & 7x_3 & = 28 \end{cases}$$

Finally, multiply equation 3 by $\frac{1}{7}$, that is, apply $\frac{1}{7}(3)$, to get

$$\begin{cases} x_1 & + 4x_2 & + 2x_3 & = -2 \\ & x_2 & + x_3 & = 1 \\ & & x_3 & = 4 \end{cases}$$

This system can be solved quickly by a process called *back substitution*. By the last equation $x_3 = 4$, so we can substitute $x_3 = 4$ in the second equation to get $x_2 = -3$. Finally, substitute $x_3 = 4$ and $x_2 = -3$ in the first equation to get $x_1 = 2$. Hence the linear system has a *unique solution*.

Example 2.1.3

$$\begin{cases} x_1 & + 3x_2 & + 3x_3 & + 2x_4 & = 1 \\ 2x_1 & + 6x_2 & + 9x_3 & + 5x_4 & = 5 \\ -x_1 & - 3x_2 & + 3x_3 & & = 5 \end{cases}$$

Apply operations $(2) - 2(1)$ and $(3) + (1)$ successively to the linear system to get

$$\begin{cases} x_1 & + 3x_2 & + 3x_3 & + 2x_4 & = 1 \\ & & 3x_3 & + x_4 & = 3 \\ & & 6x_3 & + 2x_4 & = 6 \end{cases}$$

Since x_2 has disappeared completely from the second and third equations, we move on to the next unknown x_3; applying $\frac{1}{3}(2)$, we obtain

$$\begin{cases} x_1 & + \ 3x_2 & + \ 3x_3 & + \ 2x_4 & = 1 \\ & & x_3 & + \ \frac{1}{3}x_4 & = 1 \\ & & 6x_3 & + \ 2x_4 & = 6 \end{cases}$$

Finally, operation $(3) - 6(2)$ gives

$$\begin{cases} x_1 & + \ 3x_2 & + \ 3x_3 & + \ 2x_4 & = 1 \\ & & x_3 & + \ \frac{1}{3}x_4 & = 1 \\ & & & 0 & = 0 \end{cases}$$

Here the third equation tells us nothing and can be ignored. Now observe that we can assign arbitrary values c and d to the unknowns x_4 and x_2 respectively, and then use back substitution to find x_3 and x_1. Hence the most general solution of the linear system is

$$x_1 = -2 - c - 3d, \quad x_2 = d, \quad x_3 = 1 - \frac{c}{3}, \quad x_4 = c.$$

Since c and d can be given arbitrary values, the linear system has *infinitely many solutions*.

What has been learned from these three examples? In the first place, the number of solutions of a linear system can be 0, 1 or infinity. More importantly, we have seen that there is a systematic method of eliminating some of the unknowns from all equations of the system beyond a certain point, with the result that a linear system is reached which is of such a simple form that it is possible either to conclude that no solutions exist or else to find all solutions by the process of back substitution. This systematic procedure is called *Gaussian elimination*; it is now time to give a general account of the way in which it works.

The general theory of linear systems

Consider a set of m linear equations in n unknowns $x_1, x_2,$ $\ldots, x_n$:

$$\begin{cases} a_{11}x_1 & + & a_{12}x_2 & + & \cdots & + & a_{1n}x_n & = & b_1 \\ a_{21}x_1 & + & a_{22}x_2 & + & \cdots & + & a_{2n}x_n & = & b_2 \\ & & \cdot & & \cdots & & \cdot & & \cdot \\ a_{m1}x_1 & + & a_{m2}x_2 & + & \cdots & + & a_{mn}x_n & = & b_m \end{cases}$$

By a *solution* of the linear system we shall mean an n-column vector

$$\begin{pmatrix} x_1 \\ x_2 \\ \vdots \\ x_n \end{pmatrix}$$

such that the scalars x_1, x_2, ..., x_n satisfy all the equations of the system. The set of all solutions is called the *general solution* of the linear system; this is normally given in the form of a single column vector containing a number of arbitrary quantities. A linear system with no solutions is said to be *inconsistent*.

Two linear systems which have the same sets of solutions are termed *equivalent*. Now in the examples discussed above three types of operation were applied to the linear systems:

(a) interchange of two equations;
(b) addition of a multiple of one equation to another equation;
(c) multiplication of one equation by a non-zero scalar.

Notice that each of these operations is invertible. The critical property of such operations is that, when they are applied to a linear system, the resulting system is equivalent to the original one. This fact was exploited in the three examples above. Indeed, by the very nature of these operations, any

solution of the original system is bound to be a solution of the
new system, and conversely, by invertibility of the operations,
any solution of the new system is also a solution of the original
system. Thus we can state the fundamental theorem:

Theorem 2.1.1

When an operation of one of the three types (a), (b), (c) *is
applied to a linear system, the resulting linear system is equivalent to the original one.*

We shall now exploit this result and describe the procedure known as *Gaussian elimination*. In this a sequence of
operations of types (a), (b), (c) is applied to a linear system
in such a way as to produce an equivalent linear system whose
form is so simple that we can quickly determine its solutions.

Suppose that a linear system of m equations in n unknowns x_1, x_2, ..., x_n is given. In Gaussian elimination the
following steps are to be carried out.

(i) Find an equation in which x_1 appears and, if necessary,
interchange this equation with the first equation. Thus we can
assume that x_1 appears in equation 1.

(ii) Multiply equation 1 by a suitable non-zero scalar in
such a way as to make the coefficient of x_1 equal to 1.

(iii) Subtract suitable multiples of equation 1 from equations 2 through m in order to eliminate x_1 from these equations.

(iv) Inspect equations 2 through m and find the first equation which involves one of the the unknowns x_2, ..., x_n , say
x_{i_2}. By interchanging equations once again, we can suppose
that x_{i_2} occurs in equation 2.

(v) Multiply equation 2 by a suitable non-zero scalar to
make the coefficient of x_{i_2} equal to 1.

(vi) Subtract multiples of equation 2 from equations 3
through m to eliminate x_{i_2} from these equations.

(vii) Examine equations 3 through m and find the first one that involves an unknown other than x_1 and x_{i_2}, say x_{i_3}. By interchanging equations we may assumethat x_{i_3} actually occurs in equation 3.

The next step is to make the coefficient of x_{i_3} equal to 1, and then to eliminate x_{i_3} from equations 4 through m, and so on.

The elimination procedure continues in this manner, producing the so-called *pivotal* unknowns $x_1 = x_{i_1}, x_{i_2}, ..., x_{i_r}$, until we reach a linear system in which no further unknowns occur in the equations beyond the rth. A linear system of this sort is said to be in *echelon form*; it will have the following shape.

$$
\left\{
\begin{array}{l}
x_{i_1} \;+ * x_{i_2} \;+ \quad \cdots \quad + * x_n \; = * \\
\qquad x_{i_2} \;+ \quad \cdots \quad + * x_n \; = * \\
\qquad \cdot \qquad \cdot \qquad \cdots \qquad \cdot \qquad \cdot \\
\qquad\qquad\quad x_{i_r} + \cdots \quad + * x_n \; = * \\
\qquad\qquad\qquad\qquad\qquad\qquad 0 \quad = * \\
\qquad\qquad\qquad\qquad\qquad\qquad \cdot \qquad \cdot \\
\qquad\qquad\qquad\qquad\qquad\qquad 0 \quad = *
\end{array}
\right.
$$

Here the asterisks represent certain scalars and the i_j are integers which satisfy $1 = i_1 < i_2 < ... < i_r \leq n$. The unknowns x_{i_j} for $j = 1$ to r are the *pivots*.

Once echelon form has been reached, the behavior of the linear system can be completely described and the solutions – if any – obtained by back substitution, as in the preceding examples. Consequently we have the following fundamental result which describes the possible behavior of a linear system.

Theorem 2.1.2

(i) *A linear system is consistent if and only if all the entries on the right hand sides of those equations in echelon form which contain no unknowns are zero.*

(ii) *If the system is consistent, the non-pivotal unknowns can be given arbitrary values; the general solution is then obtained by using back substitution to solve for the pivotal unknowns.*
(iii) *The system has a unique solution if and only if all the unknowns are pivotal.*

An important feature of Gaussian elimination is that it constitutes a practical algorithm for solving linear systems which can easily be implemented in one of the standard programming languages.

Gauss-Jordan elimination

Let us return to the echelon form of the linear system described above. We can further simplify the system by subtracting a multiple of equation 2 from equation 1 to eliminate x_{i_2} from that equation. Now x_{i_2} occurs only in the second equation. Similarly we can eliminate x_{i_3} from equations 1 and 2 by subtracting multiples of equation 3 from these equations. And so on. Ultimately a linear system is reached which is in *reduced echelon form.*

Here *each pivotal unknown appears in precisely one equation*; the non-pivotal unknowns may be given arbitrary values and the pivotal unknowns are then determined directly from the equations without back substitution.

The procedure for reaching reduced echelon form is called *Gauss-Jordan elimination*: while it results in a simpler type of linear system, this is accomplished at the cost of using more operations.

Example 2.1.4

In Example 2.1.3 above we obtained a linear system in echelon form

$$\begin{cases} x_1 & + 3x_2 & + 3x_3 & + 2x_4 & = 1 \\ & & x_3 & + \frac{1}{3}x_4 & = 1 \\ & & & 0 & = 0 \end{cases}$$

Here the pivots are x_1 and x_3. One further operation must be applied to put the system in reduced row echelon form, namely (1) - 3(2); this gives

$$\begin{cases} x_1 & + 3x_2 & & + x_4 & = -2 \\ & & x_3 & + \frac{1}{3}x_4 & = 1 \\ & & & 0 & = 0 \end{cases}$$

To obtain the general solution give the non-pivotal unknowns x_2 and x_4 the arbitrary values d and c respectively, and then read off directly the values $x_1 = -2 - c - 3d$ and $x_3 = 1 - c/3$.

Homogeneous linear systems

A very important type of linear system occurs when all the scalars on the right hand sides of the equations equal zero.

$$\begin{cases} a_{11}x_1 & + & a_{12}x_2 & + & \cdots & + & a_{1n}x_n & = 0 \\ a_{21}x_1 & + & a_{22}x_2 & + & \cdots & + & a_{2n}x_n & = 0 \\ \cdot & & \cdot & & \cdots & & \cdot & \cdot \\ a_{m1}x_1 & + & a_{m2}x_2 & + & \cdots & + & a_{mn}x_n & = 0 \end{cases}$$

Such a system is called *homogeneous*. It will always have *the trivial solution* $x_1 = 0$, $x_2 = 0$, ..., $x_n = 0$; thus a homogeneous linear system is always consistent. The interesting question about a homogeneous linear system is whether it has any non-trivial solutions. The answer is easily read off from the echelon form.

Theorem 2.1.3

A homogeneous linear system has a non-trivial solution if and only if the number of pivots in echelon form is less than the number of unknowns.

For if the number of unkowns is n and the number of pivots is r, the $n - r$ non-pivotal unknowns can be given arbitrary values, so there will be a non-trivial solution whenever

$n - r > 0$. On the other hand, if $n = r$, none of the unknowns can be given arbitrary values, and there is only one solution, namely the trivial one, as we see from reduced echelon form.

Corollary 2.1.4

A homogeneous linear system of m equations in n unknowns always has a non-trivial solution if $m < n$.

For if r is the number of pivots, then $r \leq m < n$.

Example 2.1.5

For which values of the parameter t does the following homogeneous linear system have non-trivial solutions?

$$\begin{cases} 6x_1 & - x_2 & + x_3 & = 0 \\ tx_1 & & + x_3 & = 0 \\ & x_2 & + tx_3 & = 0 \end{cases}$$

It suffices to find the number of pivotal unknowns. We proceed to put the linear system in echelon form by applying to it successively the operations $\frac{1}{6}(1)$, $(2) - t(1)$, $(2) \leftrightarrow (3)$ and $(3) - \frac{t}{6}(2)$:

$$\begin{cases} x_1 & - \frac{1}{6}x_2 & + & \frac{1}{6}x_3 & = 0 \\ & x_2 & + & tx_3 & = 0 \\ & & & (1 - \frac{t}{6} - \frac{t^2}{6})x_3 & = 0 \end{cases}$$

The number of pivots will be less than 3, the number of unknowns, precisely when $1 - t/6 - t^2/6$ equals zero, that is, when $t = 2$ or $t = -3$. These are the only values of t for which the linear system has non-trivial solutions.

The reader will have noticed that we deviated slightly from the procedure of Gaussian elimination; this was to avoid dividing by $t/6$, which would have necessitated a separate discussion of the case $t = 0$.

Exercises 2.1

In the first three problems find the general solution or else show that the linear system is inconsistent.

1.

$$\begin{cases} x_1 & + 2x_2 & - 3x_3 & + x_4 & = 7 \\ -x_1 & + x_2 & - x_3 & + x_4 & = 4 \end{cases}$$

2.

$$\begin{cases} x_1 & + x_2 & - x_3 & - x_4 & = 0 \\ x_1 & & + x_3 & - x_4 & = -1 \\ 2x_1 & + 2x_2 & + x_3 & - 3x_4 & = 2 \end{cases}$$

3.

$$\begin{cases} x_1 & + x_2 & + 2x_3 & = 4 \\ x_1 & - x_2 & - x_3 & = -1 \\ 2x_1 & - 4x_2 & - 5x_3 & = 1 \end{cases}$$

4. Solve the following homogeneous linear systems

(a) $\begin{cases} x_1 & + x_2 & + x_3 & + x_4 & = 0 \\ 2x_1 & + 2x_2 & + x_3 & + x_4 & = 0 \\ x_1 & + x_2 & - x_3 & + x_4 & = 0 \end{cases}$

(b) $\begin{cases} 2x_1 & - x_2 & + 3x_3 & = 0 \\ 4x_1 & + 2x_2 & + 2x_3 & = 0 \\ -2x_1 & + 5x_2 & - 4x_3 & = 0 \end{cases}$

5. For which values of t does the following homogeneous linear system have non-trivial solutions?

$$\begin{cases} 12x_1 & - x_2 & + x_3 & = 0 \\ tx_1 & & + x_3 & = 0 \\ & x_2 & + tx_3 & = 0 \end{cases}$$

6. For which values of t is the following linear system consistent?

$$\begin{cases} x_1 & + x_2 & - x_3 & + x_4 & = 12 \\ & 3x_2 & - 2x_3 & + x_4 & = 14 \\ 2x_1 & & + x_3 & + x_4 & = 10 \\ tx_1 & + 4x_2 & - 2x_3 & + x_4 & = 16 \end{cases}$$

7. How many operations of types (a), (b), (c) are needed in general to put a system of n linear equations in n unknowns in echelon form?

2.2 Elementary Row Operations

If we examine more closely the process of Gaussian elimination described in 2.1, it is apparent that much time and trouble could be saved by working directly with the augmented matrix of the linear system and applying certain operations to its rows. In this way we avoid having to write out the unknowns repeatedly.

The row operations referred to correspond to the three types of operation that may be applied to a linear system during Gaussian elimination. These are the so-called *elementary row operations* and they can be applied to any matrix. The row operations together with their symbolic representations are as follows:

(a) interchange rows i and j, $(R_i \leftrightarrow R_j)$;

(b) add c times row j to row i where c is any scalar, $(R_i + cR_j)$;

(c) multiply row i by a non-zero scalar c, (cR_i).

From the matrix point of view the essential content of Theorem 2.1.2 is that any matrix can be put in what is called row echelon form by application of a suitable finite sequence of elementary row operations. A matrix in *row echelon form* has the typical "descending staircase" form

$$
\begin{pmatrix}
0 & \cdots & 0 & \boxed{1} & * & \cdots & * & * & * & \cdots & * & * & \cdots & * & * \\
0 & \cdots & 0 & 0 & 0 & \cdots & 0 & \boxed{1} & * & \cdots & * & * & \cdots & * & * \\
\cdot & \cdots & \cdot & \cdot & \cdot & \cdots & \cdot & \cdot & \cdot & \cdots & \cdot & \cdot & \cdots & \cdot & \cdot \\
0 & \cdots & 0 & 0 & 0 & \cdots & 0 & 0 & 0 & \cdots & \boxed{1} & * & \cdots & * & * \\
0 & \cdots & 0 & 0 & 0 & \cdots & 0 & 0 & 0 & \cdots & 0 & 0 & \cdots & 0 & * \\
\cdot & \cdots & \cdot & \cdot & \cdot & \cdots & \cdot & \cdot & \cdot & \cdots & \cdot & \cdot & \cdots & \cdot & \cdot \\
0 & \cdots & 0 & 0 & 0 & \cdots & 0 & 0 & 0 & \cdots & 0 & 0 & \cdots & 0 & *
\end{pmatrix}
$$

Here the asterisks denote certain scalars.

Example 2.2.1

Put the following matrix in row echelon form by applying suitable elementary row operations:

$$
\begin{pmatrix}
1 & 3 & 3 & 2 & 1 \\
2 & 6 & 9 & 5 & 5 \\
-1 & -3 & 3 & 0 & 5
\end{pmatrix}.
$$

Applying the row operations $R_2 - 2R_1$ and $R_3 + R_1$, we obtain

$$
\begin{pmatrix}
1 & 3 & 3 & 2 & 1 \\
0 & 0 & 3 & 1 & 3 \\
0 & 0 & 6 & 2 & 6
\end{pmatrix}.
$$

Then, after applying the operations $\frac{1}{3}R_2$ and $R_3 - 6R_2$, we get

$$
\begin{pmatrix}
1 & 3 & 3 & 2 & 1 \\
0 & 0 & 1 & 1/3 & 1 \\
0 & 0 & 0 & 0 & 0
\end{pmatrix},
$$

which is in row echelon form.

Suppose now that we wish to solve the linear system with matrix form $AX = B$, using elementary row operations. The first step is to identify the augmented matrix $M = [A \mid B]$.

Then we put M in row echelon form, using row operations. From this we can determine if the original linear system is consistent; for this to be true, in the row echelon form of M the scalars in the last column which lie below the final pivot must all be zero. To find the general solution of a consistent system we convert the row echelon matrix back to a linear system and use back substitution to solve it.

Example 2.2.2

Consider once again the linear system of Example 2.1.3;

$$\begin{cases} x_1 & + 3x_2 & + 3x_3 & + 2x_4 & = 1 \\ 2x_1 & + 6x_2 & + 9x_3 & + 5x_4 & = 5 \\ -x_1 & - 3x_2 & + 3x_3 & & = 5 \end{cases}$$

The augmented matrix here is

$$\begin{pmatrix} 1 & 3 & 3 & 2 & | & 1 \\ 2 & 6 & 9 & 5 & | & 5 \\ -1 & -3 & 3 & 0 & | & 5 \end{pmatrix}.$$

Now we have just seen in Example 2.2.1 that this matrix has row echelon form

$$\begin{pmatrix} 1 & 3 & 3 & 2 & | & 1 \\ 0 & 0 & 1 & 1/3 & | & 1 \\ 0 & 0 & 0 & 0 & | & 0 \end{pmatrix}.$$

Because the lower right hand entry is 0, the linear system is consistent. The linear system corresponding to the last matrix is

$$\begin{cases} x_1 + 3x_2 + 3x_3 + 2x_4 = 1 \\ \qquad\qquad x_3 + \dfrac{1}{3}x_4 = 1 \\ \qquad\qquad\qquad\quad 0 = 0 \end{cases}$$

Hence the general solution given by back substitution is $x_1 = -2 - c - 3d$, $x_2 = d$, $x_3 = 1 - c/3$, $x_4 = c$, where c and d are arbitrary scalars.

The matrix formulation enables us to put our conclusions about linear systems in a succinct form.

Theorem 2.2.1

Let $AX = B$ be a linear system of equations in n unknowns with augmented matrix $M = [A \mid B]$.

(i) The linear system is consistent if and only if the matrices A and M have the same numbers of pivots in row echelon form.

(ii) If the linear system is consistent and r denotes the number of pivots of A in row echelon form, then the $n - r$ unknowns that correspond to columns of A not containing a pivot can be given arbitrary values. Thus the system has a unique solution if and only if $r = n$.

Proof

For the linear system to be consistent, the row echelon form of M must have only zero entries in the last column below the final pivot; but this is just the condition for A and M to have the same numbers of pivots.

Finally, if the linear system is consistent, the unknowns corresponding to columns that do not contain pivots may be given arbitrary values and the remaining unknowns found by back substitution.

Reduced row echelon form

A matrix is said to be in *reduced row echelon form* if it is in row echelon form and if in each column containing a pivot all entries other than the pivot itself are zero.

Example 2.2.3

Put the matrix
$$\begin{pmatrix} 1 & 1 & 2 & 2 \\ 4 & 4 & 9 & 10 \\ 3 & 3 & 6 & 7 \end{pmatrix}$$

in reduced row echelon form.

By applying suitable row operations we find the row echelon form to be
$$\begin{pmatrix} 1 & 1 & 2 & 2 \\ 0 & 0 & 1 & 2 \\ 0 & 0 & 0 & 1 \end{pmatrix}.$$

Notice that columns 1, 3 and 4 contain pivots. To pass to reduced row echelon form, apply the row operations $R_1 - 2R_2$, $R_1 + 2R_3$ and $R_2 - 2R_3$: the answer is

$$\begin{pmatrix} 1 & 1 & 0 & 0 \\ 0 & 0 & 1 & 0 \\ 0 & 0 & 0 & 1 \end{pmatrix}.$$

As this example illustrates, one can pass from row echelon form to reduced row echelon form by applying further row operations; notice that this will not change the number of pivots. Thus *an arbitrary matrix can be put in reduced row echelon form by applying a finite sequence of elementary row operations.* The reader should observe that this is just the matrix formulation of the Gauss-Jordan elimination procedure.

Exercises 2.2

1. Put each of the following matrices in row echelon form:

$$\text{(a)} \begin{pmatrix} 2 & -3 & -4 \\ 4 & -6 & 1 \\ 1 & 1 & 0 \end{pmatrix}, \quad \text{(b)} \begin{pmatrix} 1 & 2 & -3 \\ 3 & 1 & 2 \\ 8 & 1 & 9 \end{pmatrix},$$

$$\text{(c)} \begin{pmatrix} 1 & 2 & -3 & 1 \\ 3 & 1 & 2 & 2 \\ 8 & 1 & 9 & 1 \end{pmatrix}.$$

2. Put each of the matrices in Exercise 2.2.1 in reduced row echelon form.

3. Prove that the row operation of type (a) which interchanges rows i and j can be obtained by a combination of row operations of the other two types, that is, types (b) and (c).

4. Do Exercises 2.1.1 to 2.1.4 by applying row operations to the augmented matrices.

5. How many row operations are needed in general to put an $n \times n$ matrix in row echelon form?

6. How many row operations are needed in general to put an $n \times n$ matrix in reduced row echelon form?

7. Give an example to show that a matrix can have more than one row echelon form.

8. If A is an invertible $n \times n$ matrix, prove that the linear system $AX = B$ has a unique solution. What does this tell you about the number of pivots of A?

9. Show that each elementary row operation has an inverse which is also an elementary row operation.

2.3 Elementary Matrices

An $n \times n$ matrix is called *elementary* if it is obtained from the identity matrix I_n in one of three ways:
(a) interchange rows i and j where $i \neq j$;
(b) insert a scalar c as the (i, j) entry where $i \neq j$;
(c) put a non-zero scalar c in the (i, i) position.

Example 2.3.1

Write down all the possible types of elementary 2×2 matrices. These are the elementary matrices that arise from the matrix $I_2 = \begin{pmatrix} 1 & 0 \\ 0 & 1 \end{pmatrix}$; they are

$$E_1 = \begin{pmatrix} 0 & 1 \\ 1 & 0 \end{pmatrix}, \ E_2 = \begin{pmatrix} 1 & c \\ 0 & 1 \end{pmatrix}, \ E_3 = \begin{pmatrix} 1 & 0 \\ c & 1 \end{pmatrix}$$

and

$$E_4 = \begin{pmatrix} c & 0 \\ 0 & 1 \end{pmatrix}, \ E_5 = \begin{pmatrix} 1 & 0 \\ 0 & c \end{pmatrix}.$$

Here c is a scalar which must be non-zero in the case of E_4 and E_5.

The significance of elementary matrices from our point of view lies in the fact that when we premultiply a matrix by an elementary matrix, the effect is to perform an elementary row operation on the matrix. For example, with the matrix

$$A = \begin{pmatrix} a_{11} & a_{12} \\ a_{21} & a_{22} \end{pmatrix}$$

and elementary matrices listed in Example 2.3.1, we have

$$E_1 A = \begin{pmatrix} a_{21} & a_{22} \\ a_{11} & a_{12} \end{pmatrix}, \ E_2 A = \begin{pmatrix} a_{11} + c a_{21} & a_{12} + c a_{22} \\ a_{21} & a_{22} \end{pmatrix}$$

and

$$E_5 A = \begin{pmatrix} a_{11} & a_{12} \\ ca_{21} & ca_{22} \end{pmatrix}.$$

Thus premultiplication by E_1 interchanges rows 1 and 2; premultiplication by E_2 adds c times row 2 to row 1; premultiplication by E_5 multiplies row 2 by c. What then is the general rule?

Theorem 2.3.1

Let A be an $m \times n$ matrix and let E be an elementary $m \times m$ matrix.

(i) If E is of type (a), *then EA is the matrix obtained from A by interchanging rows i and j of A;*

(ii) if E is type (b), *then EA is the matrix obtained from A by adding c times row j to row i;*

(iii) if E is of type (c), *then EA arises from A by multiplying row i by c.*

Now recall from 2.2 that every matrix can be put in reduced row echelon form by applying elementary row operations. Combining this observation with 2.3.1, we obtain

Theorem 2.3.2

Let A be any $m \times n$ matrix. Then there exist elementary $m \times m$ matrices E_1, E_2, ..., E_k such that the matrix $E_k E_{k-1} \cdots E_1 A$ is in reduced row echelon form.

Example 2.3.2

Consider the matrix

$$A = \begin{pmatrix} 0 & 1 & 2 \\ 2 & 1 & 0 \end{pmatrix}.$$

We easily put this in reduced row echelon form B by applying successively the row operations $R_1 \leftrightarrow R_2$, $\frac{1}{2}R_1$, $R_1 - \frac{1}{2}R_2$:

$$A \to \begin{pmatrix} 2 & 1 & 0 \\ 0 & 1 & 2 \end{pmatrix} \to \begin{pmatrix} 1 & 1/2 & 0 \\ 0 & 1 & 2 \end{pmatrix} \to \begin{pmatrix} 1 & 0 & -1 \\ 0 & 1 & 2 \end{pmatrix} = B.$$

Hence $E_3 E_2 E_1 A = B$ where

$$E_1 = \begin{pmatrix} 0 & 1 \\ 1 & 0 \end{pmatrix}, \; E_2 = \begin{pmatrix} 1/2 & 0 \\ 0 & 1 \end{pmatrix}, \; E_3 = \begin{pmatrix} 1 & -1/2 \\ 0 & 1 \end{pmatrix}$$

Column operations

Just as for rows, there are three types of *elementary column operation*, namely:

(a) interchange columns i and j , ($C_i \leftrightarrow C_j$);
(b) add c times column j to column i where c is a scalar, ($C_i + cC_j$);
(c) multiply column i by a non-zero scalar c, (cC_i).

(The reader is warned, however, that column operations cannot in general be applied to the augmented matrix of a linear system without changing the solutions of the system.)

The effect of applying an elementary column operation to a matrix is simulated by right multiplication by a suitable elementary matrix. But there is one important difference from the row case. In order to perform the operation $C_i + cC_j$ to a matrix A one multiplies on the right by the elementary matrix *whose (j, i) element is c.* For example, let

$$E = \begin{pmatrix} 1 & 0 \\ c & 1 \end{pmatrix} \quad \text{and} \quad A = \begin{pmatrix} a_{11} & a_{12} \\ a_{21} & a_{22} \end{pmatrix}.$$

Then

$$AE = \begin{pmatrix} a_{11} + ca_{12} & a_{12} \\ a_{21} + ca_{22} & a_{22} \end{pmatrix}.$$

Thus E performs the column operation $C_1 + 2C_2$ and not $C_2 + 2C_1$. By multiplying a matrix on the right by suitable sequences of elementary matrices, a matrix can be put in *column echelon form* or in *reduced column echelon form*; these are just the transposes of row echelon form and reduced row echelon form respectively.

Example 2.3.3

Put the matrix $A = \begin{pmatrix} 3 & 6 & 2 \\ 1 & 2 & 7 \end{pmatrix}$ in reduced column echelon form.

Apply the column operations $\frac{1}{3}C_1$, $C_2 - 6C_1$, $C_3 - 2C_1$, $C_2 \leftrightarrow C_3$, $\frac{3}{19}C_2$, and $C_1 - \frac{1}{3}C_2$:

$$A \to \begin{pmatrix} 1 & 6 & 2 \\ 1/3 & 2 & 7 \end{pmatrix} \to \begin{pmatrix} 1 & 0 & 0 \\ 1/3 & 0 & 19/3 \end{pmatrix} \to$$

$$\begin{pmatrix} 1 & 0 & 0 \\ 1/3 & 19/3 & 0 \end{pmatrix} \to \begin{pmatrix} 1 & 0 & 0 \\ 1/3 & 1 & 0 \end{pmatrix},$$

$$\to \begin{pmatrix} 1 & 0 & 0 \\ 0 & 1 & 0 \end{pmatrix}.$$

We leave the reader to write down the elementary matrices that produce these column operations.

Now suppose we are allowed to apply *both* row and column operations to a matrix. Then we can obtain first row echelon form; subsequently column operations may be applied to give a matrix of the very simple type

$$\begin{pmatrix} I_r & 0 \\ 0 & 0 \end{pmatrix}$$

where r is the number of pivots. This is called the *normal form* of the matrix; we shall see in 5.2 that every matrix has a unique normal form. These conclusions are summed up in the following result.

Theorem 2.3.3

Let A be an $m \times n$ matrix. Then there exist elementary $m \times m$ matrices $E_1, \ldots, E_k$ and elementary $n \times n$ matrices $F_1, \ldots, F_l$ such that

$$E_k \cdots E_1 A F_1 \cdots F_l = N,$$

the normal form of A.

Proof

By applying suitable row operations to A we can find elementary matrices E_1, ..., E_k such that $B = E_k \cdots E_1 A$ is in row echelon form. Then column operations are applied to reduce B to normal form; this procedure yields elementary matrices F_1, ..., F_l such that $N = B F_1 \cdots F_l = E_k \cdots E_1 A F_1 \cdots F_l$ is the normal form of A.

Corollary 2.3.4

For any matrix A there are invertible matrices X and Y such that $N = XAY$, or equivalently $A = X^{-1} N Y^{-1}$, where N is the normal form of A.

For it is easy to see that *every elementary matrix is invertible*; indeed the inverse matrix represents the inverse of the corresponding elementary row (or column) operation. Since by 1.2.3 any product of invertible matrices is invertible, the corollary follows from 2.3.3.

Example 2.3.4

Let $A = \begin{pmatrix} 1 & 2 & 2 \\ 2 & 3 & 4 \end{pmatrix}$. Find the normal form N of A and write N as the product of A and elementary matrices as specified in 2.3.3.

All we need do is to put A in normal form, while keeping track of the elementary matrices that perform the necessary row and column operations. Thus

$$A \rightarrow \begin{pmatrix} 1 & 2 & 2 \\ 0 & -1 & 0 \end{pmatrix} \rightarrow \begin{pmatrix} 1 & 2 & 2 \\ 0 & 1 & 0 \end{pmatrix} \rightarrow \begin{pmatrix} 1 & 0 & 2 \\ 0 & 1 & 0 \end{pmatrix}$$

$$\rightarrow \begin{pmatrix} 1 & 0 & 0 \\ 0 & 1 & 0 \end{pmatrix},$$

which is the normal form of A. Here three row operations and one column operation were used to reduce A to its normal form. Therefore

$$E_3 E_2 E_1 A F_1 = N$$

where

$$E_1 = \begin{pmatrix} 1 & 0 \\ -2 & 1 \end{pmatrix}, \quad E_2 = \begin{pmatrix} 1 & 0 \\ 0 & -1 \end{pmatrix}, \quad E_3 = \begin{pmatrix} 1 & -2 \\ 0 & 1 \end{pmatrix}$$

and

$$F_1 = \begin{pmatrix} 1 & 0 & -2 \\ 0 & 1 & 0 \\ 0 & 0 & 1 \end{pmatrix}.$$

Inverses of matrices

Inverses of matrices were defined in 1.2, but we deferred the important problem of computing inverses until more was known about linear systems. It is now time to address this problem. Some initial information is given by

Theorem 2.3.5

Let A be an $n \times n$ matrix. Then the following statements about A are equivalent, that is, each one implies all of the others.

(a) *A is invertible;*

(b) *the linear system $AX = 0$ has only the trivial solution;*

(c) *the reduced row echelon form of A is I_n;*

(d) *A is a product of elementary matrices.*

Proof
We shall establish the logical implications (a) → (b), (b) → (c), (c) → (d), and (d) → (a). This will serve to establish the equivalence of the four statements.

If (a) holds, then A^{-1} exists; thus if we multiply both sides of the equation $AX = 0$ on the left by A^{-1}, we get $A^{-1}AX = A^{-1}0$, so that $X = A^{-1}0 = 0$ and the only solution of the linear system is the trivial one. Thus (b) holds.

If (b) holds, then we know from 2.1.3 that the number of pivots of A in reduced row echelon form is n. Since A is $n \times n$, this must mean that I_n is the reduced row echelon form of A, so that (c) holds.

If (c) holds, then 2.3.2 shows that there are elementary matrices $E_1, ..., E_k$ such that $E_k \cdots E_1 A = I_n$. Since elementary matrices are invertible, $E_k \cdots E_1$ is invertible, and thus $A = (E_k \cdots E_1)^{-1} = E_1^{-1} \cdots E_k^{-1}$, so that (d) is true.

Finally, (d) implies (a) since a product of elementary matrices is always invertible.

A procedure for finding the inverse of a matrix

As an application of the ideas in this section, we shall describe an efficient method of computing the inverse of an invertible matrix.

Suppose that A is an invertible $n \times n$ matrix. Then there exist elementary $n \times n$ matrices $E_1, E_2, ..., E_k$ such that $E_k \cdots E_2 E_1 A = I_n$, by 2.3.2 and 2.3.5. Therefore

$$A^{-1} = I_n A^{-1} = (E_k \cdots E_2 E_1 A)A^{-1} = (E_k \cdots E_2 E_1)I_n.$$

This means that the row operations which reduce A to its reduced row echelon form I_n will automatically transform I_n to A^{-1}. It is this crucial observation which enables us to compute A^{-1}.

The procedure for computing A^{-1} starts with the partitioned matrix

$$[A \mid I_n]$$

and then puts it in reduced row echelon form. If A is invertible, the reduced row echelon form will be

$$[I_n \mid A^{-1}],$$

as the discussion just given shows. On the other hand, if the procedure is applied to a matrix that is not invertible, it will be impossible to reach a reduced row echelon form of the above type, that is, one with I_n on the left. Thus the procedure will also detect non-invertibility of a matrix.

Example 2.3.5

Find the inverse of the matrix

$$A = \begin{pmatrix} 2 & -1 & 0 \\ -1 & 2 & -1 \\ 0 & -1 & 2 \end{pmatrix}.$$

Put the matrix $[A \mid I_3]$ in reduced row echelon form, using elementary row operations as described above:

$$\begin{pmatrix} 2 & -1 & 0 & | & 1 & 0 & 0 \\ -1 & 2 & -1 & | & 0 & 1 & 0 \\ 0 & -1 & 2 & | & 0 & 0 & 1 \end{pmatrix} \rightarrow$$

$$\begin{pmatrix} 1 & -1/2 & 0 & | & 1/2 & 0 & 0 \\ -1 & 2 & -1 & | & 0 & 1 & 0 \\ 0 & -1 & 2 & | & 0 & 0 & 1 \end{pmatrix} \rightarrow$$

$$\begin{pmatrix} 1 & -1/2 & 0 & | & 1/2 & 0 & 0 \\ 0 & 3/2 & -1 & | & 1/2 & 1 & 0 \\ 0 & -1 & 2 & | & 0 & 0 & 1 \end{pmatrix} \rightarrow$$

$$\begin{pmatrix} 1 & -1/2 & 0 & | & 1/2 & 0 & 0 \\ 0 & 1 & -2/3 & | & 1/3 & 2/3 & 0 \\ 0 & -1 & 2 & | & 0 & 0 & 1 \end{pmatrix} \rightarrow$$

$$\begin{pmatrix} 1 & 0 & -1/3 & | & 2/3 & 1/3 & 0 \\ 0 & 1 & -2/3 & | & 1/3 & 2/3 & 0 \\ 0 & 0 & 4/3 & | & 1/3 & 2/3 & 1 \end{pmatrix} \rightarrow$$

$$\begin{pmatrix} 1 & 0 & -1/3 & | & 2/3 & 1/3 & 0 \\ 0 & 1 & -2/3 & | & 1/3 & 2/3 & 0 \\ 0 & 0 & 1 & | & 1/4 & 1/2 & 3/4 \end{pmatrix} \rightarrow$$

$$\begin{pmatrix} 1 & 0 & 0 & | & 3/4 & 1/2 & 1/4 \\ 0 & 1 & 0 & | & 1/2 & 1 & 1/2 \\ 0 & 0 & 1 & | & 1/4 & 1/2 & 3/4 \end{pmatrix},$$

which is the reduced row echelon form. Therefore A is invertible and

$$A^{-1} = \begin{pmatrix} 3/4 & 1/2 & 1/4 \\ 1/2 & 1 & 1/2 \\ 1/4 & 1/2 & 3/4 \end{pmatrix}.$$

This answer can be verified by checking that $AA^{-1} = I_3 = A^{-1}A$.

As this example illustrates, the procedure for finding the inverse of a $n \times n$ matrix is an efficient one; in fact at most n^2 row operations are required to complete it (see Exercise 2.3.10).

Exercises 2.3

1. Express each of the following matrices as a product of elementary matrices and its reduced row echelon form:

(a) $\begin{pmatrix} 1 & 2 & 3 \\ 4 & 5 & 6 \end{pmatrix}$; (b) $\begin{pmatrix} 1 & -1 & 2 \\ 1 & 0 & 0 \\ 0 & -1 & 2 \end{pmatrix}$.

2. Express the second matrix in Exercise 1 as a product of elementary matrices and its reduced column echelon form.

3. Find the normal form of each matrix in Exercise 1.

4. Find the inverses of the three types of elementary matrix, and observe that each is elementary and corresponds to the inverse row operation.

5. What is the maximum number of column operations needed in general to put an $n \times n$ matrix in column echelon form and in reduced column echelon form?

6. Compute the inverses of the following matrices if they exist:

$$\text{(a)} \begin{pmatrix} 4 & 3 \\ -1 & 2 \end{pmatrix}; \quad \text{(b)} \begin{pmatrix} 2 & -3 & 1 \\ 1 & 0 & 2 \\ 0 & -1 & -3 \end{pmatrix}; \quad \text{(c)} \begin{pmatrix} 2 & 1 & 7 \\ -1 & 4 & 10 \\ 3 & 2 & 12 \end{pmatrix}.$$

7. For which values of t does the matrix

$$\begin{pmatrix} 6 & -1 & 1 \\ t & 0 & 1 \\ 0 & 1 & t \end{pmatrix}$$

not have an inverse?

8. Give necessary and sufficient conditions for an upper triangular matrix to be invertible.

9. Show by an example that if an elementary column operation is applied to the augmented matrix of a linear system, the resulting linear system need not be equivalent to the original one.

10. Prove that the number of elementary row operations needed to find the inverse of an $n \times n$ matrix is at most n^2.

Chapter Three

DETERMINANTS

Associated with every square matrix is a scalar called the determinant. Perhaps the most striking property of the determinant of a matrix is the fact that it tells us if the matrix is invertible. On the other hand, there is obviously a limit to the amount of information about a matrix which can be carried by a single scalar, and this is probably why determinants are considered less important today than, say, a hundred years ago. Nevertheless, associated with an arbitrary square matrix is an important polynomial, the characteristic polynomial, which is a determinant. As we shall see in Chapter Eight, this polynomial carries a vast amount of information about the matrix.

3.1 Permutations and the Definition of a Determinant

Let $A = (a_{ij})$ be an $n \times n$ matrix over some field of scalars (which the reader should feel free to assume is either $\mathbf{R}$ or $\mathbf{C}$). Our first task is to show how to define the determinant of A, which will be written either

$$\det(A)$$

or else in the extended form

$$\begin{vmatrix} a_{11} & a_{12} & \cdots & a_{1n} \\ a_{21} & a_{22} & \cdots & a_{2n} \\ \cdot & \cdot & \cdots & \cdot \\ a_{n1} & a_{n2} & \cdots & a_{nn} \end{vmatrix}.$$

For $n = 1$ and 2 the definition is simple enough:

$$|a_{11}| = a_{11} \text{ and } \begin{vmatrix} a_{11} & a_{12} \\ a_{21} & a_{22} \end{vmatrix} = a_{11}a_{22} - a_{12}a_{21}.$$

For example, $|6| = 6$ and $\begin{vmatrix} 2 & -3 \\ 4 & 1 \end{vmatrix} = 14.$

Where does the expression $a_{11}a_{22} - a_{12}a_{21}$ come from? The motivation is provided by linear systems. Suppose that we want to solve the linear system

$$\begin{cases} a_{11}x_1 + a_{12}x_2 = b_1 \\ a_{21}x_1 + a_{22}x_2 = b_2 \end{cases}$$

for unknowns x_1 and x_2. Eliminate x_2 by subtracting a_{12} times equation 2 from a_{22} times equation 1; in this way we obtain

$$(a_{11}a_{22} - a_{12}a_{21})x_1 = b_1a_{22} - a_{12}b_2.$$

This equation expresses x_1 as the quotient of a pair of 2×2 determinants:

$$x_1 = \frac{\begin{vmatrix} b_1 & a_{12} \\ b_2 & a_{22} \end{vmatrix}}{\begin{vmatrix} a_{11} & a_{12} \\ a_{21} & a_{22} \end{vmatrix}},$$

provided, of course, that the denominator does not vanish. There is a similar expression for x_2.

The preceding calculation indicates that 2×2 determinants are likely to be of significance for linear systems. And this is confirmed if we try the same computation for a linear system of three equations in three unknowns. While the resulting solutions are complicated, they do suggest the following definition for $\det(A)$ where $A = (a_{ij})_{3,3}$;

$$a_{11}a_{22}a_{33} + a_{12}a_{23}a_{31} + a_{13}a_{21}a_{32}$$

$$-a_{12}a_{21}a_{33} - a_{13}a_{22}a_{31} - a_{11}a_{23}a_{32}$$

What are we to make of this expression? In the first place it contains six terms, each of which is a product of three entries of A. The second subscripts in each term correspond to the six ways of ordering the integers 1, 2, 3, namely

$$1,2,3 \quad 2,3,1 \quad 3,1,2 \quad 2,1,3 \quad 3,2,1 \quad 1,3,2.$$

Also each term is a product of three entries of A, while three of the terms have positive signs and three have negative signs.

There is something of a pattern here, but how can one tell which terms are to get a plus sign and which are get a minus sign? The answer is given by permutations.

Permutations

Let n be a fixed positive integer. By a *permutation* of the integers $1, 2, \ldots, n$ we shall mean an arrangement of these integers in some definite order. For example, as has been observed, there are six permutations of the integers 1, 2, 3.

In general, a permutation of $1, 2, \ldots, n$ can be written in the form

$$i_1, i_2, \ldots, i_n$$

where $i_1, i_2, \ldots, i_n$ are the integers $1, 2, \ldots, n$ in some order. Thus to construct a permutation we have only to choose distinct integers $i_1, i_2, \ldots, i_n$ from the set $\{1, 2, \ldots, n\}$. Clearly there are n choices for i_1; once i_1 has been chosen, it cannot be chosen again, so there are just $n - 1$ choices for i_2; since i_1 and i_2 cannot be chosen again, there are $n - 2$ choices for i_3, and so on. There will be only one possible choice for i_n since $n - 1$ integers have already been selected. The number of ways of constructing a permutation is therefore equal to the *product* of these numbers

$$n(n - 1)(n - 2) \cdots 2 \cdot 1,$$

which is written

$$n!$$

and referred to as "n factorial". Thus we can state the following basic result.

Theorem 3.1.1

The number of permutations of the integers $1, 2, \ldots, n$ equals $n! = n(n-1) \cdots 2 \cdot 1$.

Even and odd permutations

A permutation of the integers $1, 2, \ldots, n$ is called *even* or *odd* according to whether the number of inversions of the natural order $1, 2, \ldots, n$ that are present in the permutation is even or odd respectively. For example, the permutation 1, 3, 2 involves a single inversion, for 3 comes before 2; so this is an odd permutation. For permutations of longer sequences of integers it is advantageous to count inversions by means of what is called a *crossover diagram*. This is best explained by an example.

Example 3.1.1

Is the permutation 8, 3, 2, 6, 5, 1, 4, 7 even or odd?

The procedure is to write the integers 1 through 8 in the natural order in a horizontal line, and then to write down the entries of the permutation in the line below. Join each integer i in the top line to the same integer i where it appears in the bottom line, taking care to avoid multiple intersections. The number of intersections or *crossovers* will be the number of inversions present in the permutation:

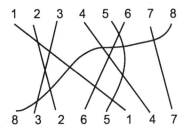

Since there are 15 crossovers in the diagram, this permutation is odd.

A *transposition* is a permutation that is obtained from $1, 2, \ldots, n$ by interchanging just two integers. Thus $2, 1, 3, 4, \ldots, n$ is an example of a transposition. An important fact about transpositions is that they are always odd.

Theorem 3.1.2

Transpositions are odd permutations.

Proof

Consider the transposition which interchanges i and j, with $i < j$ say. The crossover diagram for this transposition is

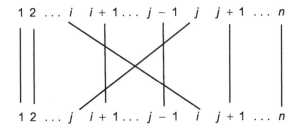

Each of the $j - i - 1$ integers $i + 1, i + 2, \ldots, j - 1$ gives rise to 2 crossovers, while i and j add one more. Hence the total number of crossovers in the diagram equals $2(j - i - 1) + 1$, which is odd.

It is important to determine the numbers of even and odd permutations.

Theorem 3.1.3

If $n > 1$, there are $\frac{1}{2}(n!)$ even permutations of $1, 2, \ldots, n$ and the same number of odd permutations.

Proof
If the first two integers are interchanged in a permutation, it is clear from the crossover diagram that an inversion is either added or removed. Thus the operation changes an even permutation to an odd permutation and an odd permutation to an even one. This makes it clear that the numbers of even and odd permutations must be equal. Since the total number of permutations is $n!$, the result follows.

Example 3.1.2

The even permutations of 1, 2, 3 are

$$1, 2, 3 \quad 2, 3, 1 \quad 3, 1, 2,$$

while the odd permutations are

$$2, 1, 3 \quad 3, 2, 1 \quad 1, 3, 2.$$

Next we define the *sign* of a permutation $i_1, i_2, \ldots, i_n$

$$\text{sign}(i_1, i_2, \ldots, i_n)$$

to be $+1$ if the permutation is even and -1 if the permutation is odd. For example, $\text{sign}(3, 2, 1) = -1$ since 3, 2, 1 is an odd permutation.

Permutation matrices

Before proceeding to the formal definition of a determinant, we pause to show how permutations can be represented by matrices. An $n \times n$ matrix is called a *permutation matrix* if it can be obtained from the identity matrix I_n by rearranging the rows or columns. For example, the permutation matrix

$$\begin{pmatrix} 0 & 1 & 0 \\ 0 & 0 & 1 \\ 1 & 0 & 0 \end{pmatrix}$$

is obtained from I_3 by cyclically permuting the columns, $C_1 \rightarrow C_2 \rightarrow C_3 \rightarrow C_1$. Permutation matrices are easy to recognize since each row and each column contains a single 1, while all other entries are zero.

Consider a permutation $i_1, i_2, \ldots, i_n$ of $1, 2, \ldots, n$, and let P be the permutation matrix which has (j, i_j) entry equal to 1 for $j = 1, 2, \ldots, n$, and all other entries zero. This means that P is obtained from I_n by rearranging the columns in the manner specified by the permutation $i_1, \ldots, i_n$, that is, $C_j \rightarrow C_{i_j}$. Then, as matrix multiplication shows,

$$P \begin{pmatrix} 1 \\ 2 \\ \vdots \\ n \end{pmatrix} = \begin{pmatrix} i_1 \\ i_2 \\ \vdots \\ i_n \end{pmatrix}.$$

Thus the effect of a permutation on the order $1, 2, \ldots, n$ is reproduced by left multiplication by the corresponding permutation matrix.

Example 3.1.3

The permutation matrix which corresponds to the permutation 4, 2, 1, 3 is obtained from I_4 by the column replacements $C_1 \rightarrow C_4$, $C_2 \rightarrow C_2$, $C_3 \rightarrow C_1$, $C_4 \rightarrow C_3$. It is

$$P = \begin{pmatrix} 0 & 0 & 0 & 1 \\ 0 & 1 & 0 & 0 \\ 1 & 0 & 0 & 0 \\ 0 & 0 & 1 & 0 \end{pmatrix}$$

and indeed

$$P \begin{pmatrix} 1 \\ 2 \\ 3 \\ 4 \end{pmatrix} = \begin{pmatrix} 4 \\ 2 \\ 1 \\ 3 \end{pmatrix}.$$

Definition of a determinant in general

We are now in a position to define the general $n \times n$ determinant. Let $A = (a_{ij})_{n,n}$ be an $n \times n$ matrix over some field of scalars. Then the *determinant* of A is the scalar defined by the equation

$$\det(A) = \sum_{i_1, i_2, \ldots, i_n} \mathrm{sign}(i_1, i_2, \ldots, i_n) a_{1i_1} a_{2i_2} \cdots a_{ni_n}$$

where the sum is taken over all permutations $i_1, i_2, \ldots, i_n$ of $1, 2, \ldots, n$.

Thus $\det(A)$ is a sum of $n!$ terms each of which involves a product of n elements of A, one from each row and one from each column. A term has a positive or negative sign according to whether the corresponding permutation is even or odd respectively. One determinant which can be immediately evaluated from the definition is that of I_n:

$$\det(I_n) = 1.$$

This is because only the permutation $1, 2, \ldots, n$ contributes a non-zero term to the sum that defines $\det(I_n)$.

If we specialise the above definition to the cases $n = 1, 2, 3$, we obtain the expressions for $\det(A)$ given at the beginning of the section. For example, let $n = 3$; the even and odd permutations are listed above in Example 3.1.2. If we write down the terms of the determinant in the same order, we obtain

$$a_{11}a_{22}a_{33} + a_{12}a_{23}a_{31} + a_{13}a_{21}a_{32}$$

$$-a_{12}a_{21}a_{33} - a_{13}a_{22}a_{31} - a_{11}a_{23}a_{32}$$

We could in a similar fashion write down the general 4×4 determinant as a sum of $4! = 24$ terms, 12 with a positive sign and 12 with a negative sign. Of course, it is clear that the definition does not provide a convenient means of computing determinants with large numbers of rows and columns; we shall shortly see that much more efficient procedures are available.

Example 3.1.4

What term in the expansion of the 8×8 determinant $\det((a_{ij}))$ corresponds to the permutation 8, 3, 2, 6, 5, 1, 4, 7 ?

We saw in Example 3.1.1 that this permutation is odd, so its sign is -1; hence the term sought is

$$-a_{18}a_{23}a_{32}a_{46}a_{55}a_{61}a_{74}a_{87}.$$

Minors and cofactors

In the theory of determinants certain subdeterminants called minors prove to be a useful tool. Let $A = (a_{ij})$ be an $n \times n$ matrix. The (i, j) *minor* M_{ij} of A is defined to be the determinant of the submatrix of A that remains when row i and column j of A are deleted.

The (i, j) *cofactor* A_{ij} of A is simply the minor with an appropriate sign:

$$A_{ij} = (-1)^{i+j} M_{ij}.$$

For example, if

$$A = \begin{pmatrix} a_{11} & a_{12} & a_{13} \\ a_{21} & a_{22} & a_{23} \\ a_{31} & a_{32} & a_{33} \end{pmatrix},$$

then

$$M_{23} = \begin{vmatrix} a_{11} & a_{12} \\ a_{31} & a_{32} \end{vmatrix} = a_{11}a_{32} - a_{12}a_{31}$$

and

$$A_{23} = (-1)^{2+3}M_{23} = a_{12}a_{31} - a_{11}a_{32}.$$

One reason for the introduction of cofactors is that they provide us with methods of calculating determinants called *row expansion* and *column expansion*. These are a great improvement on the defining sum as a means of computing determinants. The next result tells us how they operate.

Theorem 3.1.4

Let $A = (a_{ij})$ be an $n \times n$ matrix. Then

(i) $\det(A) = \sum_{k=1}^{n} a_{ik}A_{ik}$, *(expansion by row i);*

(ii) $\det(A) = \sum_{k=1}^{n} a_{kj}A_{kj}$, *(expansion by column j).*

Thus to expand by row i, we multiply each element in row i by its cofactor and add up the resulting products.

Proof of Theorem 3.1.4

We shall give the proof of (i); the proof of (ii) is similar. It is sufficient to show that the coefficient of a_{ik} in the defining expansion of $\det(A)$ equals A_{ik}. Consider first the simplest case, where $i = 1 = k$. The terms in the defining expansion of $\det(A)$ that involve a_{11} are those that appear in the sum

$$\sum \text{sign}(1, i_2, \ldots, i_n)a_{11}a_{2i_2} \cdots a_{ni_n}.$$

Here the sum is taken over all permutations of $1, 2, \ldots, n$ which have the form $1, i_2, i_3, \ldots, i_n$. This sum is clearly the same as

$$a_{11}\left(\sum \text{sign}(i_2, i_3, \ldots, i_n)a_{2i_2}a_{3i_3} \cdots a_{ni_n}\right)$$

where the summation is now over all permutations $i_2, \ldots, i_n$ of the integers $2, \ldots, n$. But the coefficient of a_{11} in this last expression is just $M_{11} = A_{11}$. Hence the coefficient of a_{11} is the same on both sides of the equation in (i).

We can deduce the corresponding statement for general i and k by means of the following device. The idea is to move a_{ik} to the $(1, 1)$ position of the matrix in such a way that it will still have the same minor M_{ik}. To do this we interchange row i of A successively with rows $i-1, i-2, \ldots, 1$, after which a_{ik} will be in the $(1, k)$ position. Then we interchange column k with the columns $k-1, k-2, \ldots, 1$ successively, until a_{ik} is in the $(1, 1)$ position. If we keep track of the determinants that arise during this process, we find that in the final determinant the minor of a_{ik} is still M_{ik}. So by the result of the first paragraph, the coefficient of a_{ik} in the new determinant is M_{ik}.

However each row and column interchange changes the sign of the determinant. For the effect of such an interchange is to switch two entries in every permutation, and, as was pointed out during the proof of 3.1.3, this changes a permutation from even to odd, or from odd to even. Thus the sign of each permutation is changed by -1. The total number of interchanges that have been applied is $(i - 1) + (k - 1) = i + k - 2$. The sign of the determinant is therefore changed by $(-1)^{i+k-2} = (-1)^{i+k}$. It follows that the coefficient of a_{ik} in $\det(A)$ is $(-1)^{i+k} M_{ik}$, which is just the definition of A_{ik}.

(It is a good idea for the reader to write out explicitly the row and column interchanges in the case $n = 3$ and $i = 2$, $k = 3$, and to verify the statement about the minor M_{23}).

The theorem provides a practical method of computing 3×3 determinants; for determinants of larger size there are more efficient methods, as we shall see.

Example 3.1.5

Compute the determinant

$$\begin{vmatrix} 1 & 2 & 0 \\ 4 & 2 & -1 \\ 6 & 2 & 2 \end{vmatrix}.$$

For example, we may expand by row 1, obtaining

$$1(-1)^2 \begin{vmatrix} 2 & -1 \\ 2 & 2 \end{vmatrix} + 2(-1)^3 \begin{vmatrix} 4 & -1 \\ 6 & 2 \end{vmatrix} + 0(-1)^4 \begin{vmatrix} 4 & 2 \\ 6 & 2 \end{vmatrix}$$

$$= 6 - 28 + 0 = -22.$$

Alternatively, we could expand by column 2:

$$2(-1)^3 \begin{vmatrix} 4 & -1 \\ 6 & 2 \end{vmatrix} + 2(-1)^4 \begin{vmatrix} 1 & 0 \\ 6 & 2 \end{vmatrix} + 2(-1)^5 \begin{vmatrix} 1 & 0 \\ 4 & -1 \end{vmatrix}$$

$$= -28 + 4 + 2 = -22.$$

However there is an obvious advantage in expanding by a row or column which contains as many zeros as possible.

The determinant of a triangular matrix can be written down at once, an observation which is used frequently in calculating determinants.

Theorem 3.1.5

The determinant of an upper or lower triangular matrix equals the product of the entries on the principal diagonal of the matrix.

Proof

Suppose that $A = (a_{ij})_{n,n}$ is, say, upper triangular, and expand $\det(A)$ by column 1. The result is the product of a_{11} and an $(n-1) \times (n-1)$ determinant which is also upper triangular. Repeat the operation until a 1×1 determinant is obtained (or use mathematical induction).

Exercises 3.1

1. Is the permutation 1, 3, 8, 5, 2, 6, 4, 7 even or odd? What is the corresponding term in the expansion of $\det((a_{ij})_{8,8})$?

2. The same questions for the permutation 8, 5, 3, 2, 1, 7, 6, 9, 4.

3. Use the definition of a determinant to compute

$$\begin{vmatrix} 1 & -3 & 0 \\ 2 & 1 & 4 \\ -1 & 0 & 1 \end{vmatrix}.$$

4. How many additions, subtractions and multiplications are needed to compute an $n \times n$ determinant by using the definition?

5. For the matrix

$$A = \begin{pmatrix} 2 & 3 & 5 \\ 4 & 3 & 1 \\ -1 & 2 & -6 \end{pmatrix}$$

find the minors M_{13}, M_{23} and M_{33}, and the corresponding cofactors A_{13}, A_{23} and A_{33}.

6. Use the cofactors found in Exercise 5 to compute the determinant of the matrix in that problem.

7. Use row or column expansion to compute the following determinants:

(a) $\begin{vmatrix} -2 & 2 & 3 \\ 2 & 2 & 1 \\ 0 & 0 & 5 \end{vmatrix}$, (b) $\begin{vmatrix} 1 & -2 & 3 & 4 \\ 2 & 0 & 1 & 3 \\ 1 & 0 & -1 & 2 \\ 0 & 1 & 1 & 3 \end{vmatrix}$,

(c) $\begin{vmatrix} 1 & -2 & 3 & 4 \\ 0 & 4 & 2 & 1 \\ 0 & 0 & -3 & 1 \\ 0 & 0 & 0 & 3 \end{vmatrix}.$

8. If A is the $n \times n$ matrix

$$\begin{pmatrix} 0 & 0 & \cdots & 0 & a_1 \\ 0 & 0 & \cdots & a_2 & 0 \\ \cdot & \cdot & \cdots & \cdot & \cdot \\ a_n & 0 & \cdots & 0 & 0 \end{pmatrix},$$

show that $\det(A) = (-1)^{n(n-1)/2} a_1 a_2 \cdots a_n$.

9. Write down the permutation matrix that represents the permutation 3, 1, 4, 5, 2.

10. Let $i_1, \ldots, i_n$ be a permutation of $1, \ldots, n$, and let P be the corresponding permutation matrix. Show that for any $n \times n$ matrix A the matrix AP is obtained from A by rearranging the columns according to the scheme $C_j \to C_{i_j}$.

11. Prove that the sign of a permutation equals the determinant of the corresponding permutation matrix.

12. Prove that every permutation matrix is expressible as a product of elementary matrices of the type that represent row or column interchanges.

13. If P is any permutation matrix, show that $P^{-1} = P^T$. [Hint: apply Exercise 10].

3.2 Basic Properties of Determinants

We now proceed to develop the theory of determinants, establishing a number of properties which will allow us to compute determinants more efficiently.

Theorem 3.2.1
If A is an $n \times n$ matrix, then

$$\det(A^T) = \det(A).$$

Proof

The proof is by mathematical induction. The statement is certainly true if $n = 1$ since then $A^T = A$. Let $n > 1$ and assume that the theorem is true for all matrices with $n - 1$ rows and columns. Expansion by row 1 gives

$$\det(A) = \sum_{j=1}^{n} a_{1j} A_{1j}.$$

Let B denote the matrix A^T. Then $a_{ij} = b_{ji}$. By induction on n, the determinant A_{1j} equals its transpose. But this is just the $(j, 1)$ cofactor B_{j1} of B. Hence $A_{1j} = B_{j1}$ and the above equation becomes

$$\det(A) = \sum_{j=1}^{n} b_{j1} B_{j1}.$$

However the right hand side of this equation is simply the expansion of $\det(B)$ by column 1; thus $\det(A) = \det(B)$.

A useful feature of this result is that it sometimes enables us to deduce that a property known to hold for the rows of a determinant also holds for the columns.

Theorem 3.2.2

A determinant with two equal rows (or two equal columns) is zero.

Proof

Suppose that the $n \times n$ matrix A has its jth and kth rows equal. We have to show that $\det(A) = 0$. Let $i_1, i_2, \ldots, i_n$ be a permutation of $1, 2, \ldots, n$; the corresponding term in the expansion of $\det(A)$ is $\text{sign}(i_1, i_2, \ldots, i_n) a_{1i_1} a_{2i_2} \cdots a_{ni_n}$. Now if we switch i_j and i_k in this product, the sign of the permutation is changed, but the product of the a's remains the same since $a_{ji_k} = a_{ki_k}$ and $a_{ki_j} = a_{ji_j}$. This means that the term under consideration occurs a second time in the defining sum for $\det(A)$, *but with the opposite sign*. Therefore all terms in the sum cancel and $\det(A)$ equals zero.

Notice that we do not need to prove the statement for columns because of the remark following 3.2.1.

The next three results describe the effect on a determinant of applying a row or column operation to the associated matrix.

Theorem 3.2.3

(i) *If a single row (or column) of a matrix A is multiplied by a scalar c, the resulting matrix has determinant equal to $c(\det(A))$.*

(ii) *If two rows (or columns) of a matrix A are interchanged, the effect is to change the sign of the determinant.*

(iii) *The determinant of a matrix A is not changed if a multiple of one row (or column) is added to another row (or column).*

Proof

(i) The effect of the operation is to multiply every term in the sum defining $\det(A)$ by c. Therefore the determinant is multiplied by c.

(ii) Here the effect of the operation is to switch two entries in each permutation of $1, 2, \ldots, n$; we have already seen that this changes the sign of a permutation, so it multiplies the determinant by -1.

(iii) Suppose that we add c times row j to row k of the matrix: here we shall assume that $j < k$. If C is the resulting matrix, then $\det(C)$ equals

$$\sum \operatorname{sign}(i_1, \ldots, i_n) a_{1i_1} \cdots a_{ji_j} \cdots (a_{ki_k} + ca_{ji_k}) \cdots a_{ni_n},$$

which is turn equals the sum of

$$\sum \text{sign}(i_1, \ldots, i_n) a_{1i_1} \cdots a_{ji_j} \cdots a_{ki_k} \cdots a_{ni_n}$$

and

$$c \sum \text{sign}(i_1, \ldots, i_n) a_{1i_1} \cdots a_{ji_j} \cdots a_{ji_k} \cdots a_{ni_n}.$$

Now the first of these sums is simply $\det(A)$, while the second sum is the determinant of a matrix in which rows j and k are identical, so it is zero by 3.2.2. Hence $\det(C) = \det(A)$.

Now let us see how use of these properties can lighten the task of evaluating a determinant. Let A be an $n \times n$ matrix whose determinant is to be computed. Then elementary row operations can be used as in Gaussian elimination to reduce A to row echelon form B. But B is an upper triangular matrix, say

$$B = \begin{pmatrix} b_{11} & b_{12} & \cdots & b_{1n} \\ 0 & b_{22} & \cdots & b_{2n} \\ . & . & \cdots & . \\ 0 & 0 & \cdots & b_{nn} \end{pmatrix},$$

so by 3.1.5 we obtain $\det(B) = b_{11}b_{22}\cdots b_{nn}$. Thus all that has to be done is to keep track, using 3.2.3, of the changes in $\det(A)$ produced by the row operations.

Example 3.2.1

Compute the determinant

$$D = \begin{vmatrix} 0 & 1 & 2 & 3 \\ 1 & 1 & 1 & 1 \\ -2 & -2 & 3 & 3 \\ 1 & -2 & -2 & -3 \end{vmatrix}.$$

Apply row operations $R_1 \leftrightarrow R_2$ and then $R_3 + 2R_1$, $R_4 - R_1$ successively to D to get:

$$D = - \begin{vmatrix} 1 & 1 & 1 & 1 \\ 0 & 1 & 2 & 3 \\ -2 & -2 & 3 & 3 \\ 1 & -2 & -2 & -3 \end{vmatrix} = - \begin{vmatrix} 1 & 1 & 1 & 1 \\ 0 & 1 & 2 & 3 \\ 0 & 0 & 5 & 5 \\ 0 & -3 & -3 & -4 \end{vmatrix}.$$

Next apply successively $R_4 + 3R_2$ and $1/5R_3$ to get

$$D = \begin{vmatrix} 1 & 1 & 1 & 1 \\ 0 & 1 & 2 & 3 \\ 0 & 0 & 5 & 5 \\ 0 & 0 & 3 & 5 \end{vmatrix} = -5 \begin{vmatrix} 1 & 1 & 1 & 1 \\ 0 & 1 & 2 & 3 \\ 0 & 0 & 1 & 1 \\ 0 & 0 & 3 & 5 \end{vmatrix}.$$

Finally, use of $R_4 - 3R_3$ yields

$$D = -5 \begin{vmatrix} 1 & 1 & 1 & 1 \\ 0 & 1 & 2 & 3 \\ 0 & 0 & 1 & 1 \\ 0 & 0 & 0 & 2 \end{vmatrix} = -10.$$

Example 3.2.2

Use row operations to show that the following determinant is identically equal to zero.

$$\begin{vmatrix} a+2 & b+2 & c+2 \\ x+1 & y+1 & z+1 \\ 2x-a & 2y-b & 2z-c \end{vmatrix}.$$

Apply row operations $R_3 + R_1$ and $2R_2$. The resulting determinant is zero since rows 2 and 3 are identical.

Example 3.2.3

Prove that the value of the $n \times n$ determinant

$$\begin{vmatrix} 2 & 1 & 0 & \cdots & 0 & 0 & 0 \\ 1 & 2 & 1 & \cdots & 0 & 0 & 0 \\ \cdot & \cdot & \cdot & \cdots & \cdot & \cdot & \cdot \\ 0 & 0 & 0 & \cdots & 1 & 2 & 1 \\ 0 & 0 & 0 & \cdots & 0 & 1 & 2 \end{vmatrix}$$

is $n + 1$.

First note the obvious equalities $D_1 = 2$ and $D_2 = 3$. Let $n \geq 3$; then, expanding by row 1, we obtain

$$
D_n = 2D_{n-1} - \begin{vmatrix}
1 & 1 & 0 & 0 & \cdots & 0 & 0 & 0 \\
0 & 2 & 1 & 0 & \cdots & 0 & 0 & 0 \\
0 & 1 & 2 & 1 & \cdots & 0 & 0 & 0 \\
\cdot & \cdot & \cdot & \cdot & \cdots & \cdot & \cdot & \cdot \\
0 & 0 & 0 & 0 & \cdots & 1 & 2 & 1 \\
0 & 0 & 0 & 0 & \cdots & 0 & 1 & 2
\end{vmatrix}.
$$

Expanding the determinant on the right by column 1, we find it to be D_{n-2}. Thus

$$
D_n = 2D_{n-1} - D_{n-2}.
$$

This is a recurrence relation which can be used to solve for successive values of D_n. Thus $D_3 = 4$, $D_4 = 5$, $D_5 = 6$, etc. In general $D_n = n + 1$. (A systematic method for solving recurrence relations of this sort will be given in 8.2.)

The next example is concerned with an important type of determinant called a *Vandermonde determinant*; these determinants occur frequently in applications.

Example 3.2.4

Establish the identity

$$
\begin{vmatrix}
1 & 1 & \cdots & 1 \\
x_1 & x_2 & \cdots & x_n \\
x_1^2 & x_2^2 & \cdots & x_n^2 \\
\cdot & \cdot & \cdots & \cdot \\
x_1^{n-1} & x_2^{n-1} & \cdots & x_n^{n-1}
\end{vmatrix}
= \prod_{i,j} (x_i - x_j),
$$

where the expression on the right is the product of all the factors $x_i - x_j$ with $i < j$ and $i, j = 1, 2, \ldots, n$.

Let D be the value of the determinant. Clearly it is a polynomial in $x_1, x_2, \ldots, x_n$. If we apply the column operation $C_i - C_j$, with $i < j$, to the determinant, its value is unchanged. On the other hand, after this operation each entry in column i will be divisible by $x_i - x_j$. Hence D is divisible by $x_i - x_j$ for all $i, j = 1, 2, \ldots, n$ and $i < j$. Thus we have located a total of $n(n-1)/2$ distinct linear polynomials which are factors of D, this being the number of pairs of distinct positive integers i, j such that $1 \le i < j \le n$. But the degree of the polynomial D is equal to

$$1 + 2 + \cdots + (n-1) = \frac{n(n-1)}{2} :$$

for each term in the defining sum has this degree. Hence D must be the product of these $n(n-1)/2$ factors and a constant c, there being no room for further factors. Thus

$$D = c \prod (x_i - x_j),$$

with $i < j = 1, 2, \ldots, n$. In fact c is equal to 1, as can be seen by looking at the coefficient of the term $1 x_2 x_3^2 \cdots x_n^{n-1}$ in the defining sum for the determinant D; this corresponds to the permutation $1, 2, \ldots, n$, and so its coefficient is $+1$. On the other hand, in the product of the $x_i - x_j$ the coefficient of the term is 1. Hence $c = 1$.

The critical property of the Vandermonde determinant D is that $D = 0$ *if and only if at least two of* $x_1, x_2, \ldots, x_n$ *are equal.*

Exercises 3.2

1. By using elementary row operations compute the following determinants:

(a) $\begin{vmatrix} 1 & 4 & 2 \\ -2 & 4 & 7 \\ 6 & 1 & 2 \end{vmatrix}$, (b) $\begin{vmatrix} 3 & 1 & -2 \\ 0 & 4 & 4 \\ 2 & -3 & 6 \end{vmatrix}$, (c) $\begin{vmatrix} 1 & 0 & 3 & 2 \\ 3 & 4 & -1 & 2 \\ 0 & 3 & 1 & 2 \\ 1 & 5 & 2 & 3 \end{vmatrix}$.

2. If one row (or column) of a determinant is a scalar multiple of another row (or column), show that the determinant is zero.

3. If A is an $n \times n$ matrix and c is a scalar, prove that $\det(cA) = c^n \det(A)$.

4. Use row operations to show that the determinant

$$\begin{vmatrix} a^2 & b^2 & c^2 \\ 1+a & 1+b & 1+c \\ 2a^2 - a - 1 & 2b^2 - b - 1 & 2c^2 - c - 1 \end{vmatrix}$$

is identically equal to zero.

5. Let A be an $n \times n$ matrix in row echelon form. Show that $\det(A)$ equals zero if and only if the number of pivots is less than n.

6. Use row and column operations to show that

$$\begin{vmatrix} a & b & c \\ b & c & a \\ c & a & b \end{vmatrix} = (a+b+c)(-a^2 - b^2 - c^2 + ab + bc + ca).$$

7. *Without expanding* the determinant, prove that

$$\begin{vmatrix} 1 & 1 & 1 \\ x & y & z \\ x^3 & y^3 & z^3 \end{vmatrix} = (x-y)(y-z)(z-x)(x+y+z).$$

[Hint: show that the determinant has factors $x - y$, $y - z$, $z - x$, and that the remaining factor must be of degree 1 and symmetric in x, y, z].

8. Let D_n denote the "bordered" $n \times n$ determinant

$$\begin{vmatrix} 0 & a & 0 & \cdots & 0 & 0 \\ b & 0 & a & \cdots & 0 & 0 \\ 0 & b & 0 & \cdots & 0 & 0 \\ . & . & . & \cdots & . & . \\ 0 & 0 & 0 & \cdots & 0 & a \\ 0 & 0 & 0 & \cdots & b & 0 \end{vmatrix}.$$

Prove that $D_{2n-1} = 0$ and $D_{2n} = (-ab)^n$.

9. Let D_n be the $n \times n$ determinant whose (i, j) entry is $i + j$. Show that $D_n = 0$ if $n > 2$. [Hint: use row operations].

10. Let u_n denote the number of additions, subtractions and multiplications needed in general to evaluate an $n \times n$ determinant by row expansion. Prove that $u_n = nu_{n-1} + 2n - 1$. Use this formula to calculate u_n for $n = 2, 3, 4$.

3.3 Determinants and Inverses of Matrices

An important property of the determinant of a square matrix is that it tells us whether the matrix is invertible.

Theorem 3.3.1
An $n \times n$ matrix A is invertible if and only if $\det(A) \neq 0$.

Proof
By 2.3.2 there are elementary matrices $E_1, E_2, \ldots, E_k$ such that the matrix $R = E_k E_{k-1} \cdots E_2 E_1 A$ is in reduced row echelon form. Now observe that if E is any elementary $n \times n$ matrix, then $\det(EA) = c \det(A)$ for some non-zero scalar c; this is because left multiplication by E performs an elementary row operation on A and we know from 3.2.3 that such an operation will, at worst, multiply the value of the determinant by a non-zero scalar. Applying this fact repeatedly, we obtain $\det(R) = \det(E_k \cdots E_2 E_1 A) = d \det(A)$ for some non-zero scalar d. Consequently $\det(A) \neq 0$ if and only if $\det(R) \neq 0$.

Now we saw in 2.3.5 that A is invertible precisely when $R = I_n$. But, remembering the form of the matrix R, we recognise that the only way that $\det(R)$ can be non-zero is if $R = I_n$. Hence the result follows.

Example 3.3.1
The Vandermonde matrix of Example 3.2.4 is invertible if and only if $x_1, \ldots, x_n$ are all different.

Corollary 3.3.2
A linear system $AX = 0$ with n equations in n unknowns has a non-trivial solution if and only if $\det(A) = 0$.

This very useful result follows directly from 2.3.5 and 3.3.1. Theorem 3.3.1 can be used to establish a basic formula for the determinant of the product of two matrices.

Theorem 3.3.3
If A and B are any $n \times n$ matrices, then

$$\det(AB) = \det(A)\det(B).$$

Proof
Consider first the case where B is *not* invertible, which by 3.3.1 means that $\det(B) = 0$. According to 2.3.5 there is a non-zero vector X such that $BX = 0$. This clearly implies that $(AB)X = 0$, and so, by 2.3.5 and 3.3.1, $\det(AB)$ must also be zero. Thus the formula certainly holds in this case.

Suppose now that B is invertible. Then B is a product of elementary matrices, say $B = E_1 E_2 \cdots E_k$; this is by 2.3.5. Now the effect of right multiplication of A by an elementary matrix E is to apply an elementary column operation to A. What is more, we can tell from 3.2.3 just what the value of $\det(AE)$ is; indeed

$$\det(AE) = \begin{cases} -\det(A) \\ \det(A) \\ c\det(A) \end{cases},$$

according to whether E represents a column operation of the types

$$\begin{cases} C_i \leftrightarrow C_j \\ C_i + cC_j \\ cC_i \end{cases}$$

Now we can see from the form of the elementary matrix E that $\det(E)$ equals -1, 1 or c, respectively, in the three cases; hence the formula $\det(AE) = \det(A)\det(E)$ is valid. In short our formula is true when B is an elementary matrix. Applying this fact repeatedly, we find that $\det(AB)$ equals

$$\det(AE_k \cdots E_2 E_1) = \det(A)\det(E_k) \cdots \det(E_2)\det(E_1),$$

which shows that

$$\det(AB) = \det(A)\det(E_k \cdots E_1) = \det(A)\det(B).$$

Corollary 3.3.4
Let A and B be $n \times n$ matrices. If $AB = I_n$, then $BA = I_n$, and thus $B = A^{-1}$.

Proof
For $1 = \det(AB) = \det(A)\det(B)$, so $\det(A) \neq 0$ and A is invertible, by 3.3.1. Therefore $BA = A^{-1}(AB)A = A^{-1}I_n A = I_n$.

Corollary 3.3.5
If A is an invertible matrix, then $\det(A^{-1}) = 1/\det(A)$.

Proof
Clearly $1 = \det(I) = \det(AA^{-1}) = \det(A)\det(A^{-1})$, from which the statement follows.

The adjoint matrix
    Let $A = (a_{ij})$ be an $n \times n$ matrix. Then the *adjoint matrix*

$$\mathrm{adj}\,(A)$$

of A is defined to be the $n \times n$ matrix whose (i, j) element is the (j, i) cofactor A_{ji} of A. Thus $\mathrm{adj}(A)$ *is the transposed matrix of cofactors of A.* For example, the adjoint of the matrix

$$\begin{pmatrix} 1 & 2 & 1 \\ 6 & -1 & 3 \\ 2 & -3 & 4 \end{pmatrix}$$

is

$$\begin{pmatrix} 5 & -11 & 7 \\ -18 & 2 & 3 \\ -16 & 7 & -13 \end{pmatrix}.$$

The significance of the adjoint matrix is made clear by the next two results.

Theorem 3.3.6
If A is any $n \times n$ matrix, then

$$A \operatorname{adj}(A) = (\det(A))I_n = \operatorname{adj}(A)A.$$

Proof
The (i, j) entry of the matrix product $A \operatorname{adj}(A)$ is

$$\sum_{k=1}^{n} a_{ik}(\operatorname{adj}(A))_{kj} = \sum_{k=1}^{n} a_{ik} A_{jk}.$$

If $i = j$, this is just the expansion of $\det(A)$ by row i; on the other hand, if $i \neq j$, the sum is also a row expansion of a determinant, but one in which rows i and j are identical. By 3.2.2 the sum will vanish in this case. This means that the off-diagonal entries of the matrix product $A \operatorname{adj}(A)$ are zero, while the entries on the diagonal all equal $\det(A)$. Therefore $A \operatorname{adj}(A)$ is the scalar matrix $(\det(A))I_n$, as claimed. The second statement can be proved in a similar fashion.

Theorem 3.3.5 leads to an attractive formula for the inverse of an invertible matrix.

Theorem 3.3.7
If A is an invertible matrix, then $A^{-1} = (1/\det(A))\operatorname{adj}(A)$.

Proof

In the first place, remember that A^{-1} exists if and only if $\det(A) \neq 0$, by 3.3.1. From $A \operatorname{adj}(A) = (\det(A))I_n$ we obtain

$$A(1/\det(A))\operatorname{adj}(A)) = 1/\det(A)(A \operatorname{adj}(A)) = I_n,$$

by 3.3.6. The result follows in view of 3.3.4.

Example 3.3.2

Let A be the matrix

$$\begin{pmatrix} 2 & -1 & 0 \\ -1 & 2 & -1 \\ 0 & -1 & 2 \end{pmatrix}.$$

The adjoint of A is

$$\begin{pmatrix} 3 & 2 & 1 \\ 2 & 4 & 2 \\ 1 & 2 & 3 \end{pmatrix}.$$

Expanding $\det(A)$ by row 1, we find that it equals 4. Thus

$$A^{-1} = \begin{pmatrix} 3/4 & 1/2 & 1/4 \\ 1/2 & 1 & 1/2 \\ 1/4 & 1/2 & 3/4 \end{pmatrix}.$$

Despite the neat formula provided by 3.3.7, for matrices with four or more rows it is usually faster to use elementary row operations to compute the inverse, as described in 2.3: for to find the adjoint of an $n \times n$ matrix one must compute n determinants each with $n - 1$ rows and columns.

Next we give an application of determinants to geometry.

Example 3.3.3

Let $P_1(x_1, y_1, z_1)$, $P_2(x_2, y_2, z_2)$ and $P_3(x_3, y_3, z_3)$ be three non-collinear points in three dimensional space. The points therefore determine a unique plane. Find the equation of the plane by using determinants.

We know from analytical geometry that the equation of the plane must be of the form $ax + by + cz + d = 0$. Here the constants a, b, c, d cannot all be zero. Let $P(x, y, z)$ be an arbitrary point in the plane. Then the coordinates of the points P, P_1, P_2, P_3 must satisfy the equation of the plane. Therefore the following equations hold:

$$\begin{cases} ax & + & by & + & cz & + & d & = 0 \\ ax_1 & + & by_1 & + & cz_1 & + & d & = 0 \\ ax_2 & + & bx_2 & + & cz_2 & + & d & = 0 \\ ax_3 & + & by_3 & + & cz_3 & + & d & = 0 \end{cases}$$

Now this is a homogeneous linear system in the unknowns a, b, c, d; by 3.3.2 the condition for there to be a non-trivial solution is that

$$\begin{vmatrix} x & y & z & 1 \\ x_1 & y_1 & z_1 & 1 \\ x_2 & y_2 & z_2 & 1 \\ x_3 & y_3 & z_3 & 1 \end{vmatrix}.$$

This is the condition for the point P to lie in the plane, so it is the equation of the plane. That it is of the form $ax + by + cz + d = 0$ may be seen by expanding the determinant by row 1.

For example, the equation of the plane which is determined by the three points $(0, 1, 1)$, $(1, 0, 1)$ and $(1, 1, 0)$ is

$$\begin{vmatrix} x & y & z & 1 \\ 0 & 1 & 1 & 1 \\ 1 & 0 & 1 & 1 \\ 1 & 1 & 0 & 1 \end{vmatrix},$$

which becomes on expansion $x + y + z - 2 = 0$.

Cramer's Rule

For a second illustration of the uses of determinants, we return to the study of linear systems. Consider a linear system of n equations in n unknowns $x_1, x_2, \ldots, x_n$

$$AX = B,$$

where the coefficient matrix A has non-zero determinant. The system has a unique solution, namely $X = A^{-1}B$. There is a simple expression for this solution in terms of determinants.

Using 3.3.7 we obtain

$$X = A^{-1}B = 1/\det(A) \; (\text{adj(A) } B).$$

From the matrix product $\text{adj}(A)B$ we can read off the ith unknown as

$$x_i = (\sum_{j=1}^{n} (\text{adj}(A))_{ij} b_j)/\det(A) = (\sum_{j=1}^{n} b_j A_{ji})/\det(A).$$

Now the second sum is a determinant; in fact it is $\det(M_i)$ where M_i is the matrix obtained from A when column i is replaced by B. Hence the solution of the linear system can be expressed in the form $x_i = \det(M_i)/\det(A)$, $i = 1, 2, \ldots, n$. Thus we have obtained the following result.

Theorem 3.3.8 (*Cramer's Rule*)
If $AX = B$ is a linear system of n equations in n unknowns and $\det(A)$ is not zero, then the unique solution of the linear system can be written in the form

$$x_i = \det(M_i)/\det(A), \; i = 1, \ldots, n,$$

where M_i is the matrix obtained from A when column i is replaced by B.

The reader should note that Cramer's Rule can only be used when the linear system has the special form indicated.

Example 3.3.4

Solve the following linear system using Cramer's Rule.

$$\begin{cases} x_1 & - & x_2 & - & x_3 & = 4 \\ x_1 & + & 2x_2 & - & x_3 & = 2 \\ 2x_1 & & & & & = 1 \end{cases}$$

Here

$$A = \begin{pmatrix} 1 & -1 & -1 \\ 1 & 2 & -1 \\ 2 & 0 & 1 \end{pmatrix} \text{ and } B = \begin{pmatrix} 4 \\ 2 \\ 1 \end{pmatrix}.$$

Thus $\det(A) = 9$, and Cramer's Rule gives the solution

$$x_1 = 1/9 \begin{vmatrix} 4 & -1 & -1 \\ 2 & 2 & -1 \\ 1 & 0 & 1 \end{vmatrix} = 13/9,$$

$$x_2 = 1/9 \begin{vmatrix} 1 & 4 & -1 \\ 1 & 2 & -1 \\ 2 & 1 & 1 \end{vmatrix} = -2/3,$$

$$x_3 = 1/9 \begin{vmatrix} 1 & -1 & 4 \\ 1 & 2 & 2 \\ 2 & 0 & 1 \end{vmatrix} = -17/9.$$

Exercises 3.3

1. For the matrices

$$A = \begin{pmatrix} 6 & 3 \\ 4 & -3 \end{pmatrix} \text{ and } B = \begin{pmatrix} 2 & 5 \\ 4 & 7 \end{pmatrix}$$

verify the identity $\det(AB) = \det(A)\det(B)$.

2. By finding the relevant adjoints, compute the inverses of the following matrices:

(a) $\begin{pmatrix} 4 & -1 \\ -2 & 3 \end{pmatrix}$, (b) $\begin{pmatrix} 2 & 3 & 1 \\ 2 & 1 & 3 \\ -1 & 4 & 6 \end{pmatrix}$, (c) $\begin{pmatrix} 1 & 1 & 1 & 1 \\ 0 & 1 & 1 & 1 \\ 0 & 0 & 1 & 1 \\ 0 & 0 & 0 & 1 \end{pmatrix}$.

3. If A is a square matrix and n is a positive integer, prove that $\det(A^n) = (\det(A))^n$.

4. Use Cramer's Rule to solve the following linear systems:

(a) $\begin{cases} 2x_1 & - & 3x_2 & + & x_3 & = -1 \\ x_1 & + & 3x_2 & + & x_3 & = 6 \\ 2x_1 & + & x_2 & + & x_3 & = 11 \end{cases}$

(b) $\begin{cases} x_1 & + & x_2 & + & x_3 & = -1 \\ 2x_1 & - & x_2 & - & x_3 & = 4 \\ x_1 & + & 2x_2 & - & 3x_3 & = 7 \end{cases}$

5. Let A be an $n \times n$ matrix. Prove that A is invertible if and only if $\mathrm{adj}(A)$ is invertible.

6. Let A be any $n \times n$ matrix where $n > 1$. Prove that $\det(\mathrm{adj}(A)) = (\det(A))^{n-1}$. [Hint: first deal with the case where $\det(A) \neq 0$, by applying det to each side of the identity of 3.3.6. Then argue that the result must still be true when $\det(A) = 0$].

7. Find the equation of the plane which contains the points $(1, 1, -2), (1, -2, 7)$ and $(0, 1, -4)$.

8. Consider the four points in three dimensional space $P_i(x_i, y_i, z_i)$, $i = 1, 2, 3, 4$. Prove that a necessary and sufficient condition for the four points to lie in a plane is

$$\begin{vmatrix} x_1 & y_1 & z_1 & 1 \\ x_2 & y_2 & z_2 & 1 \\ x_3 & y_3 & z_3 & 1 \\ x_4 & y_4 & z_4 & 1 \end{vmatrix} = 0.$$

Chapter Four

INTRODUCTION TO VECTOR SPACES

The aim of this chapter is to introduce the reader to the notion of an abstract vector space. Roughly speaking, a vector space is a set of objects called vectors which it is possible to add and multiply by scalars, subject to reasonable rules. Vector spaces occur in numerous branches of mathematics, as well as in many applications; they are therefore of great importance and utility. Rather than immediately confront the reader with an abstract definition, we prefer first to discuss some vector spaces which are familiar objects. Then we proceed to extract the common features of these examples, and use them to frame the definition of a general vector space.

4.1 Examples of Vector Spaces

The first example of a vector space has a geometrical background.

Euclidean space

Choose and fix a positive integer n, and define

$$\mathbf{R}^n$$

to be the set of all n-column vectors

$$X = \begin{pmatrix} x_1 \\ x_2 \\ \vdots \\ x_n \end{pmatrix},$$

where the entries x_i are real numbers. Of course these are special types of matrices, so rules of addition and scalar multiplication are at hand, namely

$$\begin{pmatrix} x_1 \\ x_2 \\ \vdots \\ x_n \end{pmatrix} + \begin{pmatrix} y_1 \\ y_2 \\ \vdots \\ y_n \end{pmatrix} = \begin{pmatrix} x_1 + y_1 \\ x_2 + y_2 \\ \vdots \\ x_n + y_n \end{pmatrix}$$

and

$$c \begin{pmatrix} x_1 \\ x_2 \\ \vdots \\ x_n \end{pmatrix} = \begin{pmatrix} cx_1 \\ cx_2 \\ \vdots \\ cx_n \end{pmatrix}.$$

Thus the set $\mathbf{R}^n$ is "closed" with respect to the operation of adding pairs of its elements, in the sense that one cannot escape from $\mathbf{R}^n$ by adding two of its elements; similarly $\mathbf{R}^n$ is closed with respect to multiplication of its elements by scalars. Notice also that $\mathbf{R}^n$ contains the zero column vector.

Another point to observe is that the rules of matrix algebra listed in 1.2.1 which are relevant to column vectors apply to the elements of $\mathbf{R}^n$. The set $\mathbf{R}^n$, together with the operations of addition and scalar multiplication, forms a vector space which is known as *n-dimensional Euclidean space*.

Line segments and $\mathbf{R}^3$

When n is 3 or less, the vector space $\mathbf{R}^n$ has a good geometrical interpretation. Consider the case of $\mathbf{R}^3$. A typical element of $\mathbf{R}^3$ is a 3-column

$$A = \begin{pmatrix} a_1 \\ a_2 \\ a_3 \end{pmatrix}.$$

Assume that a cartesian coordinate system has been chosen with assigned x, y and z -axes. We plan to represent the column vector A by a *directed line segment* in three-dimensional space. To achieve this, choose an arbitrary point **I** with co-ordinates (u_1, u_2, u_3) as the initial point of the line segment. The end point of the segment is the point **E** with coordinates $(u_1 + a_1, u_2 + a_2, u_3 + a_3)$. The direction of the line segment **IE** is indicated by an arrow:

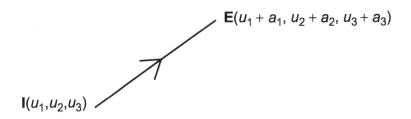

The length of **IE** equals

$$l = \sqrt{a_1^2 + a_2^2 + a_3^2}$$

and its direction is specified by the direction cosines

$$a_1/l, \ a_2/l, \ a_3/l.$$

Here the significant feature is that none of these quantities depends on the initial point **I**. Thus A is represented by in-finitely many line segments all of which have the same length and the same direction. So all the line segments which represent A are parallel and have equal length. However the zero vector is represented by a line segment of length 0 and it is not assigned a direction.

Having connected elements of $\mathbf{R}^3$ with line segments, let us see what the rule of addition in $\mathbf{R}^3$ implies about line segments. Consider two vectors in $\mathbf{R}^3$

$$A = \begin{pmatrix} a_1 \\ a_2 \\ a_3 \end{pmatrix} \text{ and } B = \begin{pmatrix} b_1 \\ b_2 \\ b_3 \end{pmatrix}$$

and their sum

$$A + B = \begin{pmatrix} a_1 + b_1 \\ a_2 + b_2 \\ a_3 + b_3 \end{pmatrix}.$$

Represent the vectors A, B and $A + B$ by line segments **IU**, **IV**, and **IW** in three dimensional space with a common initial point **I** (u_1, u_2, u_3), say. The line segments determine a figure **IUWV** as shown:

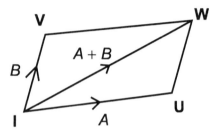

where **U**, **W** and **V** are the points

$$(u_1 + a_1,\ u_2 + a_2,\ u_3 + a_3),\ (u_1 + a_1 + b_1,\ u_2 + a_2 + b_2,\ u_3 + a_3 + b_3)$$

and

$$(u_1 + b_1,\ u_2 + b_2,\ u_3 + b_3),$$

respectively.

In fact **IUWV** is a parallelogram. To prove this, we need to find the lengths and directions of the four sides. Simple analytic geometry shows that $IU = VW = \sqrt{a_1^2 + a_2^2 + a_3^2} = l$, and that $IV = UW = \sqrt{b_1^2 + b_2^2 + b_3^2} = m$, say. Also the direction cosines of **IU** and **VW** are a_1/l, a_2/l, a_3/l, while those of **IV** and **UW** are b_1/m, b_2/m, b_3/m. It follows that opposite sides of **IUWV** are parallel and of equal length, so it is indeed a parallelogram.

These considerations show that the rule of addition for vectors in $\mathbf{R}^3$ is equivalent to the *parallelogram rule* for addition of forces, which is familiar from mechanics. To add line segments **IU** and **IV** representing the vectors A and B, complete the parallelogram formed by the lines **IU** and **IV**; the the diagonal **IW** will represent the vector $A + B$.

An equivalent formulation of this is the *triangle rule*, which is encapsulated in the diagram which follows:

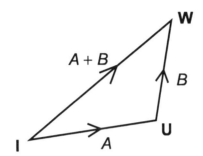

Note that this diagram is obtained from the parallelogram by deleting the upper triangle. Since **IV** and **UW** are parallel line segments of equal length, they represent the same vector B.

There is also a geometrical interpretation of the rule of scalar multiplication in $\mathbf{R}^3$. As before let A in $\mathbf{R}^3$ be represented by the line segment joining $\mathbf{I}(u_1, u_2, u_3)$ to $\mathbf{U}(u_1 + a_1, u_2 + a_2, u_3 + a_3)$. Let c be any scalar. Then cA is represented by the line segment from (u_1, u_2, u_3) to $(u_1 + ca_1, u_2 + ca_2, u_3 + ca_3)$. This line segment has length equal to $|c|$ times the length of **IU**, while its direction is the same as that of **IU** if $c > 0$, and opposite to that of **IU** if $c < 0$.

Of course, there are similar geometrical representations of vectors in $\mathbf{R}^2$ by line segments drawn in the plane, and in $\mathbf{R}^1$ by line segments drawn along a fixed line. So our first examples of vector spaces are familiar objects if $n \le 3$.

Further examples of vector spaces are obtained when the field of real numbers is replaced by the field of complex numbers $\mathbf{C}$: in this case we obtain

$$\mathbf{C}^n,$$

the vector space of all n-column vectors with entries in $\mathbf{C}$. More generally it is to carry out the same construction with an arbitrary field of scalars F, in the sense of 1.3; this yields the vector space

$$F^n$$

of all n-column vectors with entries in F, with the usual rules of matrix addition and scalar multiplication.

Vector spaces of matrices

One obvious way to extend the previous examples is by allowing matrices of arbitrary size. Let

$$M_{m,n}(\mathbf{R})$$

denote the set of all $m \times n$ matrices with real entries. This set is closed with respect to matrix addition and scalar multiplication, and it includes the zero matrix $0_{m,n}$. The rules of matrix algebra guarantee that $M_{m,n}(\mathbf{R})$ is a vector space. Of course, if $n = 1$, we recover the Euclidean space $\mathbf{R}^m$, while if $m = 1$, we obtain the vector space

$$\mathbf{R}_n$$

of all real n-row vectors. It is consistent with notation established in 1.3 if we write

$$M_n(\mathbf{R})$$

for the vector space of all real $n \times n$ matrices, instead of $M_{n,n}(\mathbf{R})$. Once again $\mathbf{R}$ can be replaced by any field of scalars F in these examples, to produce the vector spaces

$$M_{m,n}(F), \quad M_n(F) \quad \text{and} \quad F_n.$$

Vector spaces of functions

Let a and b be fixed real numbers with $a < b$, and let $C[a, b]$ denote the set of all real-valued functions of x that are continuous at each point of the closed interval $[a, b]$. If f and g are two such functions, we define their sum $f + g$ by the rule

$$f + g(x) = f(x) + g(x).$$

It is a well-known result from calculus that $f + g$ is also continuous in $[a, b]$, so that $f + g$ belongs to $C[a, b]$. Next, if c is any real number, the function cf defined by

$$cf(x) = c(f(x))$$

is continuous in $[a, b]$ and thus belongs to $C[a, b]$. The zero function, which is identically equal to zero in $[a, b]$, is also included in $C[a, b]$.

Thus once again we have a set that is closed with respect to natural operations of addition and scalar multiplication; $C[a, b]$ is the *vector space of all continuous functions on the interval* $[a, b]$. In a similar way one can form the smaller vector space $D[a, b]$ consisting of all *differentiable functions* on $[a, b]$, with the same rules of addition and scalar multiplication. A still smaller vector space is $D_\infty[a, b]$, the vector space of all functions that are infinitely differentiable in $[a, b]$

Vector spaces of polynomials

A (real) *polynomial* in an indeterminate x is an expression of the form

$$f(x) = a_0 + a_1 x + \cdots + a_n x^n$$

where the coefficients a_i are real numbers. If $a_n \neq 0$, the polynomial is said to have *degree* n. Define

$$P_n(\mathbf{R})$$

to be the set of all real polynomials in x of degree less than n.
Here we mean to include the *zero polynomial*, which has all
its coefficients equal to zero. There are natural rules of addition and scalar multiplication in $P_n(\mathbf{R})$, namely the familiar
ones of elementary algebra: to add two polynomials add corresponding coefficients; to multiply a polynomial by a scalar
c, multiply each coefficient by c. Using these operations, we
obtain the *vector space of all real polynomials of degree less
than n.*

This example could be varied by allowing polynomial of
arbitrary degree, thus yielding the *vector space of all real polynomials*

$$P(\mathbf{R}).$$

As usual $\mathbf{R}$ may be replaced by any field of scalars here.

Common features of vector spaces

The time has come to identify the common features in
the above examples: they are:

(i) a non-empty set of objects called vectors, including a
"zero" vector;

(ii) a way of adding two vectors to give another vector;

(iii) a way of multiplying a vector by a scalar to give a
vector;

(iv) a reasonable list of rules that the operations
mentioned(ii) and (iii) are required to satisfy.

We are being deliberately vague in (iv), but the rules should
correspond to properties of matrices that are known to hold
in $\mathbf{R}^n$ and $M_{m,n}(\mathbf{R})$.

Exercises 4.1

1. Give details of the geometrical interpretations of $\mathbf{R}^1$ and $\mathbf{R}^2$.

2. Which of the following might qualify as vector spaces in the sense of the examples of this section?

(a) the set of all real 3-column vectors that correspond to line segments of length 1;

(b) the set of all real polynomials of degree at least 2;

(c) the set of all line segments in $\mathbf{R}^3$ that are parallel to a given plane;

(d) the set of all continuous functions of x defined in the interval $[0, 1]$ that vanish at $x = 1/2$.

4.2 Vector Spaces and Subspaces

It is now time to give a precise formulation of the definition of a vector space.

Definition of a vector space

A *vector space V over* $\mathbf{R}$ consists of a set of objects called *vectors*, a rule for combining vectors called *addition*, and a rule for multiplying a vector by a real number to give another vector called *scalar multiplication*. If $\mathbf{u}$ and $\mathbf{v}$ are vectors, the result of adding these vectors is written $\mathbf{u} + \mathbf{v}$, the *sum* of $\mathbf{u}$ and $\mathbf{v}$; also, if c is a real number, the result of multiplying $\mathbf{v}$ by c, is written $c\mathbf{v}$, the *scalar multiple* of $\mathbf{v}$ by c.

It is understood that the following conditions must be satisfied for all vectors $\mathbf{u}$, $\mathbf{v}$, $\mathbf{w}$ and all real scalars c, d :

(i) $\mathbf{u} + \mathbf{v} = \mathbf{v} + \mathbf{u}$, (commutative law);

(ii) $(\mathbf{u} + \mathbf{v}) + \mathbf{w} = \mathbf{u} + (\mathbf{v} + \mathbf{w})$, (associative law);

(iii) there is a vector $\mathbf{0}$, called the *zero vector*, such that $\mathbf{v} + \mathbf{0} = \mathbf{v}$;

(iv) each vector $\mathbf{v}$ has a *negative*, that is, a vector $-\mathbf{v}$ such that $\mathbf{v} + (-\mathbf{v}) = \mathbf{0}$;

(v) $cd(\mathbf{v}) = c(d\mathbf{v})$;
(vi) $c(\mathbf{u} + \mathbf{v}) = c\mathbf{u} + c\mathbf{v}$: (distributive law);
(vii) $(c + d)\mathbf{v} = c\mathbf{v} + d\mathbf{v}$; (distributive law);
(viii) $1\mathbf{v} = \mathbf{v}$.

For economy of notation it is customary to use V to denote the set of vectors, as well as the vector space. Since the vector space axioms just listed hold for matrices, they are valid in $\mathbf{R}^n$; they also hold in the other examples of vector spaces described in 4.1.

More generally, we can define a *vector space over an arbitrary field of scalars F* by simply replacing $\mathbf{R}$ by F in the above axioms.

Certain simple properties of vector spaces follow easily from the axioms. Since these are used constantly, it is as well to establish them at this early stage.

Lemma 4.2.1
If $\mathbf{u}$ and $\mathbf{v}$ are vectors in a vector space, the following statements are true:
    (a) $0\mathbf{v} = \mathbf{0}$ *and* $c\,\mathbf{0} = \mathbf{0}$ *where c is a scalar;*
    (b) *if* $\mathbf{u} + \mathbf{v} = \mathbf{0}$, *then* $\mathbf{u} = -\mathbf{v}$;
    (c) $(-1)\mathbf{v} = -\mathbf{v}$.

Proof
(a) In property (vii) above put $c = 0 = d$, to get $0\mathbf{v} = 0\mathbf{v} + 0\mathbf{v}$. Add $-(0\mathbf{v})$ to both sides of this equation and use the associative law (ii) to deduce that

$$\mathbf{0} = -(0\mathbf{v}) + 0\mathbf{v} = (-(0\mathbf{v}) + 0\mathbf{v}) + 0\mathbf{v},$$

which leads to $\mathbf{0} = 0\mathbf{v}$. Proceed similarly in the second part.
(b) Add $-\mathbf{v}$ to both sides of $\mathbf{u} + \mathbf{v} = \mathbf{0}$ and use the associative law.
(c) Using (vii) and (viii), and also (a), we obtain

$$\mathbf{v} + (-1)\mathbf{v} = 1\mathbf{v} + (-1)\mathbf{v} = (1 + (-1))\mathbf{v} = 0\mathbf{v} = \mathbf{0}.$$

Hence $(-1)\mathbf{v} = -\mathbf{v}$ by (b).

Subspaces

Roughly speaking, a subspace is a vector space contained within a larger vector space; for example, the vector space $P_2(\mathbf{R})$ is a subspace of $P_3(\mathbf{R})$. More precisely, a subset S of a vector space V is called a *subspace* of V if the following statements are true:

(i) S contains the zero vector $\mathbf{0}$;
(ii) if $\mathbf{v}$ belongs to S, then so does $c\mathbf{v}$ for every scalar c, that is, S is *closed under scalar multiplication*;
(iii) if $\mathbf{u}$ and $\mathbf{v}$ belong to S, then so does $\mathbf{u} + \mathbf{v}$ that is, S is *closed under addition*.

Thus a subspace of V is a subset S which is itself a vector space with respect to the same rules of addition and scalar multiplication as V. Of course, the vector space axioms hold in S since they are already valid in V.

Examples of subspaces

If V is any vector space, then V itself is a subspace, for trivial reasons. It is often called the *improper subspace*. At the other extreme is the *zero subspace*, written 0 or 0_V, which contains only the zero vector $\mathbf{0}$. This is the smallest subspace of V. (In general a vector space that contains only the zero vector is called a *zero space*). The zero subspace and the improper subspace are present in every vector space. We move on now to some more interesting examples of subspaces.

Example 4.2.1

Let S be the subset of $\mathbf{R}^2$ consisting of all columns of the form

$$\begin{pmatrix} 2t \\ -3t \end{pmatrix}$$

where t is an arbitrary real number. Since

$$\begin{pmatrix} 2t_1 \\ -3t_1 \end{pmatrix} + \begin{pmatrix} 2t_2 \\ -3t_2 \end{pmatrix} = \begin{pmatrix} 2(t_1 + t_2) \\ -3(t_1 + t_2) \end{pmatrix}$$

and

$$c\begin{pmatrix} 2t \\ -3t \end{pmatrix} = \begin{pmatrix} 2ct \\ -3ct \end{pmatrix},$$

S is closed under addition and scalar multiplication; also S contains the zero vector $\begin{pmatrix} 0 \\ 0 \end{pmatrix}$, as may be seen by taking t to be to 0. Hence S is a subspace of $\mathbf{R}^2$.

In fact this subspace has geometrical significance. For an arbitrary vector $\begin{pmatrix} 2t \\ -3t \end{pmatrix}$ of S may be represented by a line segment in the plane with initial point the origin and end point $(2t, -3t)$. But the latter is a general point on the line with equation $3x + 2y = 0$. Therefore the subspace S corresponds to the set of line segments drawn from the origin along the line $3x + 2y = 0$.

Example 4.2.2

This example is an important one. Consider the homogeneous linear system

$$AX = 0$$

in n unknowns over some field of scalars F and let S denote the set of all solutions of the linear system, that is, all the n-column vectors X over F that satisfy $AX = 0$. Then S is a subset of F^n and it certainly contains the zero vector. Now if X and Y are solutions of the linear system and c is any scalar, then

$$A(X + Y) = AX + AY = 0 \quad \text{and} \quad A(cX) = c(AX) = 0.$$

Thus $X + Y$ and cX belong to S and it follows that S is a subspace of the vector space $\mathbf{R}^n$. This subspace is called the *solution space* of the homogeneous linear system $AX = 0$; it is also known as the *null space* of the matrix A. (Question: why is it necessary to have a *homogeneous* linear system here?)

Example 4.2.3

Let S denote the set of all real solutions $y = y(x)$ of the homogeneous linear differential equation

$$y'' + 5y' + 6y = 0$$

defined in some interval $[a, b]$. Thus S is a subset of the vector space $C[a, b]$ of continuous functions on $[a, b]$. It is easy to verify that S contains the zero function and that S is closed with respect to addition and scalar multiplication; in other words S is a subspace of $C[a, b]$.

The subspace S in this example is called *the solution space* of the differential equation. More generally, one can define the solution space of an arbitrary homogeneous linear differential equation, or even of a system of such differential equations. Systems of homogeneous linear differential equations are studied in Chapter Eight.

Linear combinations of vectors

Let $\mathbf{v}_1, \mathbf{v}_2, \ldots, \mathbf{v}_k$ be vectors in a vector space V. If c_1, $c_2, \ldots, c_k$ are any scalars, the vector

$$c_1\mathbf{v}_1 + c_2\mathbf{v}_2 + \cdots + c_k\mathbf{v}_k$$

is called a *linear combination* of $\mathbf{v}_1, \mathbf{v}_2, \ldots, \mathbf{v}_k$.

For example, consider two vectors in $\mathbf{R}^2$

$$X_1 = \begin{pmatrix} 2 \\ 1 \end{pmatrix}, \quad X_2 = \begin{pmatrix} -3 \\ 4 \end{pmatrix}.$$

The most general linear combination of X_1 and X_2 is

$$c_1 X_1 + c_2 X_2 = \begin{pmatrix} 2c_1 - 3c_2 \\ c_1 + 4c_2 \end{pmatrix}.$$

In general let X be any non-empty subset of a vector space V and denote by

$$< X >$$

the set of all linear combinations of vectors in X. Thus a typical element of $< X >$ is a vector of the form

$$c_1\mathbf{x}_1 + c_2\mathbf{x}_2 + \cdots + c_k\mathbf{x}_k$$

where $\mathbf{x}_1, \mathbf{x}_2, \ldots, \mathbf{x}_k$ are vectors belonging to X and $c_1, c_2, \ldots,$ c_k are scalars. From this formula it is clear that the sum of any two elements of $< X >$ is still in $< X >$ and that a scalar multiple of an element of $< X >$ is in $< X >$. Thus we have the following important result.

Theorem 4.2.2

If X is a non-empty subset of a vector space V, then $< X >$, the set of all linear combinations of elements of X, is a subspace of V.

We refer to $< X >$ as the subspace of V *generated* (or *spanned*) by X. A good way to think of $< X >$ is as the smallest subspace of V that contains X. For any subspace of V that contains X will necessarily contain all linear combinations of vectors in X and so must contain $< X >$ as a subset. In particular, a subset X is a subspace if and only if $X = < X >$. In the case of a finite set $X = \{\mathbf{x}_1, \mathbf{x}_2, \ldots, \mathbf{x}_k\}$, we shall write

$$< \mathbf{x}_1, \mathbf{x}_2, \ldots, \mathbf{x}_k >$$

for $< X >$.

Example 4.2.4

For the three vectors of $\mathbf{R}^3$ given below, determine whether C belongs to the subspace generated by A and B:

$$A = \begin{pmatrix} 1 \\ 1 \\ 4 \end{pmatrix}, \quad B = \begin{pmatrix} -1 \\ 2 \\ 1 \end{pmatrix}, \quad C = \begin{pmatrix} -1 \\ 5 \\ 6 \end{pmatrix}.$$

We have to decide if there are real numbers c and d such that $cA + dB = C$. To see what this entails, equate corresponding vector entries on both sides of the equation to obtain

$$\begin{cases} c & - & d & = -1 \\ c & + & 2d & = 5 \\ 4c & + & d & = 6 \end{cases}$$

Thus C belongs to $< A, B >$ if and only if this linear system is consistent. It is quickly seen that the linear system has the (unique) solution $c = 1, d = 2$. Hence $C = A + 2B$, so that C does belong to the subspace $< A, B >$.

What is the geometrical meaning of this conclusion? Recall that A, B and C can be represented by line segments in 3-dimensional space with a common initial point $\mathbf{I}$, say $\mathbf{IP}$, $\mathbf{IQ}$ and $\mathbf{IR}$. A typical vector in $< A, B >$ can be expressed in the form $sA + tB$ with real numbers s and t . Now sA and tB are representable by line segments parallel to $\mathbf{IP}$ and $\mathbf{IQ}$ respectively. We obtain a line segment that represents $sA+tB$ by applying the parallelogram law; clearly the resulting line segment will lie in *the plane determined by* $\mathbf{IP}$ *and* $\mathbf{IQ}$. Conversely, it is not difficult to see that any line segment lying in this plane represents a vector of the form $sA + tB$. Therefore the vectors in the subspace $< A, B >$ are those that can be represented by line segments drawn from $\mathbf{I}$ lying in the plane determined by $\mathbf{IP}$ and $\mathbf{IQ}$. What we have shown is that $\mathbf{IR}$ lies in this plane.

Finitely generated vector spaces

A vector space V is said to be *finitely generated* if there is a finite subset $\{\mathbf{v}_1, \mathbf{v}_2, \ldots, \mathbf{v}_k\}$ of V such that

$$V =< \mathbf{v}_1, \mathbf{v}_2, \ldots, \mathbf{v}_k >,$$

that is to say, every vector in V is a linear combination of the vectors $\mathbf{v}_1, \mathbf{v}_2, \ldots, \mathbf{v}_k$, and so has the form

$$c_1\mathbf{v}_1 + c_2\mathbf{v}_2 + \cdots + c_k\mathbf{v}_k$$

for some scalars c_i. If, on the other hand, no finite subset generates V, then V is said to be *infinitely generated*.

Example 4.2.5

Show that the Euclidean space $\mathbf{R}^n$ is finitely generated.

Let $X_1, X_2, \ldots, X_n$ be the columns of the identity matrix I_n. If

$$A = \begin{pmatrix} a_1 \\ a_2 \\ \vdots \\ a_n \end{pmatrix}$$

is any vector in $\mathbf{R}^n$, then $A = a_1 X_1 + a_2 X_2 + \cdots + a_n X_n$; therefore $X_1, X_2, \ldots, X_n$ generate $\mathbf{R}^n$ and consequently this vector space is finitely generated.

On the other hand, one does not have to look far to find infinitely generated vector spaces.

Example 4.2.6

Show that the vector space $P(\mathbf{R})$ of all real polynomials in x is infinitely generated.

To prove this we adopt the method of *proof by contradiction*. Assume that $P(\mathbf{R})$ is finitely generated, say by polynomials $p_1, p_2, \ldots, p_k$, and look for a contradiction. Clearly we may assume that all of these polynomials are non-zero; let m be the largest of their degrees. Then the degree of any linear combination of $p_1, p_2 \ldots, p_k$ certainly cannot exceed m. But this means that x^{m+1}, for example, is not such a linear combination. Consequently $p_1, p_2, \ldots, p_k$ do not generate $P(\mathbf{R})$, and we have reached a contradiction. This establishes the truth of the claim.

Exercises 4.2

1. Which of the following are vector spaces? The operations of addition and scalar multiplication are the natural ones:

(a) the set of all 2×2 real matrices with determinant equal to zero;

(b) the set of all solutions X of a linear system $AX = B$ where $B \neq 0$;

(c) the set of all functions $y = y(x)$ that are solutions of the homogeneous linear differential equation

$$a_n(x)y^{(n)} + a_{n-1}(x)y^{(n-1)} + \cdots + a_1(x)y' + a_0(x)y = 0.$$

2. In the following examples say whether S is a subspace of the vector space V :

(a) $V = \mathbf{R}^2$ and S is the subset of all matrices of the form $\begin{pmatrix} a^2 \\ a \end{pmatrix}$ where a is an arbitrary real number;

(b) $V = C[0,1]$ and S is the set of all infinitely differentiable functions in V.

(c) $V = P(\mathbf{R})$ and S is the set of all polynomials p such that $p(1) = 0$.

3. Does the polynomial $1 - 2x + x^2$ belong to the subspace of $P_3(\mathbf{R})$ generated by the polynomials $1 + x^2$, $x^2 - x$ and $3 - 2x$?

4. Determine if the matrix $\begin{pmatrix} 4 & 3 \\ 1 & -2 \end{pmatrix}$ is in the subspace of $M_2(\mathbf{R})$ generated by the following matrices:

$$\begin{pmatrix} 3 & 4 \\ 1 & 2 \end{pmatrix}, \quad \begin{pmatrix} 0 & 2 \\ -1/3 & 4 \end{pmatrix}, \quad \begin{pmatrix} 0 & 2 \\ 6 & 1 \end{pmatrix}.$$

5. Prove that the vector spaces $M_{m,n}(F)$ and $P_n(F)$ are finitely generated where F is an arbitrary field.

6. Prove that the vector spaces $C[0, 1]$ and $P(F)$ are infinitely generated, where F is any field.

7. Let A and B be vectors in $\mathbf{R}^2$. Show that A and B generate $\mathbf{R}^2$ if and only if neither is a scalar multiple of the other. Interpret this result geometrically.

4.3 Linear Independence in Vector Spaces

We begin with the crucial definition. Let V be a vector space and let X be a non-empty subset of V. Then X is said to be *linearly dependent* if there are distinct vectors $\mathbf{v}_1, \mathbf{v}_2, \ldots, \mathbf{v}_k$ in X, and scalars $c_1, c_2, \ldots, c_k$, *not all of them zero*, such that

$$c_1\mathbf{v}_1 + c_2\mathbf{v}_2 + \cdots + c_k\mathbf{v}_k = \mathbf{0}.$$

This amounts to saying that *at least one of the vectors $\mathbf{v}_i$ can be expressed as a linear combination of the others*. Indeed, if say $c_i \neq 0$, then we can solve the equation for $\mathbf{v}_i$, obtaining

$$\mathbf{v}_i = \sum_{j=1,\ j\neq i}^{n} (-c_i^{-1}c_j)\mathbf{v}_j.$$

For example, a one-element set $\{\mathbf{v}\}$ is linearly dependent if and only if $\mathbf{v} = \mathbf{0}$. A set with two elements is linearly dependent if and only if one of the elements is a scalar multiple of the other.

A subset which is not linearly dependent is said to be *linearly independent*. Thus a set of distinct vectors $\{\mathbf{v}_1, \ldots, \mathbf{v}_k\}$ is linearly independent if and only if an equation of the form $c_1\mathbf{v}_1 + \cdots + c_k\mathbf{v}_k = 0$ always implies that $c_1 = c_2 = \cdots = c_k = 0$.

We shall often say that vectors $\mathbf{v}_1, \ldots, \mathbf{v}_k$ are linearly dependent or independent, meaning that the subset $\{\mathbf{v}_1, \ldots, \mathbf{v}_k\}$ has this property.

Linear dependence in $\mathbf{R}^3$

Consider three vectors A, B, C in Euclidean space $\mathbf{R}^3$, and represent them by line segments in 3-dimensional space with a common initial point. If these vectors form a linearly dependent set, then one of them, say A, can be expressed as a linear combination of the other two, $A = uB + vC$; this equation says that the line segment representing A lies in the same plane as the line segments that represent B and C. Thus, if the three vectors form a linearly dependent set, their line segments must be coplanar.

Conversely, assume that A, B, C are vectors in $\mathbf{R}^3$ which are represented by line segments drawn from the origin, all of which lie in a plane. We claim that the vectors will then be linearly dependent. To see this, let the equation of the plane be $ux + vy + wz = 0$; keep in mind that the plane passes through the origin. Let the entries of A be written a_1, a_2, a_3, with a similar notation for B and C. Then the respective end points of the line segments have coordinates (a_1, a_2, a_3), (b_1, b_2, b_3), (c_1, c_2, c_3). Since these points lie on the plane, we have the equations

$$\begin{cases} ua_1 + va_2 + wa_3 & = 0 \\ ub_1 + vb_2 + wb_3 & = 0 \\ uc_1 + vc_2 + wc_3 & = 0 \end{cases}$$

This homogeneous linear system has a non-trivial solution for u, v, w, so the determinant of its coefficient matrix is zero by 3.3.2. Now the coefficient matrix of the linear system

$$\begin{cases} ua_1 + vb_1 + wc_1 & = 0 \\ ua_2 + vb_2 + wc_2 & = 0 \\ ua_3 + vb_3 + wc_3 & = 0 \end{cases}$$

is the transpose of the previous one, so by 3.2.1 it has the same determinant. It follows that the second linear system also has a non-trivial solution u, v, w. But then $uA + vB + wC = 0$, which shows that the vectors A, B, C are linearly independent.

Thus there is a natural geometrical interpretation of linear dependence in the Euclidean space $\mathbf{R}^3$: *three vectors are linearly dependent if and only if they are represented by line segments lying in the same plane.* There is a corresponding interpretation of linear dependence in $\mathbf{R}^2$ (see Exercise 4.3.11).

Example 4.3.1

Are the polynomials $x + 1$, $x + 2$, $x^2 - 1$ linearly dependent in the vector space $P_3(\mathbf{R})$?

To answer this, suppose that c_1, c_2, c_3 are scalars satisfying

$$c_1(x + 1) + c_2(x + 2) + c_3(x^2 - 1) = 0.$$

Equating to zero the coefficients of 1, x, x^2, we obtain the homogeneous linear system

$$\begin{cases} c_1 & + 2c_2 & - c_3 & = 0 \\ c_1 & + c_2 & & = 0 \\ & & c_3 & = 0 \end{cases}$$

This has only the trivial solution $c_1 = c_2 = c_3 = 0$; hence the polynomials are linearly independent.

Example 4.3.2

Show that the vectors

$$\begin{pmatrix} -1 \\ 2 \end{pmatrix}, \begin{pmatrix} 1 \\ 2 \end{pmatrix}, \begin{pmatrix} 2 \\ -4 \end{pmatrix}$$

are linearly dependent in $\mathbf{R}^2$.

Proceeding as in the last example, we let c_1, c_2, c_3 be scalars such that

$$c_1 \begin{pmatrix} -1 \\ 2 \end{pmatrix} + c_2 \begin{pmatrix} 1 \\ 2 \end{pmatrix} + c_3 \begin{pmatrix} 2 \\ -4 \end{pmatrix} = \begin{pmatrix} 0 \\ 0 \end{pmatrix}.$$

This is equivalent to the homogeneous linear system

$$\begin{cases} -c_1 & + & c_2 & + & 2c_3 & = 0 \\ 2c_1 & + & 2c_2 & - & 4c_3 & = 0 \end{cases}$$

Since the number of unknowns is greater than the number of equations, this system has a non-trivial solution by 2.1.4. Hence the vectors are linearly dependent.

These examples suggest that the question of deciding whether a set of vectors is linearly dependent is equivalent to asking if a certain homogeneous linear system has non-trivial solutions. Further evidence for this is provided by the proof of the next result.

Theorem 4.3.1
Let $A_1, A_2, \ldots, A_m$ be vectors in the vector space F^n where F is some field. Put $A = [A_1|A_2|\ldots|A_m]$, an $n \times m$ matrix. Then $A_1, A_2, \ldots, A_m$ are linearly dependent if and only if the number of pivots of A in row echelon form is less than m.

Proof
Consider the equation $c_1 A_1 + c_2 A_2 + \cdots + c_m A_m = 0$ where $c_1, c_2, \ldots, c_m$ are scalars. Equating entries of the vector on the left side of the equation to zero, we find that this equation is equivalent to the homogeneous linear system

$$A \begin{pmatrix} c_1 \\ c_2 \\ \vdots \\ c_m \end{pmatrix} = 0.$$

By 2.1.3 the condition for this linear system to have a non-trivial solution $c_1, c_2, \ldots, c_m$ is that the number of pivots be less than m . Hence this is the condition for the set of column vectors to be linearly dependent.

In 5.1 we shall learn how to tell if a set of vectors in an arbitrary finitely generated vector space is linearly dependent.

An application to differential equations

In the theory of linear differential equations it is an important problem to decide if a given set of functions in the vector space $C[a, b]$ is linearly dependent. These functions will normally be solutions of a homogeneous linear differential equation. There is a useful way to test such a set of functions for linear independence using a determinant called the Wronskian.

Suppose that $f_1, f_2, \ldots, f_n$ are functions whose first $n - 1$ derivatives exist at all points of the interval $[a, b]$. In particular this means that the functions will be continuous throughout the interval, so they belong to $C[a, b]$. Assume that $c_1, c_2, \ldots,$ c_n are real numbers such that $c_1 f_1 + c_2 f_2 + \cdots + c_n f_n = 0$, the zero function on $[a, b]$. Now differentiate this equation $n - 1$ times, keeping in mind that the c_i are constants. This results in a set of n equations for $c_1, c_2, \ldots, c_n$

$$
\begin{cases}
c_1 f_1 + c_2 f_2 + \cdots + c_n f_n = 0 \\
c_1 f_1' + c_2 f_2' + \cdots + c_n f_n' = 0 \\
\quad \vdots \\
c_1 f_1^{(n-1)} + c_2 f_2^{(n-1)} + \cdots + c_n f_n^{(n-1)} = 0
\end{cases}
$$

This linear system can be written in matrix form:

$$
\begin{pmatrix}
f_1 & f_2 & \cdots & f_n \\
f_1' & f_2' & \cdots & f_n' \\
\vdots & \vdots & & \vdots \\
f_1^{(n-1)} & f_2^{(n-1)} & \cdots & f_n^{(n-1)}
\end{pmatrix}
\begin{pmatrix}
c_1 \\ c_2 \\ \vdots \\ c_n
\end{pmatrix} = 0.
$$

By 3.3.2, if the determinant of the coefficient matrix of the linear system is not identically equal to zero in $[a, b]$, the linear system has only the trivial solution and the functions $f_1, f_2, \ldots, f_n$ will be linearly independent. Define

$$
W(f_1, f_2, \ldots, f_n) = \begin{vmatrix} f_1 & f_2 & \cdots & f_n \\ f_1' & f_2' & \cdots & f_n' \\ \cdot & \cdot & \cdots & \cdot \\ f_1^{(n-1)} & f_2^{(n-1)} & \cdots & f_n^{(n-1)} \end{vmatrix}.
$$

This determinant is called the *Wronskian* of the functions $f_1, f_2, \ldots, f_n$. Then our discussion shows that the following is true.

Theorem 4.3.2 *Suppose that $f_1, f_2, \ldots, f_n$ are functions whose first $n - 1$ derivatives exist in the interval $[a, b]$. If $W(f_1, f_2, \ldots, f_n)$ is not identically equal to zero in this interval, then $f_1, f_2, \ldots, f_n$ are linearly independent in $[a, b]$.*

The converse of 4.3.2 is false. In general one cannot conclude that if $f_1, f_2, \ldots, f_n$ are linearly independent, then $W(f_1, f_2, \ldots, f_n)$ is not the zero function. However, it turns out that if the functions $f_1, f_2, \ldots, f_n$ are solutions of a homogeneous linear differential equation of order n, then the Wronskian can never vanish. Hence *a necessary and sufficient condition for a set of solutions of a homogeneous linear differential equation to be linearly independent is that their Wronskian should not be the zero function.* For a detailed account of this topic the reader should consult a book on differential equations such as [16].

Example 4.3.3

Show that the functions x, e^x, e^{-2x} are linearly independent in the vector space $C[0, 1]$.

The Wronskian is

$$
W(x, e^x, e^{-2x}) = \begin{vmatrix} x & e^x & e^{-2x} \\ 1 & e^x & -2e^{-2x} \\ 0 & e^x & 4e^{-2x} \end{vmatrix} = 3(2x - 1)e^{-x},
$$

which is not identically equal to zero in $[0, 1]$.

Exercises 4.3

1. In each of the following cases determine if the subset S of the vector space V is linearly dependent or linearly independent:

(a) $V = \mathbf{C}^3$ and S consists of the column vectors

$$\begin{pmatrix} -1 \\ 2 \\ \sqrt{-1} \end{pmatrix}, \quad \begin{pmatrix} 1 \\ 3 + \sqrt{-1} \\ -2 \end{pmatrix}, \quad \begin{pmatrix} -5 \\ 3\sqrt{-1} \\ -6 + 2\sqrt{-1} \end{pmatrix};$$

(b) $V = P(\mathbf{R})$ and $S = \{x - 1, \ x^2 + 1, \ x^3 - x^2 - x + 3\}$;
(c) $V = M(2, \mathbf{R})$ and S consists of the matrices

$$\begin{pmatrix} 2 & -3 \\ 6 & 4 \end{pmatrix}, \quad \begin{pmatrix} 3 & 1 \\ -1/2 & -3 \end{pmatrix}, \quad \begin{pmatrix} 12 & -7 \\ 17 & 6 \end{pmatrix}.$$

2. A subset of a vector space that contains the zero vector is linearly dependent: true or false?

3. If X is a linearly independent subset of a vector space, every non-empty subset of X is also linearly independent: true or false?

4. If X is a linearly dependent subset of a vector space, every non-empty subset of X is also linearly dependent: true or false?

5. Prove that any three vectors in $\mathbf{R}^2$ are linearly dependent. Generalize this result to $\mathbf{R}^n$.

6. Find a set of n linearly independent vectors in $\mathbf{R}^n$.

7. Find a set of mn linearly independent vectors in the vector space $M_{m,n}(\mathbf{R})$.

8. Show that the functions x, $e^x \sin x$, $e^x \cos x$ form a linearly independent subset of the vector space $C[0, \pi]$.

9. The union of two linearly independent subsets of a vector space is linearly independent: true or false?

10. If $\{\mathbf{u}, \mathbf{v}\}, \{\mathbf{v}, \mathbf{w}\}$ and $\{\mathbf{w}, \mathbf{u}\}$ are linearly independent subsets of a vector space, is the subset $\{\mathbf{u}, \mathbf{v}, \mathbf{w}\}$ necessarily linearly independent?

11. Show that two non-zero vectors in $\mathbf{R}^2$ are linearly dependent precisely when they are represented by parallel line segments in the plane.

Chapter Five

BASIS AND DIMENSION

We now specialize our study of vector spaces to finitely generated vector spaces, that is, to those that can be generated by finite subsets. The essential fact to be established is that in any non-zero vector space there is a basis, that is to say, a set of vectors in terms of which every vector of the space can be written in a unique manner. This allows the representation of vectors in abstract vector spaces by column vectors.

5.1 The Existence of a Basis

The following theorem on linear dependence is fundamental for everything in this chapter.

Theorem 5.1.1

Let $\mathbf{v}_1, \mathbf{v}_2, \ldots, \mathbf{v}_m$ be vectors in a vector space V and let $S = \langle \mathbf{v}_1, \mathbf{v}_2, \ldots, \mathbf{v}_m \rangle$, the subspace generated by these vectors. Then any subset of S containing $m + 1$ or more elements is linearly dependent.

Proof

To prove the theorem it suffices to show that if $\mathbf{u}_1, \mathbf{u}_2, \ldots, \mathbf{u}_{m+1}$ are any $m + 1$ vectors of the subspace S, then these vectors are linearly dependent. This amounts to finding scalars $c_1, c_2, \ldots, c_m$, not all of them zero, such that

$$c_1 \mathbf{u}_1 + c_2 \mathbf{u}_2 + \cdots + c_{m+1} \mathbf{u}_{m+1} = \mathbf{0}.$$

Now, because $\mathbf{u}_i$ belongs to S, there is an expression

$$\mathbf{u}_i = d_{1i} \mathbf{v}_1 + d_{2i} \mathbf{v}_2 + \cdots + d_{mi} \mathbf{v}_m$$

112

where the d_{ji} are certain scalars. On substituting for the $\mathbf{u}_i$, we obtain

$$c_1\mathbf{u}_1 + c_2\mathbf{u}_2 + \cdots + c_{m+1}\mathbf{u}_{m+1} = \sum_{i=1}^{m+1} c_i \left(\sum_{j=1}^{m} d_{ji}\mathbf{v}_j\right)$$

$$= \sum_{j=1}^{m}\left(\sum_{i=1}^{m+1} d_{ji}c_i\right)\mathbf{v}_j.$$

Here we have interchanged the summations over i and j. This is permissible since it corresponds to adding up the vectors $c_i d_{ji}\mathbf{v}_j$ in a different order, which is possible in a vector space because of the commutative law for addition.

We deduce from the last equation that the vector $c_1\mathbf{u}_1 + c_2\mathbf{u}_2 + \cdots + c_{m+1}\mathbf{u}_{m+1}$ will equal $\mathbf{0}$ provided that all the expressions $\sum d_{ji}c_i$ equal zero, that is to say, $c_1, c_2, \ldots, c_{m+1}$ form a non-trivial solution of the homogeneous linear system $DC = 0$ where D is the $m \times (m+1)$ matrix whose (j, i) entry is d_{ji} and C is the column consisting of $c_1, c_2, \ldots, c_{m+1}$. But this linear system has $m+1$ unknowns and m equations; therefore, by 2.1.4, there is a non-trivial solution C. In consequence there are indeed scalars $c_1, c_2, \ldots, c_{m+1}$, not all zero, which make the vector $c_1\mathbf{u}_1 + c_2\mathbf{u}_2 + \cdots + c_{m+1}\mathbf{u}_{m+1}$ zero.

Corollary 5.1.2

If V is a vector space which can be generated by m elements, then every subset of V with $m+1$ or more vectors is linearly dependent.

Thus the number of elements in a linearly independent subset of a finitely generated vector space cannot exceed the number of generators. On the other hand, if a subset is to generate a vector space, it surely cannot be too small. We unite these two contrasting requirements in the definition of a basis.

Bases

Let X be a non-empty subset of a vector space V. Then X is called a *basis* of V if both of the following are true:

(i) X is linearly independent;
(ii) X generates V.

Example 5.1.1

As a first example of a basis, consider the columns of the identity $n \times n$ matrix I_n:

$$E_1 = \begin{pmatrix} 1 \\ 0 \\ \vdots \\ 0 \end{pmatrix}, \quad E_2 = \begin{pmatrix} 0 \\ 1 \\ \vdots \\ 0 \end{pmatrix}, \quad \dots, \quad E_n = \begin{pmatrix} 0 \\ 0 \\ \vdots \\ 1 \end{pmatrix}.$$

From the equation

$$c_1 E_1 + c_2 E_2 + \cdots + c_n E_n = \begin{pmatrix} c_1 \\ c_2 \\ \vdots \\ c_n \end{pmatrix}$$

it follows that $E_1, E_2, \dots, E_n$ generate $\mathbf{R}^n$. But these vectors are also linearly independent; for the equation also shows that $c_1 E_1 + c_2 E_2 + \cdots + c_n E_n$ cannot equal zero unless all the c_i are zero. Therefore *the vectors $E_1, E_2, \dots, E_n$ form a basis of the Euclidean space $\mathbf{R}^n$*. This is called the *standard basis* of $\mathbf{R}^n$.

An important property of bases is uniqueness of expressibility of vectors.

Theorem 5.1.3
If $\{v_1, v_2, \ldots, v_n\}$ is a basis of a vector space V, then each vector v in V has a unique expression of the form

$$v = c_1 v_1 + c_2 v_2 + \cdots + c_n v_n$$

for certain scalars c_i.

Proof
If there are two such expressions for v, say $c_1 v_1 + \cdots + c_n v_n$ and $d_1 v_1 + \cdots + d_n v_n$, then, by equating these, we arrive at the equation

$$(c_1 - d_1)v_1 + \cdots + (c_n - d_n)v_n = 0.$$

By linear independence of the v_i this can only mean that $c_i = d_i$ for all i, so the expression is unique as claimed.

Naturally the question arises: does every vector space have a basis? The answer is negative in general. Since a zero space has 0 as its only vector, it has no linearly independent subsets at all; thus a zero space cannot have a basis. However, apart from this uninteresting case, every finitely generated vector space has a basis, a fundamental result that will now be proved. Notice that such a basis must be finite by 5.1.2.

Theorem 5.1.4
Let V be a finitely generated vector space and suppose that X_0 is a linearly independent subset of V. Then X_0 is contained in some basis X of V.

Proof
Suppose that V is generated by m elements. Then by 5.1.2 no linearly independent subset of V can contain more than m elements. From this it follows that there exists a subset X of V containing X_0 which is as large as possible subject to being linearly independent. For if this were false, it would be

possible to find arbitrarily large linearly independent subsets of V.

We will prove the theorem by showing that the subset X is a basis of V. Write $X = \{\mathbf{v}_1, \mathbf{v}_2, \ldots, \mathbf{v}_n\}$. Suppose that $\mathbf{u}$ is a vector in V which does not belong to X. Then the subset $\{\mathbf{v}_1, \mathbf{v}_2, \ldots, \mathbf{v}_n, \mathbf{u}\}$ must be linearly dependent since it properly contains X. Hence there is a linear relation of the form

$$c_1\mathbf{v}_1 + c_2\mathbf{v}_2 + \cdots + c_n\mathbf{v}_n + d\mathbf{u} = \mathbf{0}$$

where not all of the scalars $c_1, c_2, \ldots, c_n, d$ are zero. Now if the scalar d were zero, it would follow that $c_1\mathbf{v}_1 + c_2\mathbf{v}_2 + \cdots + c_n\mathbf{v}_n = 0$, which, in view of the linear independence of $\mathbf{v}_1, \mathbf{v}_2, \ldots, \mathbf{v}_n$, could only mean that $c_1 = c_2 = \cdots = c_n = 0$. But now *all* the scalars are zero, which is not true. Therefore $d \neq 0$. Consequently we can solve the above equation for $\mathbf{u}$ to obtain

$$\mathbf{u} = (-d^{-1}c_1)\mathbf{v}_1 + (-d^{-1}c_2)\mathbf{v}_2 + \cdots + (-d^{-1}c_n)\mathbf{v}_n.$$

Hence $\mathbf{u}$ belongs $to < \mathbf{v}_1, \ldots, \mathbf{v}_n >$. From this it follows that the vectors $\mathbf{v}_1, \ldots, \mathbf{v}_n$ generate V; since these are also linearly independent, they form a basis of V.

Corollary 5.1.5
Every non-zero finitely generated vector space V has a basis.

Indeed by hypothesis V contains a non-zero vector, say $\mathbf{v}$. Then $\{\mathbf{v}\}$ is linearly independent and by 5.1.4 it is contained in a basis of V.

Usually a vector space will have many bases. For example, the vector space $\mathbf{R}^2$ has the basis

$$\begin{pmatrix} 1 \\ -1 \end{pmatrix}, \begin{pmatrix} 2 \\ -1 \end{pmatrix},$$

as well as the standard basis

$$\begin{pmatrix} 1 \\ 0 \end{pmatrix}, \begin{pmatrix} 0 \\ 1 \end{pmatrix}.$$

And one can easily think of other examples. It is therefore a very significant fact that all bases of a finitely generated vector space have the same number of elements.

Theorem 5.1.6
Let V be a non-zero finitely generated vector space. Then any two bases of V have equal numbers of elements.

Proof
Let $\{\mathbf{u}_1, \mathbf{u}_2, \ldots, \mathbf{u}_m\}$ and $\{\mathbf{v}_1, \mathbf{v}_2, \ldots, \mathbf{v}_n\}$ be two bases of V. Then

$$V = < \mathbf{u}_1, \mathbf{u}_2, \ldots, \mathbf{u}_m >$$

and it follows from 5.1.2 that no linearly independent subset of V can have more than m elements; hence $n \leq m$. In the same fashion we argue that $m \leq n$. Therefore $m = n$.

Dimension

Let V be a finitely generated vector space. If V is non-zero, define the *dimension of V* to be the number of elements in a basis of V; this definition makes sense because 5.1.6 guarantees that all bases of V have the same number of elements. Of course, a zero space does not have a basis; however it is convenient to define the dimension of a zero space to be 0, so that *every finitely generated vector space has a dimension*. The dimension of a finitely generated vector space V is denoted by

$$\dim(V).$$

In fact infinitely generated vector spaces also have bases, and it is even possible to assign a dimension to such a space, namely a cardinal number, which is a sort of infinite analog of a positive integer. However this goes well beyond our brief, so we shall say no more about it.

Example 5.1.2

The dimension of $\mathbf{R}^n$ is n; indeed it has already been shown in Example 5.1.1 that the columns of the identity matrix I_n form a basis of $\mathbf{R}^n$.

Example 5.1.3

The dimension of $P_n(\mathbf{R})$ is n. In this case the polynomials $1, x, x^2, \ldots, x^{n-1}$ form a basis (called the *standard basis*) of $P_n(\mathbf{R})$.

Example 5.1.4

Find a basis for the null space of the matrix

$$A = \begin{pmatrix} 1 & -1 & 1 & 2 \\ 2 & 1 & 3 & 2 \\ 1 & 5 & 3 & -2 \end{pmatrix}.$$

Recall that the null space of A is the subspace of $\mathbf{R}^4$ consisting of all solutions X of the linear system $AX = 0$. To solve this system, put A in reduced row echelon form using row operations:

$$\begin{pmatrix} 1 & 0 & 4/3 & 4/3 \\ 0 & 1 & 1/3 & -2/3 \\ 0 & 0 & 0 & 0 \end{pmatrix}.$$

From this we read off the general solution in the usual way:

$$X = \begin{pmatrix} -4c/3 - 4d/3 \\ -c/3 + 2d/3 \\ c \\ d \end{pmatrix}.$$

Now X can be written in the form

$$X = c \begin{pmatrix} -4/3 \\ -1/3 \\ 1 \\ 0 \end{pmatrix} + d \begin{pmatrix} -4/3 \\ 2/3 \\ 0 \\ 1 \end{pmatrix},$$

where c and d are arbitrary scalars. Hence the null space of A is generated by the vectors

$$X_1 = \begin{pmatrix} -4/3 \\ -1/3 \\ 1 \\ 0 \end{pmatrix}, \quad X_2 = \begin{pmatrix} -4/3 \\ 2/3 \\ 0 \\ 1 \end{pmatrix}.$$

Notice that these vectors are obtained from the general solution X by putting $c = 1, d = 0$, and then $c = 0, d = 1$. Now X_1 and X_2 are linearly independent. Indeed, if we assume that some linear combination of them is zero, then, because of the configuration of 0's and 1's, the scalars are forced to be be zero. It follows that X_1 and X_2 form a basis of the null space of A, which therefore has dimension equal to 2.

It should be clear to the reader that this example describes a general method for finding a basis, and hence the dimension, of the null space of an arbitrary $m \times n$ matrix A. The procedure goes as follows. Using elementary row operations, put A in reduced row echelon form, with say r pivots. Then the general solution of the linear system $AX = 0$ will contain $n - r$ arbitrary scalars, say $c_1, c_2, \ldots, c_{n-r}$. The method of solving linear systems by elementary row operations shows that the general solution can be written in the form

$$X = c_1 X_1 + c_2 X_2 + \cdots + c_{n-r} X_{n-r}$$

where $X_1, \ldots, X_{n-r}$ are particular solutions. In fact the solution X_i arises from X when we put $c_i = 1$ and all other c_j's equal to 0. The vectors $X_1, X_2, \ldots, X_{n-r}$ are linearly independent, just as in the example, because of the arrangement of 0's and 1's among their entries. It follows that a basis of the null space of A is $\{X_1, X_2, \ldots, X_{n-r}\}$. We can therefore state:

Theorem 5.1.7

Let A be a matrix with n columns and suppose that the number of pivots in the reduced row echelon form of A is r. Then the null space of A has dimension $n - r$.

Coordinate column vectors

Let V be a vector space with an *ordered basis* $\{v_1, \ldots, v_n\}$; this means that the basis vectors are to be written in the prescribed order. We have seen in 5.1.3 that each vector v of V has a unique expression in terms of the basis,

$$v = c_1 v_1 + \cdots + c_n v_n$$

say. Thus v is completely determined by the scalars $c_1, \ldots, c_n$. We call the column

$$\begin{pmatrix} c_1 \\ c_2 \\ \vdots \\ c_n \end{pmatrix}$$

the *coordinate vector* of v with respect to the ordered basis $\{v_1, \ldots, v_n\}$. Thus each vector in the abstract vector space V is represented by an n-column vector. This provides us with a concrete way of representing abstract vectors.

Example 5.1.5

Find the coordinate vector of $\begin{pmatrix} 2 \\ 3 \end{pmatrix}$ with respect to the ordered basis of $\mathbf{R}^2$ consisting of the vectors

$$\begin{pmatrix} 1 \\ 1 \end{pmatrix}, \begin{pmatrix} 3 \\ 4 \end{pmatrix}.$$

First notice that these two vectors are linearly independent and generate $\mathbf{R}^2$, so that they form a basis. We need to
find scalars c and d such that

$$\begin{pmatrix} 2 \\ 3 \end{pmatrix} = c \begin{pmatrix} 1 \\ 1 \end{pmatrix} + d \begin{pmatrix} 3 \\ 4 \end{pmatrix}.$$

This amounts to solving the linear system

$$\begin{cases} c + 3d & = 2 \\ c + 4d & = 3 \end{cases}$$

The unique solution is $c = -1$, $d = 1$, and hence the coordinate vector is $\begin{pmatrix} -1 \\ 1 \end{pmatrix}$.

Coordinate vectors provide us with a method of testing a
subset of an arbitrary finitely generated vector space for linear
dependence.

Theorem 5.1.8
*Let $\{\mathbf{v}_1, \ldots, \mathbf{v}_n\}$ be an ordered basis of a vector space V. Let
$\mathbf{u}_1, \ldots, \mathbf{u}_m$ be a set of vectors in V whose coordinate vectors
with respect to the given ordered basis are $X_1, \ldots, X_m$ respectively. Then $\{\mathbf{u}_1, \ldots, \mathbf{u}_m\}$ is linearly dependent if and only
if the number of pivots of the matrix $A = [X_1 | X_2 | \ldots | X_m]$ is
less than m.*

Proof
Write $\mathbf{u}_i = \sum_{i=1}^{n} a_{ji} \mathbf{v}_j$; then the entries of X_i are $a_{1i}, \ldots, a_{ni}$,
so the (j, i) entry of A is a_{ji}. If $c_1, \ldots, c_m$ are any scalars,
then

$$c_1 \mathbf{u}_1 + \cdots + c_m \mathbf{u}_m = \sum_{i=1}^{m} c_i \left(\sum_{j=1}^{n} a_{ji} \mathbf{v}_j \right) = \sum_{j=1}^{n} \left(\sum_{i=1}^{m} a_{ji} c_i \right) \mathbf{v}_j.$$

Since $\mathbf{v}_1, \ldots, \mathbf{v}_n$ are linearly independent, the only way that $c_1 \mathbf{u}_1 + \cdots + c_m \mathbf{u}_m$ can be zero is if the sums $\sum_{i=1}^m a_{ji} c_i$ vanish for $j = 1, \ldots, n$. This amounts to requiring that $AC = 0$ where C is the column consisting of $c_1, \ldots, c_m$. We know from 2.1.3 that there is such a C different from 0 precisely when the number of pivots of A is less than m. So this is the condition for $\mathbf{u}_1, \ldots, \mathbf{u}_m$ to be linearly dependent.

Example 5.1.6
Are the polynomials $1 - x + 2x^2 - x^3$, $x + x^3$, $2 + x + 4x^2 + x^3$ linearly independent in $P_4(\mathbf{R})$?

Use the standard ordered basis $\{1, x, x^2, x^3\}$ of $P_4(\mathbf{R})$. Then the coordinate columns of the given polynomials are the columns of the matrix

$$\begin{pmatrix} 1 & 0 & 2 \\ -1 & 1 & 1 \\ 2 & 0 & 4 \\ -1 & 1 & 1 \end{pmatrix}.$$

Using row operations, we see that the number of pivots of the matrix is 2, which is less than the number of vectors. Therefore the given polynomials are linearly dependent.

The next theorem lessens the work needed to show that a particular set is a basis.

Theorem 5.1.9
Let V be a finitely generated vector space with positive dimension n. Then

(i) *any set of n linearly independent vectors of V is a basis;*

(ii) *any set of n vectors that generates V is a basis.*

Proof
Assume first that the vectors $\mathbf{v}_1, \mathbf{v}_2, \ldots, \mathbf{v}_n$ are linearly independent. Then by 5.1.4 the set $\{\mathbf{v}_1, \mathbf{v}_2, \ldots, \mathbf{v}_n\}$ is contained

in a basis of V. But the latter must have n elements by 5.1.6, and so it coincides with the set of $\mathbf{v}_i$'s.

Now assume that the vectors $\mathbf{v}_1, \mathbf{v}_2, \ldots, \mathbf{v}_n$ generate V. If these vectors are linearly dependent, then one of them, say $\mathbf{v}_i$, can be expressed as a linear combination of the others. But this means that we can dispense with $\mathbf{v}_i$ completely and generate V using only the $\mathbf{v}_j$'s for $j \neq i$, of which there are $n-1$. Therefore $\dim(V) \leq n-1$ by 5.1.2. By this contradiction $\mathbf{v}_1, \mathbf{v}_2, \ldots, \mathbf{v}_n$ are linearly independent, so they form a basis of V.

Example 5.1.7

The vectors
$$\begin{pmatrix} 1 \\ 1 \\ -1 \end{pmatrix}, \begin{pmatrix} 3 \\ 4 \\ 2 \end{pmatrix}, \begin{pmatrix} 0 \\ 1 \\ 2 \end{pmatrix}$$
are linearly independent since the matrix which they form has three pivots; therefore these vectors constitute a basis of $\mathbf{R}^3$.

We conclude with an application of the ideas of this section to accounting systems.

Example 5.1.8 (*Transactions on an accounting system*)

Consider an accounting system with n accounts, say $\alpha_1, \alpha_2, \ldots, \alpha_n$. At any instant each account has a balance which can be a credit (positive), a debit (negative), or zero. Since the accounting system must at all times be in balance, the sum of the balances of all the accounts will always be zero. Now suppose that a *transaction* is applied to the system. By this we mean that there is a flow of funds between accounts of the system. If as a result of the transaction the balance of account α_i changes by an amount t_i, then the transaction can be represented by an n-column vector with entries $t_1, t_2, \ldots, t_n$. Since the accounting system must still be in balance after the transaction has been applied, the sum of the t_i will be zero.

Hence the transactions correspond to column vectors

$$\begin{pmatrix} t_1 \\ \vdots \\ t_n \end{pmatrix}$$

such that $t_1 + \cdots + t_n = 0$. Now vectors of this form are easily seen to constitute a subspace T of the vector space $\mathbf{R}^n$; this is called the *transaction space*. Evidently T is just the null space of the matrix

$$A = \begin{pmatrix} 1 & 1 & 1 & \cdots & 1 \\ 0 & 0 & 0 & \cdots & 0 \\ \cdot & \cdot & \cdot & \cdots & \cdot \\ 0 & 0 & 0 & \cdots & 0 \end{pmatrix}.$$

Now A is already in reduced row echelon form, so we can read off at once the general solution of the linear system $AX = 0$:

$$X = \begin{pmatrix} -c_2 - c_3 - \cdots - c_n \\ c_2 \\ c_3 \\ \vdots \\ c_n \end{pmatrix}$$

with arbitrary real scalars $c_2, c_3, \ldots, c_n$. Now we can find a basis of the null space in the usual way. For $i = 2, \ldots, n$ define T_i to be the n-column vector with first entry -1, ith entry 1, and all other entries zero. Then

$$X = c_2 T_2 + c_3 T_3 + \cdots + c_n T_n$$

and $\{T_2, T_3, \ldots, T_n\}$ is a basis of the transaction space T. Thus $\dim(T) = n - 1$. Observe that T_i corresponds to a *simple transaction*, in which there is a flow of funds amounting to one unit from account α_1 to account α_i and which does not affect other accounts.

Exercises 5.1

1. Show that the following sets of vectors form bases of $\mathbf{R}^3$, and then express the vectors E_1, E_2, E_3 of the standard basis in terms of these:

$$\text{(a) } X_1 = \begin{pmatrix} 4 \\ 2 \\ 1 \end{pmatrix}, \quad X_2 = \begin{pmatrix} -5 \\ 2 \\ -3 \end{pmatrix}, \quad X_3 = \begin{pmatrix} 1 \\ 3 \\ 0 \end{pmatrix};$$

$$\text{(b) } Y_1 = \begin{pmatrix} 1 \\ 1 \\ -1 \end{pmatrix}, \quad Y_2 = \begin{pmatrix} 2 \\ 1 \\ 0 \end{pmatrix}, \quad Y_3 = \begin{pmatrix} 5 \\ 2 \\ 2 \end{pmatrix}.$$

2. Find a basis for the null space of each of the following matrices:

$$\text{(a) } \begin{pmatrix} 1 & -5 & 3 \\ -4 & 2 & -6 \\ 3 & 1 & 7 \end{pmatrix}; \quad \text{(b) } \begin{pmatrix} 2 & 3 & 1 & 1 \\ -3 & 1 & 4 & -7 \\ 1 & 2 & 1 & 0 \end{pmatrix}.$$

3. What is the dimension of the vector space $M_{m,n}(F)$ where F is an arbitrary field of scalars?

4. Let V be a vector space containing vectors $\mathbf{v}_1, \mathbf{v}_2, \ldots, \mathbf{v}_n$ and suppose that each vector of V has a *unique* expression as a linear combination of $\mathbf{v}_1, \mathbf{v}_2, \ldots, \mathbf{v}_n$. Prove that the $\mathbf{v}_i$'s form a basis of V.

5. If S is a subspace of a finitely generated vector space V, establish the inequality $\dim(S) \leq \dim(V)$.

6. If in the last problem $\dim(S) = \dim(V)$, show that $S = V$.

7. If V is a vector space of dimension n, show that for each integer i satisfying $0 \leq i \leq n$ there is a subspace of V which has dimension i.

8. Write the transaction $\begin{pmatrix} 6 \\ -4 \\ -2 \end{pmatrix}$ as a linear combination of simple transactions.

9. Prove that vectors A, B, C generate $\mathbf{R}^3$ if and only if none of these vectors belongs to the subspace generated by the other two. Interpret this result geometrically.

10. If V is a vector space with dimension n over the field of two elements, prove that V contains exactly 2^n vectors.

5.2 The Row and Column Spaces of a Matrix

Let A be an $m \times n$ matrix over some field of scalars F. Then the columns of A are m-column vectors, so they belong to the vector space F^m, while the rows of A are n-row vectors and belong to the vector space F_n. Thus there are two natural subspaces associated with A, the *row space*, which is generated by the rows of A and is a subspace of F_n, and the *column space*, generated by the columns of A, which is a subspace of F^m.

We begin the study of these important subspaces by investigating the effect upon them of applying row and column operations to the matrix.

Theorem 5.2.1

Let A be any matrix.

(i) *The row space is unchanged when an elementary row operation is applied to A.*

(ii) *The column space is unchanged when an elementary column operation is applied to A.*

Proof
Let B arise from A when an elementary row operation is applied. Then by 2.3.1 there is an elementary matrix E such that $B = EA$. The row-times-column rule of matrix multiplication shows that each row of B is a linear combination of the rows of A. Hence the row space of B is contained in the row space of A. But $A = E^{-1}B$, since elementary matrices are invertible, so the same argument shows that the row space of A is contained in the row space of B. Therefore the row spaces of A and B are identical. Of course, the argument for column spaces is analogous.

There are simple procedures available for finding bases for the row and column spaces of a matrix.

(I) To find a basis of the row space of a matrix A, use elementary row operations to put A in reduced row echelon form. Discard any zero rows; then the remaining rows will form a basis of the row space of A.

(II) To find a basis of the column space of a matrix A, use elementary column operations to put A in reduced column echelon form. Discard any zero columns; then the remaining columns will form a basis of the column space of A.

Why do these procedures work? By 5.2.1 the row space of A equals the row space of R, its reduced row echelon form, and this is certainly generated by the non-zero rows of R. Also the non-zero rows of R are linearly independent because of the arrangement of 0's and 1's in R ; therefore these rows form a basis of the row space of A. Again the argument for columns is similar. This discussion makes the following result obvious.

Corollary 5.2.2

For any matrix the dimension of the row space equals the number of pivots in reduced row echelon form, with a like statement for columns.

Example 5.2.1

Consider the matrix

$$A = \begin{pmatrix} 2 & 1 & 1 & 3 & 2 \\ -1 & 2 & 1 & 1 & 3 \\ 0 & 0 & 1 & 0 & 1 \\ 0 & 1 & 0 & 1 & 1 \end{pmatrix}.$$

The reduced row echelon form of A is found to be

$$\begin{pmatrix} 1 & 0 & 0 & 1 & 0 \\ 0 & 1 & 0 & 1 & 1 \\ 0 & 0 & 1 & 0 & 1 \\ 0 & 0 & 0 & 0 & 0 \end{pmatrix}.$$

Hence the row vectors $[1\ 0\ 0\ 1\ 0]$, $[0\ 1\ 0\ 1\ 1]$, $[0\ 0\ 1\ 0\ 1]$ form a basis of the row space of A and the dimension of this space is 3.

In general elementary row operations change the column space of a matrix, and column operations change the row space. However it is an important fact that such operations do not change the dimension.

Theorem 5.2.3

For any matrix, elementary row operations do not change the dimension of the column space and elementary column operations do not change the dimension of the row space.

Proof
Take the case of row operations first. Let A be a matrix with n columns and suppose that $B = EA$ where E is an elementary matrix. We have to show that the column spaces of A and B have the same dimension. Denote the columns of A by $A_1, A_2, \ldots, A_n$. If some of these columns are linearly dependent, then there are integers $i_1 < i_2 < \cdots < i_r$ and non-zero scalars $c_{i_1}, c_{i_2}, \ldots, c_{i_r}$ such that

$$c_{i_1} A_{i_1} + c_{i_2} A_{i_2} + \cdots + c_{i_r} A_{i_r} = 0.$$

Consequently there is a non-trivial solution C of the linear system $AC = 0$ such that $c_j \neq 0$ for $j = i_1, \ldots, i_r$. Using the equation $B = EA$, we find that $BC = EAC = E0 = 0$. This means that columns $i_1, \ldots, i_r$ of B are also linearly dependent. Therefore, if columns $j_1, \ldots, j_s$ of B are linearly independent, then so are columns $j_1, \ldots, j_s$ of A. Hence the dimension of the column space of B does not exceed the dimension of the column space of A.

Since $A = E^{-1}B$, this argument can be applied equally well to show that the dimension of the column space of A does not exceed that of B. Therefore these dimensions are equal.

The truth of the corresponding statement for row spaces can be quickly deduced from what has just been proved. Let $B = AE$ where E is an elementary matrix. Then $B^T = (AE)^T = E^T A^T$. Now E^T is also an elementary matrix, so by the last paragraph the column spaces of A^T and B^T have the same dimension. But obviously the column space of A^T and the row space of A have the same dimension, and there is a similar statement for B: the required result follows at once.

We are now in a position to connect row and column spaces with normal form and at the same time to clarify a point left open in Chapter Two.

Theorem 5.2.4

If A is any matrix, then the following integers are equal:
  (i)   *the dimension of the row space of A;*
  (ii) *the dimension of the column space of A;*
  (iii) *the number of 1's in a normal form of A.*

Proof

By applying elementary row and column operations to A, we can reduce it to normal form, say

$$N = \begin{pmatrix} I_r & 0 \\ 0 & 0 \end{pmatrix}.$$

Now by 5.2.1 and 5.2.3 the row spaces of A and N have the same dimension, with a like statement for column spaces. But it is clear from the form of N that the dimensions of its row and column spaces are both equal to r, so the result follows.

It is a consequence of 5.2.4 that *every matrix has a unique normal form*; for the normal form is completely determined by the number of 1's on the diagonal.

The rank of a matrix

The *rank* of a matrix is defined to be the dimension of the row or column space. With this definition we can reformulate the condition for a linear system to be consistent.

Theorem 5.2.5

A linear system is consistent if and only if the ranks of the coefficient matrix and the augmented matrix are equal.

This is an immediate consequence of 2.2.1, and 5.2.2.

Finding a basis for a subspace

Suppose that $X_1, X_2, \ldots, X_k$ are vectors in F^n where F is a field. In effect we already know how to find a basis for the subspace generated by these vectors; for this subspace is simply the column space of the matrix $[X_1|X_2|\ldots|X_k]$. But what about subspaces of vector spaces other than F^n? It turns out that use of coordinate vectors allows us to reduce the problem to the case of F^n.

Let V be a vector space over F with a given ordered basis $\mathbf{v}_1, \mathbf{v}_2, \ldots, \mathbf{v}_n$, and suppose that S is the subspace of V generated by some given set of vectors $\mathbf{w}_1, \mathbf{w}_2, \ldots, \mathbf{w}_m$. The problem is to find a basis of S. Recall that each vector in V has a unique expression as a linear combination of the basis vectors $\mathbf{v}_1, \ldots, \mathbf{v}_n$ and hence has a unique coordinate column vector, as described in 5.1. Let $\mathbf{w}_i$ have coordinate column vector X_i with respect to the given basis. Then the coordinate column vector of the linear combination

$c_1 \mathbf{w}_1 + c_2 \mathbf{w}_2 + \cdots + c_k \mathbf{w}_k$ is surely $c_1 X_1 + c_2 X_2 + \cdots + c_k X_k$. Hence the set of all coordinate column vectors of elements of S equals the subspace T of F^n which is generated by $X_1, \ldots, X_k$. Moreover $\mathbf{w}_1, \mathbf{w}_2, \ldots, \mathbf{w}_k$ will be linearly independent if and only if $X_1, X_2, \ldots, X_k$ are. In short $\mathbf{w}_1, \mathbf{w}_2, \ldots, \mathbf{w}_k$ form a basis of S if and only if $X_1, X_2, \ldots, X_k$ form a basis of T; thus our problem is solved.

Example 5.2.2

Find a basis for the subspace of $P_4(\mathbf{R})$ generated by the polynomials $1 - x - 2x^3$, $1 + x^3$, $1 + x + 4x^3$, x^2.

Of course we will use the standard ordered basis for $P_4(\mathbf{R})$ consisting of $1, x, x^2, x^3$. The first step is to write down the coordinate vectors of the given polynomials with respect to the standard basis and arrange them as the columns of a matrix A; thus

$$A = \begin{pmatrix} 1 & 1 & 1 & 0 \\ -1 & 0 & 1 & 0 \\ 0 & 0 & 0 & 1 \\ -2 & 1 & 4 & 0 \end{pmatrix}.$$

To find a basis for the column space of A, use column operations to put it in reduced column echelon form:

$$\begin{pmatrix} 1 & 0 & 0 & 0 \\ 0 & 1 & 0 & 0 \\ 0 & 0 & 1 & 0 \\ 1 & 3 & 0 & 0 \end{pmatrix}.$$

The first three columns form a basis for the column space of A. Therefore we get a basis for the subspace of $P_4(\mathbf{R})$ generated by the given polynomials by simply writing down the polynomials that have these columns as their coordinate column vectors; in this way we arrive at the basis

$$1 + x^3, \quad x + 3x^3, \quad x^2.$$

Hence the subspace generated by the given polynomials has dimension 3.

Exercises 5.2

1. Find bases for the row and column spaces of the following matrices:

$$\text{(a)} \begin{pmatrix} 2 & -3 & 9 \\ 4 & -5 & 36 \end{pmatrix}, \quad \text{(b)} \begin{pmatrix} -1 & 6 & 1 & 5 \\ 3 & 1 & 1 & 5 \\ 1 & 13 & 3 & 15 \end{pmatrix}.$$

2. Find bases for the subspaces generated by the given vectors in the vector spaces indicated:

(a) $1 - 2x - x^3$, $3x - x^2$, $1 + x + x^2 + x^3$, $4 + 7x + x^2 + 2x^3$ in $P_4(\mathbf{R})$;

(b) $\begin{pmatrix} 3 & 4 \\ 1 & 2 \end{pmatrix}$, $\begin{pmatrix} 2 & 5 \\ 1 & 1 \end{pmatrix}$, $\begin{pmatrix} 0 & -1 \\ -1 & 1 \end{pmatrix}$ in $M_2(\mathbf{R})$.

3. Let A be a matrix and let N, R and C be the null space, row space and column space of A respectively. Prove that

$$\dim(R) + \dim(N) = \dim(C) + \dim(N) = n$$

where n is the number of columns of A.

4. If A is any matrix, show that A and A^T have the same rank.

5. Suppose that A is an $m \times n$ matrix with rank r. What is the dimension of the null space of A^T?

6. Let A and B be $m \times n$ and $n \times p$ matrices respectively. Prove that the row space of AB is contained in the row space of B, and the column space of AB is contained in the the column space of A. What can one conclude about the ranks of AB and BA ?

7. The rank of a matrix can be defined as the maximum number of rows in an invertible submatrix: justify this statement.

5.3 Operations with Subspaces

If U and W are subspaces of a vector space V, there are two natural ways of combining U and W to form new subspaces of V. The first of these subspaces is the *intersection*

$$U \cap W,$$

which is the set of all vectors that belong to both U and V.

The second subspace that can be formed from U and W is not, as one might perhaps expect, their union $U \cup W$; for this is not in general closed under addition, so it may not be a subspace. The subspace we are looking for is the *sum*

$$U + W,$$

which is defined to be the set of all vectors of the form $\mathbf{u} + \mathbf{w}$ where $\mathbf{u}$ belongs to U and $\mathbf{w}$ to W.

The first point to note is that these are indeed subspaces.

Theorem 5.3.1
If U and W are subspaces of a vector space V, then $U \cap W$ and $U + W$ are subspaces of V.

Proof
Certainly $U \cap W$ contains the zero vector and it is closed with respect to addition and scalar multiplication since both U and W are; therefore $U \cap W$ is a subspace.

The same method applies to $U + W$. Clearly this contains $\mathbf{0} + \mathbf{0} = \mathbf{0}$. Also, if $\mathbf{u}_1$, $\mathbf{u}_2$ and $\mathbf{w}_1$, $\mathbf{w}_2$ are vectors in U and W respectively, and c is a scalar, then

$$(\mathbf{u}_1 + \mathbf{w}_1) + (\mathbf{u}_2 + \mathbf{w}_2) = (\mathbf{u}_1 + \mathbf{u}_2) + (\mathbf{w}_1 + \mathbf{w}_2)$$

and

$$c(\mathbf{u}_1 + \mathbf{w}_1) = c\mathbf{u}_1 + c\mathbf{w}_1,$$

both of which belong to $U + W$. Thus $U + W$ is closed with respect to addition and scalar multiplication and so it is a subspace.

Example 5.3.1

Consider the subspaces U and W of $\mathbf{R}^4$ consisting of all vectors of the forms

$$\begin{pmatrix} a \\ b \\ c \\ 0 \end{pmatrix} \quad \text{and} \quad \begin{pmatrix} 0 \\ d \\ e \\ f \end{pmatrix}$$

respectively, where a, b, c, d, e are arbitrary scalars. Then $U \cap W$ consists of all vectors of the form

$$\begin{pmatrix} 0 \\ b \\ c \\ 0 \end{pmatrix},$$

while $U + W$ equals $\mathbf{R}^4$ since every vector in $\mathbf{R}^4$ can be expressed as the sum of a vector in U and a vector in W.

For subspaces of a finitely generated vector space there is an important formula connecting the dimensions of their sum and intersection.

Theorem 5.3.2

Let U and W be subspaces of a finitely generated vector space V. Then

$$\dim(U + W) + \dim(U \cap W) = \dim(U) + \dim(W).$$

Proof

If $U = 0$, then obviously $U + W = W$ and $U \cap W = 0$; in this case the formula is certainly true, as it is when $W = 0$.

Assume therefore that $U \neq 0$ and $W \neq 0$, and put $m = \dim(U)$ and $n = \dim(W)$. Consider first the case where $U \cap W = 0$. Let $\{\mathbf{u}_1, \mathbf{u}_2, \ldots, \mathbf{u}_m\}$ and $\{\mathbf{w}_1, \mathbf{w}_2, \ldots, \mathbf{w}_n\}$ be bases of U and W respectively. Then the vectors $\mathbf{u}_1, \ldots, \mathbf{u}_m$ and $\mathbf{w}_1, \ldots, \mathbf{w}_n$ surely generate $U + W$. In fact these vectors are also linearly independent: for if there is a linear relation between them, say

$$c_1 \mathbf{u}_1 + \cdots + c_m \mathbf{u}_m + d_1 \mathbf{w}_1 + \cdots + d_n \mathbf{w}_n = \mathbf{0},$$

then

$$c_1 \mathbf{u}_1 + \cdots + c_m \mathbf{u}_m = (-d_1)\mathbf{w}_1 + \cdots + (-d_n)\mathbf{w}_n,$$

a vector which belongs to both U and W, and so to $U \cap W$, which is the zero subspace. Consequently this vector must be the zero vector. Therefore all the c_i and d_j must be zero since the $\mathbf{u}_i$ are linearly independent, as are the $\mathbf{w}_j$. Consequently the vectors $\mathbf{u}_1, \ldots, \mathbf{u}_m, \mathbf{w}_1, \ldots, \mathbf{w}_n$ form a basis of $U + W$, so that $\dim(U + W) = m + n = \dim(U) + \dim(W)$, the correct formula since $U \cap W = 0$ in the case under consideration.

Now we tackle the more difficult case where $U \cap W \neq 0$. First choose a basis for $U \cap W$, say $\{\mathbf{z}_1, \ldots, \mathbf{z}_r\}$. By 5.1.4 this may be extended to bases of U and of W, say

$$\{\mathbf{z}_1, \ldots, \mathbf{z}_r, \mathbf{u}_{r+1}, \ldots, \mathbf{u}_m\}$$

and

$$\{\mathbf{z}_1, \ldots, \mathbf{z}_r, \mathbf{w}_{r+1}, \ldots, \mathbf{w}_n\}$$

respectively. Now the vectors

$$\mathbf{z}_1 \ldots, \mathbf{z}_r, \mathbf{u}_{r+1}, \ldots, \mathbf{u}_m, \mathbf{w}_{r+1}, \ldots, \mathbf{w}_n$$

generate $U + W$: for we can express any vector of U or W in terms of them. What still needs to be proved is that they are linearly independent. Suppose that in fact there is a linear relation

$$\sum_{i=1}^{r} e_i \mathbf{z}_i + \sum_{j=r+1}^{m} c_j \mathbf{u}_j + \sum_{k=r+1}^{n} d_k \mathbf{w}_k = \mathbf{0}$$

where the e_i, c_j, d_k are scalars. Then

$$\sum_{k=r+1}^{n} d_k \mathbf{w}_k = \sum_{i=1}^{r} (-e_i)\mathbf{z}_i + \sum_{j=r+1}^{m} (-c_j)\mathbf{u}_j,$$

which belongs to both U and W and so to $U \cap W$. The vector $\sum d_k \mathbf{w}_k$ is therefore expressible as a linear combination of the $\mathbf{z}_i$ since these vectors are known to form a basis of the subspace $U \cap W$. However $\mathbf{z}_1, \ldots, \mathbf{z}_r, \mathbf{w}_{r+1}, \ldots, \mathbf{w}_n$ are definitely linearly independent. Therefore all the d_j are zero and our linear relation becomes

$$\sum_{i=1}^{r} e_i \mathbf{z}_i + \sum_{j=r+1}^{m} c_j \mathbf{u}_j = \mathbf{0}.$$

But $\mathbf{z}_1, \ldots, \mathbf{z}_r, \mathbf{u}_{r+1}, , \mathbf{u}_m$ are linearly independent, so it follows that the c_j and the e_i are also zero, which establishes linear independence.

We conclude that the vectors $\mathbf{z}_1, \ldots, \mathbf{z}_r, \mathbf{u}_{r+1}, \ldots, \mathbf{u}_m,$ $\mathbf{w}_{r+1}, \ldots, \mathbf{w}_n$ form a basis of $U + W$. A count of the basis vectors reveals that $\dim(U + W)$ equals

$$r + (m - r) + (n - r) = m + n - r$$
$$= \dim(U) + \dim(W) - \dim(U \cap W).$$

Example 5.3.2

Suppose that U and W are subspaces of $\mathbf{R}^{10}$ with dimensions 6 and 8 respectively. Find the smallest possible dimension for $U \cap W$.

Of course $\dim(\mathbf{R}^{10}) = 10$ and, since $U + W$ is a subspace of $\mathbf{R}^{10}$, its dimension cannot exceed 10. Therefore by 5.3.2

$$\dim(U \cap W) = \dim(U) + \dim(W) - \dim(U + W)$$
$$\geq 6 + 8 - 10 = 4.$$

So the dimension of the intersection is at least 4. The reader is challenged to think of an example which shows that the intersection really can have dimension 4.

Direct sums of subspaces

Let U and W be two subspaces of a vector space V. Then V is said to be the *direct sum* of U and W if

$$V = U + W \quad \text{and} \quad U \cap W = 0.$$

The notation for the direct sum is

$$V = U \oplus W.$$

Notice the consequence of the definition: each vector $\mathbf{v}$ of V has a *unique* expression of the form $\mathbf{v} = \mathbf{u} + \mathbf{w}$ where $\mathbf{u}$ belongs to U and $\mathbf{w}$ to W. Indeed, if there are two such expressions $\mathbf{v} = \mathbf{u}_1 + \mathbf{w}_1 = \mathbf{u}_2 + \mathbf{w}_2$ with $\mathbf{u}_i$ in U and $\mathbf{w}_i$ in W, then $\mathbf{u}_1 - \mathbf{u}_2 = \mathbf{w}_2 - \mathbf{w}_1$, which belongs to $U \cap W = 0$; hence $\mathbf{u}_1 = \mathbf{u}_2$ and $\mathbf{w}_1 = \mathbf{w}_2$.

Example 5.3.3

Let U denote the subset of $\mathbf{R}^3$ consisting of all vectors of the form

$$\begin{pmatrix} a \\ b \\ 0 \end{pmatrix}$$

and let W be the subset of all vectors of the form

$$\begin{pmatrix} 0 \\ 0 \\ c \end{pmatrix}$$

where a, b, c are arbitrary scalars. Then U and W are subspaces of $\mathbf{R}^3$. In addition $U + W = \mathbf{R}^3$ and $U \cap W = 0$. Hence $\mathbf{R}^3 = U \oplus W$.

Theorem 5.3.3
If V is a finitely generated vector space and U and W are subspaces of V such that $V = U \oplus W$, then

$$\dim(V) = \dim(U) + \dim(W).$$

This follows at once from 5.3.2 since $\dim(U \cap W) = 0$.

Direct sums of more than two subspaces

The concept of a direct sum can be extended to any finite set of subspaces. Let $U_1, U_2, \ldots, U_k$ be subspaces of a vector space V. First of all define the *sum* of these subspaces

$$U_1 + \cdots + U_k$$

to be the set of all vectors of the form $\mathbf{u}_1 + \cdots + \mathbf{u}_k$ where $\mathbf{u}_i$ belongs to U_i. This is clearly a subspace of V. The vector space V is said to be the *direct sum* of the subspaces $U_1, \ldots U_k$, in symbols

$$V = U_1 \oplus U_2 \oplus \cdots \oplus U_k,$$

if the following hold:
    (i) $V = U_1 + \cdots + U_k$;
    (ii) for each $i = 1, 2, \ldots, k$ the intersection of U_i with the sum of all the other subspaces U_j, $j \neq i$, equals zero.

In fact these are equivalent to requiring that every element of V be expressible in a unique fashion as a sum of the form $\mathbf{u}_1 + \cdots + \mathbf{u}_k$ where $\mathbf{u}_i$ belongs to U_i.

The concept of a direct sum is a useful one since it often allows us to express a vector space as a direct sum of subspaces that are in some sense simpler.

Example 5.3.4

Let U_1, U_2, U_3 be the subspaces of $\mathbf{R}^5$ which consist of all vectors of the forms

$$
\begin{pmatrix} 0 \\ 0 \\ a \\ 0 \\ 0 \end{pmatrix}, \quad
\begin{pmatrix} 0 \\ b \\ 0 \\ c \\ 0 \end{pmatrix}, \quad
\begin{pmatrix} d \\ 0 \\ 0 \\ 0 \\ e \end{pmatrix}
$$

respectively, where a, b, c, d, e are arbitrary scalars. Then $\mathbf{R}^5 = U_1 \oplus U_2 \oplus U_3$.

Bases for the sum and intersection of subspaces

Suppose that V is a vector space over a field F with positive dimension n and let there be given a specific ordered basis. Assume that we have vectors $\mathbf{u}_1, \ldots, \mathbf{u}_r$ and $\mathbf{w}_1, \ldots, \mathbf{w}_s$, generating subspaces U and W respectively. How can we find bases for the subspaces $U + W$ and $U \cap W$ and hence compute their dimensions?

The first step in the solution is to translate the problem to the vector space F^n. Associate with each $\mathbf{u}_i$ and $\mathbf{w}_j$ its coordinate column vector X_i and Y_j with respect to the given ordered basis of V. Then $X_1, \ldots, X_r$ and $Y_1, \ldots, Y_s$ generate respective subspaces U^* and W^* of F^n. It is sufficient if we can find bases for $U^* + W^*$ and $U^* \cap W^*$ since from these bases for $U + W$ and $U \cap W$ can be read off. So assume from now on that V equals F^n.

Take the case of $U + W$ first – it is the easier one. Let A be the matrix whose columns are $\mathbf{u}_1, \ldots, \mathbf{u}_r$: remember that these are now n-column vectors. Also let B be the matrix whose columns are $\mathbf{w}_1, \ldots, \mathbf{w}_s$. Then $U+W$ is just the column space of the matrix $M = [A \mid B]$. A basis for $U + W$ can therefore be found by putting M in reduced column echelon form and deleting the zero columns.

Turning now to $U \cap W$, we look for scalars c_i and d_j such that

$$c_1 \mathbf{u}_1 + \cdots + c_r \mathbf{u}_r = d_1 \mathbf{w}_1 + \cdots + d_s \mathbf{w}_s :$$

for every element of $U \cap W$ is of this form. Equivalently

$$c_1 \mathbf{u}_1 + \cdots + c_r \mathbf{u}_r + (-d_1) \mathbf{w}_1 + \cdots + (-d_s) \mathbf{w}_s = \mathbf{0}.$$

Now this equation asserts that the vector

$$\begin{pmatrix} c_1 \\ \vdots \\ c_r \\ -d_1 \\ \vdots \\ -d_s \end{pmatrix}$$

belongs to the null space of $[A \mid B]$. A method for finding a basis for the null space of a matrix was described in 5.1. To complete the process, read off the the first r entries of each vector in the basis of the null space of $[A \mid B]$, and take these entries to be $c_1, \ldots, c_r$. The resulting vectors form a basis of $U \cap W$.

Example 5.3.5
Let

$$M = \begin{pmatrix} 1 & 0 & 2 & 1 \\ 2 & 1 & 5 & 2 \\ 2 & -2 & -1 & -1 \\ 1 & 1 & 5 & 3 \end{pmatrix}$$

and denote by U and W the subspaces of $\mathbf{R}^4$ generated by columns 1 and 2, and by columns 3 and 4 of M respectively. Find a basis for $U + W$.

Apply the procedure for finding a basis of the column space of M. Putting M in reduced column echelon form, we obtain

$$\begin{pmatrix} 1 & 0 & 0 & 0 \\ 0 & 1 & 0 & 0 \\ 0 & 0 & 1 & 0 \\ 3 & -1/3 & -2/3 & 0 \end{pmatrix}.$$

The first three columns of this matrix form a basis of $U + W$; hence $\dim(U + W) = 3$.

Example 5.3.6

Find a basis of $U \cap W$ where U and W are the subspaces of Example 5.3.5.

Following the procedure indicated above, we put the matrix M in reduced row echelon form:

$$\begin{pmatrix} 1 & 0 & 0 & -1 \\ 0 & 1 & 0 & -1 \\ 0 & 0 & 1 & 1 \\ 0 & 0 & 0 & 0 \end{pmatrix}.$$

From this a basis for the null space of M can be read off, as described in the paragraph preceding 5.1.7; in this case the basis has the single element

$$\begin{pmatrix} 1 \\ 1 \\ -1 \\ 1 \end{pmatrix}.$$

Therefore a basis for $U \cap W$ is obtained by taking the linear combination of the generating vectors of U corresponding to the scalars in the first two rows of this vector, that is to say

$$
1 \cdot \begin{pmatrix} 1 \\ 2 \\ 2 \\ 1 \end{pmatrix} + 1 \cdot \begin{pmatrix} 0 \\ 1 \\ -2 \\ 1 \end{pmatrix} = \begin{pmatrix} 1 \\ 3 \\ 0 \\ 2 \end{pmatrix}.
$$

Thus $\dim(U \cap W) = 1$.

Example 5.3.7

Find bases for the sum and intersection of the subspaces U and W of $P_4(\mathbf{R})$ generated by the respective sets of polynomials

$$
\{1 + 2x + x^3,\ 1 - x - x^2\} \text{ and } \{x + x^2 - 3x^3,\ 2 + 2x - 2x^3\}.
$$

The first step is to translate the problem to $\mathbf{R}^4$ by writing down the coordinate columns of the given polynomials with respect to the standard ordered basis $1,\ x,\ x^2,\ x^3$ of $P_4(\mathbf{R})$. Arranged as the columns of a matrix, these are

$$
A = \begin{pmatrix} 1 & 1 & 0 & 2 \\ 2 & -1 & 1 & 2 \\ 0 & -1 & 1 & 0 \\ 1 & 0 & -3 & -2 \end{pmatrix}.
$$

Let U^* and W^* be the subspaces of $\mathbf{R}^4$ generated by the coordinate columns of the polynomials that generate U and W, that is, by columns 1 and 2, and by columns 3 and 4 of A respectively. Now find bases for $U^* + W^*$ and $U^* \cap W^*$, just as in Examples 5.3.5 and 5.3.6. It emerges that $U^* + W^*$, which is just the column space of A, has a basis

$$
\begin{pmatrix} 1 \\ 0 \\ 0 \\ -3 \end{pmatrix}, \begin{pmatrix} 0 \\ 1 \\ 0 \\ 2 \end{pmatrix}, \begin{pmatrix} 0 \\ 0 \\ 1 \\ -5 \end{pmatrix}.
$$

On writing down the polynomials with these coordinate vectors, we obtain the basis $1 - 3x^3$, $x + 2x^3$, $x^2 - 5x^3$ for $U^* + W^*$.

In the case of $U \cap W$ the procedure is to find a basis for $U^* \cap W^*$. This turns out to consist of the single vector

$$\begin{pmatrix} 1 \\ 1 \\ 1 \\ -1 \end{pmatrix}.$$

Finally, read off that the polynomial

$$1 \cdot (1 + 2x + x^3) + 1 \cdot (1 - x - x^2) = 2 + x - x^2 + x^3$$

forms a basis of $U \cap W$.

Quotient Spaces

We conclude the section by describing another subspace operation, the formation of the quotient space of a vector space with respect to a subspace. This new vector space is formed by identifying the vectors in certain subsets of the given vector space, which is a construction found throughout algebra.

Proceeding now to the details, let us consider a vector space V with a fixed subspace U. The first step is to define certain subsets called cosets: the *coset* of U containing a given vector $\mathbf{v}$ is the subset of V

$$\mathbf{v} + U = \{\mathbf{v} + \mathbf{u} \mid \mathbf{u} \in U\}.$$

Notice that the coset $\mathbf{v} + U$ really does contain the vector $\mathbf{v}$ since $\mathbf{v} = \mathbf{v} + \mathbf{0} \in \mathbf{v} + U$. Observe also that the coset $\mathbf{v} + U$ can be represented by any one of its elements in the sense that $(\mathbf{v} + \mathbf{u}) + U = \mathbf{v} + U$ for all $\mathbf{u} \in U$.

An important feature of the cosets of a given subspace is that they are disjoint, i.e., they do not overlap.

Lemma 5.3.4

If U is a subspace of a vector space V, then distinct cosets of U are disjoint. Thus V is the disjoint union of all the distinct cosets of U.

Proof

Suppose that cosets $\mathbf{v} + U$ and $\mathbf{w} + U$ both contain a vector $\mathbf{x}$: we will show that these cosets are the same. By hypothesis there are vectors $\mathbf{u}_1$, $\mathbf{u}_2$ in U such that

$$\mathbf{x} = \mathbf{v} + \mathbf{u}_1 = \mathbf{w} + \mathbf{u}_2.$$

Hence $\mathbf{v} = \mathbf{w} + \mathbf{u}$ where $\mathbf{u} = \mathbf{u}_2 - \mathbf{u}_1 \in U$, and consequently $\mathbf{v} + U = (\mathbf{w} + \mathbf{u}) + U = \mathbf{w} + U$, since $\mathbf{u} + U = U$, as claimed. Finally, V is the union of all the cosets of U since $\mathbf{v} \in \mathbf{v} + U$.

The set of all cosets of U in V is written

$$V/U.$$

A good way to think about V/U is that its elements arise by identifying all the elements in a coset, so that each coset has been "compressed" to a single vector.

The next step in the construction is to turn V/U into a vector space by defining addition and scalar multiplication on it. There are natural definitions for these operations, namely

$$\begin{cases} (\mathbf{v} + U) + (\mathbf{w} + U) = (\mathbf{v} + \mathbf{w}) + U \\ \qquad c(\mathbf{v} + U) = (c\mathbf{v}) + U \end{cases}$$

where $\mathbf{v}$, $\mathbf{w} \in V$ and c is a scalar.

Although these definitions look natural, some care must be exercised. For a coset can be represented by any of its vectors, so we must make certain that the definitions just given do not depend on the choice of $\mathbf{v}$ and $\mathbf{w}$ in the cosets $\mathbf{v} + U$ and $\mathbf{w} + U$.

To verify this, suppose we had chosen different representatives, say $\mathbf{v}'$ for $\mathbf{v} + U$ and $\mathbf{w}'$ for $\mathbf{w} + U$. Then $\mathbf{v}' = \mathbf{v} + \mathbf{u}_1$ and $\mathbf{w}' = \mathbf{w} + \mathbf{u}_2$ where $\mathbf{u}_1, \mathbf{u}_2 \in U$. Therefore

$$\mathbf{v}' + \mathbf{w}' = (\mathbf{v} + \mathbf{w}) + (\mathbf{u}_1 + \mathbf{u}_2) \in (\mathbf{v} + \mathbf{w}) + U,$$

so that $(\mathbf{v}' + \mathbf{w}') + U = (\mathbf{v} + \mathbf{w}) + U$. Also $c\mathbf{v}' = c\mathbf{v} + c\mathbf{u}_1 \in (c\mathbf{u}) + U$ and hence $c\mathbf{v}' + U = c\mathbf{v} + U$. These arguments show that our definitions are free from dependency on the choice of coset representatives.

Theorem 5.3.5

If U is a subspace of a vector space V over a field F, then V/U is a vector space over F where sum and scalar multiplication are defined above: also the zero vector is $\mathbf{0} + U = U$ and the negative of $\mathbf{v} + U$ is $(-\mathbf{v}) + U$.

Proof
We have to check that the vector space axioms hold for V/U, which is an entirely routine task. As an example, let us verify one of the distributive laws. Let $\mathbf{v}, \mathbf{w} \in V$ and let $c \in F$. Then by definition

$$\begin{aligned} c((\mathbf{v} + U) + (\mathbf{w} + U)) = c((\mathbf{v} + \mathbf{w}) + U) &= c(\mathbf{v} + \mathbf{w}) + U \\ &= (c\mathbf{v} + c\mathbf{w}) + U, \end{aligned}$$

which by definition equals $(c\mathbf{v}+U)+(c\mathbf{w}+U)$. This establishes the distributive law. Verification of the other axioms is left to the reader as an exercise. It also is easy to check that $\mathbf{0}$ is the zero vector and $(-\mathbf{v}) + U$ the negative of $\mathbf{v} + U$.

Example 5.3.8

Suppose we take U to be the zero subspace of the vector space V: then $V/0$ consists of all $\mathbf{v} + 0 = \{\mathbf{v}\}$, i.e., the one-element subsets of V. While $V/0$ is not the same vector space as V, the two spaces are clearly very much alike: this can be made precise by saying that they are isomorphic (see 6.3).

At the opposite extreme, we could take $U = V$. Now V/V consists of the cosets $\mathbf{v} + V = V$, i.e., there is just one element. So V/V is a zero vector space.

We move on to more interesting examples of coset formation.

Example 5.3.9

Let S be the set of all solutions of a consistent linear system $AX = B$ of m equations in n unknowns over a field F. If $B = 0$, then S is a subspace of F^n, namely, the solution space U of the associated homogeneous linear system $AX = 0$. However, if $B \neq 0$, then S is not a subspace: but we will see that it is a coset of the subspace U.

Since the system is consistent, there is at least one solution, say X_1. Suppose X is another solution. Then we have $AX_1 = B$ and $AX = B$. Subtracting the first of these equations from the second, we find that

$$0 = AX - AX_1 = A(X - X_1),$$

so that $X - X_1 \in U$ and $X \in X_1 + U$, where U is the solution space of the system $AX = 0$. Hence every solution of $AX = B$ belongs to the coset $X_1 + U$ and thus $S \subseteq X_1 + U$.

Conversely, consider any $Y \in X_1 + U$, say $Y = X_1 + Z$ where $Z \in U$. Then $AY = AX_1 + AZ = B + 0 = B$. Therefore $Y \in S$ and $S = X_1 + U$.

These considerations have established the following result.

Theorem 5.3.6

Let $AX = B$ be a consistent linear system. Let X_1 be any fixed solution of the system and let U be the solution space of the associated homogeneous linear system $AX = 0$. Then the set of all solutions of the linear system $AX = B$ is the coset $X_1 + U$.

Our last example of coset formation is a geometric one.

Example 5.3.10

Let A and B be vectors in $\mathbf{R}^3$ representing non-parallel line segments in 3-dimensional space. Then the subspace

$$U = \, <A, \ B>$$

has dimension 2 and consists of all $cA + dB$, $(c, d \in \mathbf{R})$. The vectors in U are represented by line segments, drawn from the origin, which lie in a plane P. Now choose $X \in \mathbf{R}^3$, with $X = (x_1, \ x_2, \ x_3)^T$.

A typical vector in the coset $X + U$ has the form $X + cA + dB$, with $c, d \in \mathbf{R}$, i.e.,

$$(x_1 + ca_1 + db_1 \quad x_2 + ca_2 + db_2 \quad x_3 + ca_3 + db_3)^T.$$

Now the points $(x_1 + ca_1 + db_1, \ x_2 + ca_2 + db_2, \ x_3 + ca_3 + db_3)$ lie in the plane P_1 passing through the point (x_1, x_2, x_3), which is parallel to the plane P. This is seen by forming the line segment joining two such points. The elements of $X + U$ correspond to the points in the plane P_1: the latter is called a *translate* of the plane P.

Dimension of a Quotient Space

We conclude the discussion by noting a simple formula for the dimension of a quotient space of a finite dimensional vector space.

Theorem 5.3.7
Let U be a subspace of a finite dimensional vector space V. Then

$$\dim(V/U) \ = \ \dim(V) \ - \ \dim(U).$$

Proof

If $U = 0$, then $\dim(U) = 0$ and $V/0 = \{\{\mathbf{v}\} \mid \mathbf{v} \in V\}$, which clearly has the same dimension as V. Thus the formula is valid in this case.

Now let $U \neq 0$ and choose a basis $\{\mathbf{u}_1, \ldots, \mathbf{u}_m\}$ of U. By 5.1.4 we may extend this to a basis

$$\{\mathbf{u}_1, \ldots, \mathbf{u}_m, \mathbf{u}_{m+1}, \ldots, \mathbf{u}_n\}$$

of V. Here of course $m = \dim(U)$ and $n = \dim(V)$. A typical element $\mathbf{v}$ of V has the form $\mathbf{v} = \sum\limits_{i=1}^{n} c_i \mathbf{u}_i$, where the c_i are scalars. Next

$$\mathbf{v} + U = \Big(\sum_{i=m+1}^{n} c_i \mathbf{u}_i \Big) + U = \sum_{i=m+1}^{n} c_i (\mathbf{u}_i + U),$$

since $\sum\limits_{i=1}^{m} c_i \mathbf{u}_i \in U$. Hence $\mathbf{u}_{m+1} + U, \ldots, \mathbf{u}_n + U$ generate the quotient space V/U.

On the other hand, if $\sum\limits_{i=m+1}^{n} c_i (\mathbf{u}_i + U) = \mathbf{0}_{V/U} = U$, then $\sum\limits_{i=m+1}^{n} c_i \mathbf{u}_i \in U$, so that this vector is a linear combination of $\mathbf{u}_1, \ldots, \mathbf{u}_m$. Since the $\mathbf{u}_i$ are linearly independent, it follows that $c_{m+1} = \cdots = c_n = 0$. Therefore $\mathbf{u}_{m+1} + U, \ldots, \mathbf{u}_n + U$ form a basis of V/U and hence

$$\dim(V/U) = n - m = \dim(V) - \dim(U).$$

Exercises 5.3

1. Find three distinct subspaces U, V, W of $\mathbf{R}^2$ such that $\mathbf{R}^2 = U \oplus V = V \oplus W = W \oplus U$.

2. Let U and W denote the sets of all $n \times n$ real symmetric and skew-symmetric matrices respectively. Show that these are subspaces of $M_n(\mathbf{R})$, and that $M_n(\mathbf{R})$ is the direct sum of U and W. Find $\dim(U)$ and $\dim(W)$.

3. Let U and W be subspaces of a vector space V and suppose that each vector $\mathbf{v}$ in V has a *unique* expression of the form $\mathbf{v} = \mathbf{u} + \mathbf{w}$ where $\mathbf{u}$ belongs to U and $\mathbf{w}$ to W. Prove that $V = U \oplus W$.

4. Let U, V, W be subspaces of some vector space and suppose that $U \subseteq W$. Prove that

$$(U + V) \cap W = U + (V \cap W).$$

5. Prove or disprove the following statement: if U, V, W are subspaces of a vector space, then $(U + V) \cap W = (U \cap W) + (V \cap W)$.

6. Suppose that U and W are subspaces of $P_{14}(\mathbf{R})$ with $\dim(U) = 7$ and $\dim(W) = 11$. Show that $\dim(U \cap W) \geq 4$. Give an example to show that this minimum dimension can occur.

7. Let M be the matrix

$$\begin{pmatrix} 3 & 3 & 2 & 8 \\ 1 & 1 & -1 & 1 \\ 1 & 1 & 3 & 5 \\ -2 & 4 & 6 & 8 \end{pmatrix}$$

and let U and W be the subspaces of $\mathbf{R}^4$ generated by rows 1 and 2 of M, and by rows 3 and 4 of M respectively. Find the dimensions of $U + W$ and $U \cap W$.

8. Define polynomials

$$f_1 = 1 - 2x + x^3, \quad f_2 = x + x^2 - x^3.$$

and

$$g_1 = 2 + 2x - 4x^2 + x^3, \ g_2 = 1 - x + x^2, \ g_3 = 2 + 3x - x^2.$$

Let U be the subspace of $P_4(\mathbf{R})$ generated by $\{f_1, \ f_2\}$ and let W be the subspace generated by $\{g_1, \ g_2, \ g_3\}$. Find bases for the subspaces $U + W$ and $U \cap W$.

9. Let $U_1, \ldots, U_k$ be subspaces of a vector space V. Prove that $V = U_1 \oplus \cdots \oplus U_k$ if and only if each element of V has a unique expression of the form $\mathbf{u}_1 + \cdots + \mathbf{u}_k$ where $\mathbf{u}_i$ belongs to U_i.

10. Every vector space of dimension n is a direct sum of n subspaces each of which has dimension 1. Explain why this true.

11. If $U_1, \ldots, U_k$ are subspaces of a finitely generated vector space whose sum is the direct sum, find the dimension of $U_1 \oplus \cdots \oplus U_k$.

12. Let U_1, U_2, U_3 be subspaces of a vector space such that $U_1 \cap U_2 = U_2 \cap U_3 = U_3 \cap U_1 = 0$. Does it follow that $U_1 + U_2 + U_3 = U_1 \oplus U_2 \oplus U_3$? Justify your answer.

13. Verify that all the vector space axioms hold for a quotient space V/U.

14. Consider the linear system of Exercise 2.1.1,

$$\begin{cases} x_1 & + \ 2x_2 & - \ 3x_3 & + \ x_4 & = 7 \\ -x_1 & + \ x_2 & - \ x_3 & + \ x_4 & = 4 \end{cases}$$

(a) Write the general solution of the system in the form $X_0 + Y$, where X_0 is a particular solution and Y is the general solution of the associated homogeneous system.

(b) Identify the set of all solutions of the given linear system as a coset of the solution space of the associated homogeneous linear system.

15. Find the dimension of the quotient space $P_n(\mathbf{R})/U$ where U is the subspace of all real constant polynomials.

16. Let V be an n-dimensional vector space over an arbitrary field. Prove that there exists a quotient space of V of each dimension i where $0 \leq i \leq n$.

17. Let V be a finite-dimensional vector space and let U and W be two subspaces of V. Prove that

$$\dim((U+W)/W) = \dim(U/(U \cap W)).$$

Chapter Six

LINEAR TRANSFORMATIONS

A linear transformation is a function between two vector spaces which relates the structures of the spaces. Linear transformations include operations as diverse as multiplication of column vectors by matrices and differentiation of functions of a real variable. Despite their diversity, linear transformations have many common properties which can be exploited in different contexts. This is a good reason for studying linear transformations and indeed much else in linear algebra.

In order to establish notation and basic ideas, we begin with a brief discussion of functions defined on arbitrary sets. Readers who are familiar with this elementary material may wish to skip 6.1.

6.1 Functions Defined on Sets

If X and Y are two non-empty sets, a *function* or *mapping* from X to Y,

$$F : X \to Y,$$

is a rule that assigns to each element x of X a unique element $F(x)$ of Y, called the *image of x under F*. The sets X and Y are called the *domain* and *codomain* of the function F respectively. The set of all images of elements of X is called the *image of the function F*; it is written

$$\operatorname{Im}(F).$$

Examples of functions abound; the most familiar are quite likely the functions that arise in calculus, namely functions whose domain and codomain are subsets of the set of real numbers **R**. An example of a function which has the flavor of linear algebra is $F : M_{m,n}(\mathbf{R}) \to \mathbf{R}$ defined by $F(A) = \det(A)$, that is, the determinant function.

A very simple, but nonetheless important, example of a function is the *identity function* on a set X; this is the function

$$1_X : X \to X$$

which leaves every element of the set X fixed, that is, $1_X(x) = x$ for all elements x of X.

Next, three important special types of function will be introduced. A function $F : X \to Y$ is said to be *injective* (or *one-one*) if distinct elements of X always have distinct images under F, that is, if the equation $F(x_1) = F(x_2)$ implies that $x_1 = x_2$. On the other hand, F is said to be *surjective* (or *onto*) if every element y of Y is the image under F of at least one element of X, that is, if $y = F(x)$ for some x in X. Finally, F is said to be *bijective* (or a *one-one correspondence*) if it is both injective and surjective.

We need to give some examples to illustrate these concepts. For convenience these will be real-valued functions of a real variable x.

Example 6.1.1

Define $F_1 : \mathbf{R} \to \mathbf{R}$ by the rule $F_1(x) = 2^x$. Then F_1 is injective since $2^x = 2^y$ clearly implies that $x = y$. But F_1 cannot be surjective since 2^x is always positive and so, for example, 0 is not the image of any element under F.

Example 6.1.2

Define a function $F_2 : \mathbf{R} \to \mathbf{R}$ by $F_2(x) = x^2(x - 1)$. Here F_2 is not injective; indeed $F_2(0) = 0 = F_2(1)$. However F_2 is surjective since the expression $x^2(x-1)$ assumes all real values as x varies. The best way to see this is to draw the graph of the function $y = x^2(x - 1)$ and observe that it extends over the entire y-axis.

Example 6.1.3

Define $F_3 : \mathbf{R} \to \mathbf{R}$ by $F_2(x) = 2x - 1$. This function is both injective and surjective, so it is bijective. (The reader should supply the proof.)

Composition of functions

Consider two functions $F : X \to Y$ and $G : U \to V$ such that the image of G is a subset of X. Then it is possible to combine the functions to produce a new function called the *composite* of F and G

$$F \circ G \; : \; U \to Y,$$

by applying first G and then F; thus the image of an element x of U is given by the formula

$$F \circ G(x) = F(G(x)).$$

Here it is necessary to know that $\text{Im}(G)$ is contained in X, since otherwise the expression $F(G(x))$ might be meaningless.

Example 6.1.4

Consider the functions $F : \mathbf{R}^2 \to \mathbf{R}$ and $G : \mathbf{C} \to \mathbf{R}^2$ defined by the rules

$$F(\begin{pmatrix} a \\ b \end{pmatrix}) = \sqrt{a^2 + b^2} \text{ and } G(a + \sqrt{-1}b) = \begin{pmatrix} 2a \\ 2b \end{pmatrix}.$$

Here a and b are arbitrary real numbers. Then $F \circ G : \mathbf{C} \to \mathbf{R}$ exists and its effect is described by

$$F \circ G(a + \sqrt{-1}b) = F(\begin{pmatrix} 2a \\ 2b \end{pmatrix}) = \sqrt{4a^2 + 4b^2}.$$

A basic fact about functional composition is that it satisfies the associative law. First let us agree that two functions

F and G are to be considered *equal* – in symbols $F = G$ – if they have the same domain and codomain and if $F(x) = G(x)$ for all x.

Theorem 6.1.1
Let $F : X \rightarrow Y$, $G : U \rightarrow V$ and $H : R \rightarrow S$ be functions such that $\operatorname{Im}(H)$ is contained in U and $\operatorname{Im}(G)$ is contained in X. Then $F \circ (G \circ H) = (F \circ G) \circ H$.

Proof
First observe that the various composites mentioned in the formula make sense: this is because of the assumptions about $\operatorname{Im}(H)$ and $\operatorname{Im}(G)$. Let x be an element of X. Then, by the definition of a composite,

$$F \circ (G \circ H)(x) = F((G \circ H)(x)) = F(G(H(x))).$$

In a similar manner we find that $(F \circ G) \circ H(x)$ is also equal to this element. Therefore $F \circ (G \circ H) = (F \circ G) \circ H$, as claimed.

Another basic result asserts that a function is unchanged when it is composed with an identity function.

Theorem 6.1.2
If $F : X \rightarrow Y$ is any function, then $F \circ 1_X = F = 1_Y \circ F$.

The very easy proof is left to the reader as an exercise.

Inverses of functions
Suppose that $F : X \rightarrow Y$ is a function. An *inverse* of F is a function of the form $G : Y \rightarrow X$ such that $F \circ G$ and $G \circ F$ are the identity functions on Y and on X respectively, that is,
$$F(G(y)) = y \quad \text{and} \quad G(F(x)) = x$$

for all x in X and y in Y. A function which has an inverse is said to be *invertible*.

Example 6.1.5

Consider the functions F and G with domain and codomain **R** which are defined by $F(x) = 2x - 1$ and $G(x) = (x+1)/2$. Then G is an inverse of F since $F \circ G$ and $G \circ F$ are both equal to $1_{\mathbf{R}}$. Indeed

$$F \circ G(x) = F(G(x)) = F((x+1)/2) = 2((x+1)/2) - 1 = x,$$

with a similar computation for $G \circ F(x)$.

Not every function has an inverse; in fact a basic theorem asserts that only the bijective ones do.

Theorem 6.1.3

A function $F : X \to Y$ has an inverse if and only if it is bijective.

Proof

Suppose first that F has an inverse function $G : Y \to X$. If $F(x_1) = F(x_2)$, then, on applying G to both sides, we obtain $G \circ F(x_1) = G \circ F(x_2)$. But $G \circ F$ is the identity function on X, so $x_1 = x_2$. Hence F is injective. Next let y be any element of Y; then, since $F \circ G$ is the identity function, $y = F \circ G(y) = F(G(y))$, which shows that y belongs to the image of F and F is surjective. Therefore F is bijective.

Conversely, assume that F is a bijective function. We need to find an inverse function $G : Y \to X$ for F. To this end let y belong to Y; then, since F is surjective, $y = F(x)$ for some x in X; moreover x is uniquely determined by y since F is injective. This allows us to define $G(y)$ to be x. Then $G(F(x)) = G(y) = x$ and $F(G(y)) = F(x) = y$. Here it is necessary to observe that every element of X is of the form $G(y)$ for some y in Y, so that $G(F(x))$ equals x for *all* elements x of X. Therefore G is an inverse function for F.

The next observation is that when inverse functions do exist, they are unique.

Theorem 6.1.4
Every bijective function $F : X \to Y$ *has a unique inverse function.*

Proof
Suppose that F has two inverse functions, say G_1 and G_2. Then $(G_1 \circ F) \circ G_2 = 1_X \circ G_2 = G_2$ by 6.1.2. On the other hand, by 6.1.1 this function is also equal to $G_1 \circ (F \circ G_2) = G_1 \circ 1_Y = G_1$. Thus $G_1 = G_2$.

Because of this result it is unambiguous to denote the inverse of a bijective function $F : X \to Y$ by

$$F^{-1} : Y \to X.$$

To conclude this brief account of the elementary theory of functions, we record two frequently used results about inverse functions.

Theorem 6.1.5
(a) *If* $F : X \to Y$ *is an invertible function, then* F^{-1} *is invertible with inverse* F.
(b) *If* $F : X \to Y$ *and* $G : U \to X$ *are invertible functions, then the function* $F \circ G : U \to Y$ *is invertible and its inverse is* $G^{-1} \circ F^{-1}$.

Proof
Since $F \circ F^{-1} = 1_Y$ and $F^{-1} \circ F = 1_X$, it follows that F is the inverse of F^{-1}. For the second statement it is enough to check that when $G^{-1} \circ F^{-1}$ is composed with $F \circ G$ on both sides, identity functions result. To prove this simply apply the associative law twice.

Exercises 6.1

1. Label each of the following functions $F : \mathbf{R} \to \mathbf{R}$ injective, surjective or bijective, as is most appropriate. (You may wish to draw the graph of the function in some cases):

(a) $F(x) = x^2$; (b) $F(x) = x^3/(x^2 + 1)$;

(c) $F(x) = x(x - 1)(x - 2)$; (d) $F(x) = e^x + 2$.

2. Let functions F and G from $\mathbf{R}$ to $\mathbf{R}$ be defined by $F(x) = 2x - 3$, and $G(x) = (x^2 - 1)/(x^2 + 1)$. Show that the composite functions $F \circ G$ and $G \circ F$ are different.

3. Verify that the following functions from $\mathbf{R}$ to $\mathbf{R}$ are mutually inverse: $F(x) = 3x - 5$ and $G(x) = (x + 5)/3$.

4. Find the inverse of the bijective function $F : \mathbf{R} \to \mathbf{R}$ defined by $F(x) = 2x^3 - 5$.

5. Let $G : Y \to X$ be an injective function. Construct a function $F : X \to Y$ such that $F \circ G$ is the identity function on Y. Then use this result to show that there exist functions $F, G : \mathbf{R} \to \mathbf{R}$ such that $F \circ G = 1_{\mathbf{R}}$ but $G \circ F \neq 1_{\mathbf{R}}$.

6. Prove 6.1.2.

7. Complete the proof of part (b) of 6.1.5.

6.2 Linear Transformations and Matrices

After the preliminaries on functions, we proceed at once to the fundamental definition of the chapter, that of a linear transformation. Let V and W be two vector spaces over the same field of scalars F. A *linear transformation* (or *linear mapping*) from V to W is a function

$$T : V \to W$$

with the properties

$$T(\mathbf{v}_1 + \mathbf{v}_2) = T(\mathbf{v}_1) + T(\mathbf{v}_2) \text{ and } T(c\mathbf{v}) = cT(\mathbf{v})$$

for all vectors $\mathbf{v}$, $\mathbf{v}_1$, $\mathbf{v}_2$ in V and all scalars c in F. In short the function T is required to act in a "linear" fashion on sums and scalar multiples of vectors in V. In the case where T is a linear transformation from V to V, we say that T is a *linear operator on* V.

Of course we need some examples of linear transformations, but these are not hard to find.

Example 6.2.1

Let the function $T : \mathbf{R}^3 \to \mathbf{R}^2$ be defined by the rule

$$T(\begin{pmatrix} a \\ b \\ c \end{pmatrix}) = \begin{pmatrix} a \\ b \end{pmatrix}.$$

Thus T simply "forgets" the third entry of a vector. From this definition it is obvious that T is a linear transformation.

Now recall from Chapter Four the geometrical interpretation of the column vector with entries a, b, c as the line segment joining the origin to the point with coordinates (a, b, c). Then the linear transformation T projects the line segment onto the xy-plane. Consequently projection of a line in 3-dimensional space which passes through the origin onto the xy-plane is a linear transformation from $\mathbf{R}^3$ to $\mathbf{R}^2$.

The next example of a linear transformation is also of a geometrical nature.

Example 6.2.2

Suppose that an anti-clockwise rotation through angle θ about the origin O is applied to the xy-plane. Since vectors in $\mathbf{R}^2$ are represented by line segments in the plane drawn from the origin, such a rotation determines a function $T : \mathbf{R}^2 \to \mathbf{R}^2$; here the line segment representing $T(X)$ is obtained by rotating the line segment that represents X.

To show that T is a linear operator on $\mathbf{R}^2$, we suppose that Y is another vector in $\mathbf{R}^2$.

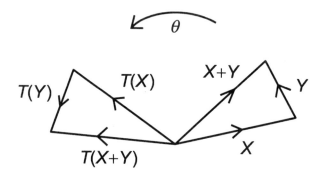

Referring to the diagram above, we know from the triangle rule that $X+Y$ is represented by the third side of the triangle formed by the line segments representing X and Y. When the rotation is applied to this triangle, the sides of the resulting triangle represent the vectors $T(X)$, $T(Y)$, $T(X)+T(Y)$, as shown in the diagram. The triangle rule then shows that $T(X+Y) = T(X) + T(Y)$.

In a similar way we can see from the geometrical interpretation of scalar multiples in $\mathbf{R}^2$ that $T(cX) = cT(X)$ for any scalar c. It follows that T is a linear operator on $\mathbf{R}^2$.

Example 6.2.3

Define $T : D_\infty[a, b] \to D_\infty[a, b]$ to be differentiation, that is,

$$T(f(x)) = f'(x).$$

Here $D_\infty[a, b]$ denotes the vector space of all functions of x that are infinitely differentiable in the interval $[a, b]$. Then well-known facts from calculus guarantee that T is a linear operator on $D_\infty[a, b]$.

This example can be generalized in a significant fashion as follows. Let $a_1, a_2, \ldots, a_n$ be functions in $D_\infty[a, b]$. For any

f in $D_\infty[a, b]$, define $T(f)$ to be

$$a_n f^{(n)} + a_{n-1} f^{(n-1)} + \cdots + a_1 f' + a_0 f.$$

Then T is a linear operator on $D_\infty[a, b]$, once again by elementary results from calculus. Here one can think of T as a sort of generalized differential operator that can be applied to functions in $D_\infty[a, b]$.

Our next example of a linear transformation involves quotient spaces, which were defined in 5.3.

Example 6.2.4

Let U be a subspace of a vector space V and define a function $T : V \to V/U$ by the rule $T(\mathbf{v}) = \mathbf{v} + U$. It is simple to verify that T is a linear transformation: indeed,

$$T(\mathbf{v}_1 + \mathbf{v}_2) = (\mathbf{v}_1 + \mathbf{v}_2) + U = (\mathbf{v}_1 + U) + (\mathbf{v}_2 + U)$$
$$= T(\mathbf{v}_1) + T(\mathbf{v}_2)$$

by definition of the sum of two vectors in a quotient space. In a similar way one can show that $T(c\mathbf{v}) = c(T(\mathbf{v}))$.

The function just defined is often called the *canonical linear transformation associated with the subspace U*.

Finally, we record two very simple examples of linear transformations.

Example 6.2.5

(a) Let V and W be two vector spaces over the same field. The function which sends every vector in V to the zero vector of W is a linear transformation called the *zero linear transformation* from V to W; it is written

$$0_{V,W} \quad \text{or} \quad \text{simply} \quad 0.$$

(b) The identity function $1_V : V \to V$ is a linear operator on V.

After these examples it is time to present some elementary properties of linear transformations.

Theorem 6.2.1
Let $T : V \to W$ be a linear transformation. Then

$$T(\mathbf{0}_V) = \mathbf{0}_W$$

and

$$T(c_1\mathbf{v}_1 + c_2\mathbf{v}_2 + \cdots + c_k\mathbf{v}_k) = c_1 T(\mathbf{v}_1) + c_2 T(\mathbf{v}_2) + \cdots + c_k T(\mathbf{v}_k)$$

for all vectors $\mathbf{v}_i$ and scalars c_i.

Thus a linear transformation always sends a zero vector to a zero vector; it also sends a linear combination of vectors to the corresponding linear combination of the images of the vectors.

Proof
In the first place we have

$$T(\mathbf{0}_V) = T(\mathbf{0}_V + \mathbf{0}_V) = T(\mathbf{0}_V) + T(\mathbf{0}_V)$$

by the first defining property of linear transformations. Addition of $-T(\mathbf{0}_V)$ to both sides gives $\mathbf{0}_W = T(\mathbf{0}_V)$, as required.
Next, use of both parts of the definition shows that

$$T(c_1\mathbf{v}_1 + \cdots + c_{k-1}\mathbf{v}_{k-1} + c_k\mathbf{v}_k)$$

is equal to the vector

$$T(c_1\mathbf{v}_1 + \cdots + c_{k-1}\mathbf{v}_{k-1}) + c_k T(\mathbf{v}_k).$$

By repeated application of this procedure, or more properly induction on k, we obtain the second result.

Representing linear transformations by matrices

We now specialize the discussion to linear transformations of the type

$$T : F^n \to F^m$$

where F is some field of scalars. Let $\{E_1, E_2, ..., E_n\}$ be the standard basis of F^n written in the usual order, that of the columns of the identity matrix 1_n. Also let $\{D_1, D_2, ..., D_m\}$ be the corresponding ordered basis of F^m. Since $T(E_j)$ is a vector in F^m, it can be written in the form

$$T(E_j) = \begin{pmatrix} a_{1j} \\ \vdots \\ a_{mj} \end{pmatrix} = a_{1j}D_1 + \cdots + a_{mj}D_m = \sum_{i=1}^{m} a_{ij}D_i.$$

Put $A = [a_{ij}]_{m,n}$, so that the columns of the matrix A are the vectors $T(E_1), ..., T(E_n)$. We show that T is completely determined by the matrix A.

Take an arbitrary vector in F^n, say

$$X = \begin{pmatrix} x_1 \\ \vdots \\ x_n \end{pmatrix} = x_1E_1 + \cdots + x_nE_n = \sum_{j=1}^{n} x_jE_j.$$

Then, using 6.2.1 together with the expression for $T(E_j)$, we obtain

$$T(X) = T(\sum_{j=1}^{n} x_jE_j) = \sum_{j=1}^{n} x_jT(E_j) = \sum_{j=1}^{n} x_j(\sum_{i=1}^{m} a_{ij}D_i)$$
$$= \sum_{i=1}^{m}(\sum_{j=1}^{n} a_{ij}x_j)D_i.$$

Therefore the ith entry of $T(X)$ equals the ith entry of the matrix product AX. Thus we have shown that

$$T(X) = AX,$$

which means that the effect of T on a vector in F^n is to multiply it on the left by the matrix A. Thus A determines T completely.

Conversely, suppose that we start with an $m \times n$ matrix A over F; then we can define a function $T : F^n \to F^m$ by the rule $T(X) = AX$. The laws of matrix algebra guarantee that T is a linear transformation; for by 1.2.1

$$A(X_1 + X_2) = AX_1 + AX_2 \text{ and } A(cX) = c(AX).$$

We have now established a fundamental connection between matrices and linear transformations.

Theorem 6.2.2

(i) *Let* $T : F^n \to F^m$ *be a linear transformation. Then* $T(X) = AX$ *for all* X *in* F^n *where* A *is the* $m \times n$ *matrix whose columns are the images under* T *of the standard basis vectors of* F^n.

(ii) *Conversely, if* A *is any* $m \times n$ *matrix over the field* F, *the function* $T : F^n \to F^m$ *defined by* $T(X) = AX$ *is a linear transformation.*

Example 6.2.6

Define $T : \mathbf{R}^3 \to \mathbf{R}^2$ by the rule

$$T\left(\begin{pmatrix} x_1 \\ x_2 \\ x_3 \end{pmatrix}\right) = \begin{pmatrix} x_2 - x_1 \\ 3x_3 - x_2 \end{pmatrix}.$$

One quickly checks that T is a linear transformation. The images under T of the standard basis vectors E_1, E_2, E_3 are

$$\begin{pmatrix} -1 \\ 0 \end{pmatrix}, \quad \begin{pmatrix} 1 \\ -1 \end{pmatrix}, \quad \begin{pmatrix} 0 \\ 3 \end{pmatrix}$$

respectively. It follows that T is represented by the matrix

$$A = \begin{pmatrix} -1 & 1 & 0 \\ 0 & -1 & 3 \end{pmatrix}.$$

Consequently

$$T\left(\begin{pmatrix} x_1 \\ x_2 \\ x_3 \end{pmatrix}\right) = A \begin{pmatrix} x_1 \\ x_2 \\ x_3 \end{pmatrix},$$

as can be verified directly by matrix multiplication.

Example 6.2.7

Consider the linear operator $T : \mathbf{R}^2 \to \mathbf{R}^2$ which arises from an anti-clockwise rotation in the xy-plane through an angle θ (see Example 6.2.2.) The problem is to write down the matrix which represents T.

All that need be done is to identify the vectors $T(E_1)$ and $T(E_2)$ where E_1 and E_2 are the vectors of the standard ordered basis.

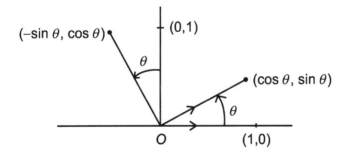

The line segment representing E_1 is drawn from the origin O to the point $(1, 0)$, and after rotation it becomes the line segment from O to the point $(\cos\,\theta,\,\sin\,\theta)$; thus $T(E_1) = \begin{pmatrix} \cos\,\theta \\ \sin\,\theta \end{pmatrix}$. Similarly $T(E_2) = \begin{pmatrix} -\sin\,\theta \\ \cos\,\theta \end{pmatrix}$. It follows that the matrix which represents the rotation T is

$$\begin{pmatrix} \cos\,\theta & -\sin\,\theta \\ \sin\,\theta & \cos\,\theta \end{pmatrix}.$$

Representing linear transformations by matrices: The general case

We turn now to the problem of representing by matrices linear transformations between arbitrary finite-dimensional vector spaces.

Let V and W be two non-zero finite-dimensional vector spaces over the same field of scalars F. Consider a linear transformation $T : V \to W$. The first thing to do is to choose and fix ordered bases for V and W, say

$$\mathcal{B} = \{\mathbf{v}_1, \mathbf{v}_2 \ldots, \mathbf{v}_n\} \text{ and } \mathcal{C} = \{\mathbf{w}_1, \mathbf{w}_2 \ldots, \mathbf{w}_m\}$$

respectively. We saw in 5.1 how any vector $\mathbf{v}$ of V can be represented by a unique coordinate vector with respect to the ordered basis $\mathcal{B}$. If $\mathbf{v} = c_1\mathbf{v}_1 + \cdots + c_n\mathbf{v}_n$, this coordinate vector is

$$[\mathbf{v}]_\mathcal{B} = \begin{pmatrix} c_1 \\ \vdots \\ c_n \end{pmatrix}.$$

Similarly each $\mathbf{w}$ in W may be represented by a coordinate vector $[\mathbf{w}]_\mathcal{C}$ with respect to $\mathcal{C}$.

To represent T by a matrix with respect to these chosen ordered bases, we first express the image under T of each vector in $\mathcal{B}$ as a linear combination of the vectors of $\mathcal{C}$, say

$$T(\mathbf{v}_j) = a_{1j}\mathbf{w}_1 + \cdots + a_{mj}\mathbf{w}_m = \sum_{i=1}^{m} a_{ij}\mathbf{w}_i$$

where the a_{ij} are scalars. Thus $[T(\mathbf{v}_j)]_\mathcal{C}$ is the column vector with entries $a_{1j}, \ldots, a_{mj}$. Let A be the $m \times n$ matrix whose (i, j) entry is a_{ij}. Thus the columns of A are just the coordinate vectors of $T(\mathbf{v}_1), \ldots, T(\mathbf{v}_n)$ with respect to $\mathcal{C}$.

Now consider the effect of T on an arbitrary vector of V, say $\mathbf{v} = c_1\mathbf{v}_1 + \cdots + c_n\mathbf{v}_n$. This is computed by using the expression for $T(\mathbf{v}_j)$ given above:

$$T(\mathbf{v}) = T(\sum_{j=1}^{n} c_j\mathbf{v}_j) = \sum_{j=1}^{n} c_j T(\mathbf{v}_j) = \sum_{j=1}^{n} c_j(\sum_{i=1}^{m} a_{ij}\mathbf{w}_i).$$

On interchanging the order of summations, this becomes

$$T(\mathbf{v}) = \sum_{i=1}^{m}(\sum_{j=1}^{n} a_{ij}c_j)\mathbf{w}_i.$$

Hence the coordinate vector of $T(\mathbf{v})$ with respect to the ordered basis C has entries $\sum_{j=1}^{n} a_{ij}c_j$ for $i = 1, 2, \ldots, m$. This means that
$$[T(\mathbf{v})]_C = A[\mathbf{v}]_B.$$

The conclusions of this discussion can be summed up as follows.

Theorem 6.2.3
Let $T : V \to W$ be a linear transformation between two non-zero finite-dimensional vector spaces V and W over the same field. Suppose that B and C are ordered bases for V and W respectively. If $\mathbf{v}$ is any vector of V, then

$$[T(\mathbf{v})]_C = A[v]_B$$

where A is the $m \times n$ matrix whose jth column is the coordinate vector of the image under T of the jth vector of B, taken with respect to the basis C.

What this result means is that a linear transformation between non-zero finite-dimensional vector spaces can always be represented by left multiplication by a suitable matrix. At this point the reader may wonder if it is worth the trouble of introducing linear transformations, given that they can be described by matrices. The answer is that there are situations where the functional nature of a linear transformation is a decided advantage. In addition there is the fact that a given linear transformation can be represented by a host of different matrices, depending on which ordered bases are used. The real object of interest is the linear transformation, not the representing matrix, which is dependent on the choice of bases.

Example 6.2.8
Define $T : P_{n+1}(\mathbf{R}) \rightarrow P_n(\mathbf{R})$ by the rule $T(f) = f'$, the derivative. Let us use the standard bases $\mathcal{B} = \{1, x, x^2, ..., x^n\}$ and $\mathcal{C} = \{1, x, x^2, ..., x^{n-1}\}$ for the two vector spaces. Here $T(x^i) = ix^{i-1}$, so $[T(x^i)]_{\mathcal{C}}$ is the vector whose ith entry is i and whose other entries are zero. Therefore T is represented by the $n \times (n+1)$ matrix

$$A = \begin{pmatrix} 0 & 1 & 0 & \cdots & 0 \\ 0 & 0 & 2 & \cdots & 0 \\ 0 & 0 & 0 & \cdots & 0 \\ . & . & . & \cdots & . \\ 0 & 0 & 0 & \cdots & n \\ 0 & 0 & 0 & \cdots & 0 \end{pmatrix}.$$

For example,

$$A \begin{pmatrix} 2 \\ -1 \\ 3 \\ 0 \\ \vdots \\ 0 \end{pmatrix} = \begin{pmatrix} -1 \\ 6 \\ 0 \\ \vdots \\ 0 \end{pmatrix},$$

which corresponds to the differentiation

$$T(2 - x + 3x^2) = (2 - x + 3x^2)' = 6x - 1.$$

Change of basis

Being aware of a dependence on the choice of bases, we wish to determine the effect on the matrix representing a linear transformation when the ordered bases are changed. The first step is to find a matrix that describes the change of basis.

Let $\mathcal{B} = \{\mathbf{v}_1, \ldots, \mathbf{v}_n\}$ and $\mathcal{B}' = \{\mathbf{v}'_1, \ldots, \mathbf{v}'_n\}$ be two ordered bases of a finite-dimensional vector space V. Then each $\mathbf{v}'_i$ can be expressed as a linear combination of $\mathbf{v}_1, \ldots, \mathbf{v}_n$, say

$$\mathbf{v}'_i = \sum_{j=1}^{n} s_{ji}\mathbf{v}_j$$

for certain scalars s_{ji}. The change of basis $\mathcal{B}' \to \mathcal{B}$ is determined by the $n \times n$ matrix $S = [s_{ij}]$. To see how this works we take an arbitrary vector $\mathbf{v}$ in V and write it in the form

$$\mathbf{v} = \sum_{i=1}^{n} c_i'\mathbf{v}_i',$$

where, of course, $c_1', \ldots, c_n'$ are the entries of the coordinate vector $[\mathbf{v}]_{\mathcal{B}'}$. Replace each $\mathbf{v}_i'$ by its expression in terms of the $\mathbf{v}_j$ to get

$$\mathbf{v} = \sum_{i=1}^{n} c_i' \left(\sum_{j=1}^{n} s_{ji}\mathbf{v}_j\right) = \sum_{j=1}^{n}\left(\sum_{i=1}^{n} s_{ji}c_i'\right)\mathbf{v}_j.$$

From this one sees that the entries of the coordinate vector $[\mathbf{v}]_{\mathcal{B}}$ are just the scalars $\sum_{i=1}^{n} s_{ji}c_i'$, for $j = 1, 2, \ldots, n$. But the latter are the entries of the product

$$S \begin{pmatrix} c_1' \\ \vdots \\ c_n' \end{pmatrix}.$$

Therefore we obtain the fundamental relation

$$[\mathbf{v}]_{\mathcal{B}} = S[\mathbf{v}]_{\mathcal{B}'}.$$

Thus left multiplication by the change of basis matrix S transforms coordinate vectors with respect to $\mathcal{B}'$ into coordinate vectors with respect to $\mathcal{B}$. It is in this sense that the matrix S describes the basis change $\mathcal{B}' \to \mathcal{B}$. Here it is important to observe how S is formed: its ith column is the coordinate vector of $\mathbf{v}_i'$, the ith vector of $\mathcal{B}'$, with respect to the basis $\mathcal{B}$.

It is a crucial remark that *the change of basis matrix S is always invertible*. Indeed, if this were false, there would by 2.3.5 be a non-zero n-column vector X such that $SX = 0$. However, if $\mathbf{u}$ denotes the vector in V whose coordinate vector with respect to basis $\mathcal{B}'$ is X, then $[\mathbf{u}]_{\mathcal{B}} = SX = 0$, which can only mean that $\mathbf{u} = \mathbf{0}$ and $X = 0$, a contradiction.

As one would expect, the matrix S^{-1} represents the inverse change of basis $\mathcal{B} \to \mathcal{B}'$: for the equation $[\mathbf{v}]_{\mathcal{B}} = S[\mathbf{v}]_{\mathcal{B}'}$ implies that

$$[\mathbf{v}]_{\mathcal{B}'} = S^{-1}[\mathbf{v}]_{\mathcal{B}}.$$

These conclusions can be summed up in the following form.

Theorem 6.2.4

Let $\mathcal{B}$ and $\mathcal{B}'$ be two ordered bases of an n-dimensional vector space V. Define S to be the $n \times n$ matrix whose ith column is the coordinate vector of the ith vector of $\mathcal{B}'$ with respect to the basis $\mathcal{B}$. Then S is invertible and, if $\mathbf{v}$ is any vector of V,

$$[\mathbf{v}]_{\mathcal{B}} = S[\mathbf{v}]_{\mathcal{B}'} \quad \text{and} \quad [\mathbf{v}]_{\mathcal{B}'} = S^{-1}[\mathbf{v}]_{\mathcal{B}}.$$

Example 6.2.9

Consider two ordered bases of the vector space $P_3(\mathbf{R})$:

$$\mathcal{B} = \{1, x, x^2\} \quad \text{and} \quad \mathcal{B}' = \{1,\ 2x,\ 4x^2 - 2\}.$$

In order to find the matrix S which describes the change of basis $\mathcal{B}' \to \mathcal{B}$, we must write down the coordinate vectors of the elements of $\mathcal{B}'$ with respect to the standard basis $\mathcal{B}$: these are

$$[1]_\mathcal{B} = \begin{pmatrix} 1 \\ 0 \\ 0 \end{pmatrix}, \quad [2x]_\mathcal{B} = \begin{pmatrix} 0 \\ 2 \\ 0 \end{pmatrix}, \quad [4x^2 - 2]_\mathcal{B} = \begin{pmatrix} -2 \\ 0 \\ 4 \end{pmatrix}.$$

Therefore

$$S = \begin{pmatrix} 1 & 0 & -2 \\ 0 & 2 & 0 \\ 0 & 0 & 4 \end{pmatrix}.$$

The matrix which describes the change of basis $\mathcal{B} \to \mathcal{B}'$ is

$$S^{-1} = \begin{pmatrix} 1 & 0 & 1/2 \\ 0 & 1/2 & 0 \\ 0 & 0 & 1/4 \end{pmatrix}.$$

For example, to express $f = a + bx + cx^2$ in terms of the basis $\mathcal{B}'$, we compute

$$[f]_{\mathcal{B}'} = S^{-1}[f]_\mathcal{B} = S^{-1} \begin{pmatrix} a \\ b \\ c \end{pmatrix} = \begin{pmatrix} a + c/2 \\ b/2 \\ c/4 \end{pmatrix}.$$

Thus $f = (a + c/2)1 + (b/2)2x + (c/4)(4x^2 - 2)$, which is of course easy to verify.

Example 6.2.10

Consider the change of basis in $\mathbf{R}^2$ which arises when the x- and y-axes are rotated through angle θ in an anticlockwise direction. As was noted in Example 6.2.6, the effect of this rotation is to replace the standard ordered basis $\mathcal{B} = \{E_1, E_2\}$ by the basis $\mathcal{B}'$ consisting of

$$\begin{pmatrix} \cos\theta \\ \sin\theta \end{pmatrix} \quad \text{and} \quad \begin{pmatrix} -\sin\theta \\ \cos\theta \end{pmatrix}.$$

The matrix which describes the change of basis $\mathcal{B}' \to \mathcal{B}$ is

$$S = \begin{pmatrix} \cos\theta & -\sin\theta \\ \sin\theta & \cos\theta \end{pmatrix},$$

so the change of basis $\mathcal{B} \to \mathcal{B}'$ is described by

$$S^{-1} = \begin{pmatrix} \cos\theta & \sin\theta \\ -\sin\theta & \cos\theta \end{pmatrix}.$$

Hence, if $X = \begin{pmatrix} a \\ b \end{pmatrix}$, the coordinate of vector of X with respect to the basis $\mathcal{B}'$ is

$$[X]_{\mathcal{B}'} = S^{-1}X = \begin{pmatrix} a\,\cos\theta + b\,\sin\theta \\ -a\,\sin\theta + b\,\cos\theta \end{pmatrix}.$$

This means that the coordinates of the point (a, b) with respect to the rotated axes are

$$a' = a\,\cos\theta + b\,\sin\theta \quad \text{and} \quad b' = -a\,\sin\theta + b\,\cos\theta,$$

respectively.

Change of basis and linear transformations

We are now in a position to calculate the effect of change of bases on the matrix representing a linear transformation.

Let $\mathcal{B}$ and $\mathcal{C}$ be ordered bases of finite-dimensional vector spaces V and W over the same field, and let $T : V \to W$ be a linear transformation. Then T is represented by a matrix A with respect to these bases.

Suppose now that we select new bases $\mathcal{B}'$ and $\mathcal{C}'$ for V and W respectively. Then T will be represented with respect to these bases by another matrix, say A'. The question before us is: what is the relation between A and A'?

Let X and Y be the invertible matrices that represent the changes of bases $\mathcal{B} \to \mathcal{B}'$ and $\mathcal{C} \to \mathcal{C}'$ respectively. Then, for any vectors $\mathbf{v}$ of V and $\mathbf{w}$ of W, we have

$$[\mathbf{v}]_{\mathcal{B}'} = X[\mathbf{v}]_{\mathcal{B}} \ \text{ and } \ [\mathbf{w}]_{\mathcal{C}'} = Y[\mathbf{w}]_{\mathcal{C}}.$$

Now by 6.2.3

$$[T(\mathbf{v})]_{\mathcal{C}} = A[\mathbf{v}]_{\mathcal{B}} \ \text{ and } \ [T(\mathbf{v})]_{\mathcal{C}'} = A'[\mathbf{v}]_{\mathcal{B}'}.$$

On combining these equations, we obtain

$$[T(\mathbf{v})]_{\mathcal{C}'} = Y[T(\mathbf{v})]_{\mathcal{C}} = YA[\mathbf{v}]_{\mathcal{B}} = YAX^{-1}[\mathbf{v}]_{\mathcal{B}'}.$$

But this means that the matrix YAX^{-1} describes the linear transformation T with respect to the bases $\mathcal{B}'$ and $\mathcal{C}'$ of V and W respectively. Hence $A' = YAX^{-1}$.

We summarise these conclusions in

Theorem 6.2.5

Let V and W be non-zero finite-dimensional vector spaces over the same field. Let $\mathcal{B}$ and $\mathcal{B}'$ be ordered bases of V, and $\mathcal{C}$ and $\mathcal{C}'$ ordered bases of W. Suppose that matrices X and Y describe the respective changes of bases $\mathcal{B} \to \mathcal{B}'$ and $\mathcal{C} \to \mathcal{C}'$. If the linear transformation $T : V \to W$ is represented by a *matrix A with respect to $\mathcal{B}$ and $\mathcal{C}$, and by a matrix A' with respect to $\mathcal{B}'$ and $\mathcal{C}'$, then*

$$A' = YAX^{-1}.$$

The most important case is that of a linear operator $T : V \rightarrow V$, when the ordered basis $\mathcal{B}$ is used for both domain and codomain.

Theorem 6.2.6
Let $\mathcal{B}$ and $\mathcal{B}'$ be two ordered bases of a finite-dimensional vector space V and let T be a linear operator on V. If T is represented by matrices A and A' with respect to $\mathcal{B}$ and $\mathcal{B}'$ respectively, then

$$A' = SAS^{-1}$$

where S is the matrix representing the change of basis $\mathcal{B} \rightarrow \mathcal{B}'$.

Example 6.2.11
Let T be the linear transformation on $P_3(\mathbf{R})$ defined by $T(f) = f'$. Consider the ordered bases of $P_3(\mathbf{R})$

$$\mathcal{B} = \{1, x, x^2\} \quad \text{and} \quad \mathcal{B}' = \{1, 2x, 4x^2 - 2\}.$$

We saw in Example 6.2.9 that the change of basis $\mathcal{B} \rightarrow \mathcal{B}'$ is represented by the matrix

$$U = \begin{pmatrix} 1 & 0 & 1/2 \\ 0 & 1/2 & 0 \\ 0 & 0 & 1/4 \end{pmatrix}.$$

Now T is represented with respect to $\mathcal{B}$ by the matrix

$$A = \begin{pmatrix} 0 & 1 & 0 \\ 0 & 0 & 2 \\ 0 & 0 & 0 \end{pmatrix}.$$

Hence T is represented with respect to $\mathcal{B}'$ by

$$UAU^{-1} = \begin{pmatrix} 0 & 2 & 0 \\ 0 & 0 & 4 \\ 0 & 0 & 0 \end{pmatrix}.$$

This conclusion is easily checked. An arbitrary element of $P_3(\mathbf{R})$ can be written in the form $f = a(1) + b(2x) + c(4x^2 - 2)$. Then it is claimed that the coordinate vector of $T(f)$ with respect to the basis $\mathcal{B}'$ is

$$\begin{pmatrix} 0 & 2 & 0 \\ 0 & 0 & 4 \\ 0 & 0 & 0 \end{pmatrix} \begin{pmatrix} a \\ b \\ c \end{pmatrix} = \begin{pmatrix} 2b \\ 4c \\ 0 \end{pmatrix}.$$

This is correct since

$$2b(1) + 4c(2x) + 0(4x^2 - 2) = 2b + 8cx$$
$$= (a(1) + b(2x) + c(4x^2 - 2))'.$$

Similar matrices

Let A and B be two $n \times n$ matrices over a field F; then B is said to be *similar to* A *over* F if there is an invertible $n \times n$ matrix S with entries in F such that

$$B = SAS^{-1}.$$

Thus the essential content of 6.2.6 is that two matrices which represent the same linear operator on a finite-dimensional vector space are similar. Because of this fact it is to be expected that similar matrices will have many properties in common: for example, *similar matrices have the same determinant*. Indeed if $B = SAS^{-1}$, then by 3.3.3 and 3.3.5

$$\det(B) = \det(S)\det(A)\det(S)^{-1} = \det(A).$$

We shall encounter other common properties of similar matrices in Chapter Eight.

Exercises 6.2

1. Which of the following functions are linear transformations?

    (a) $T_1 : \mathbf{R}_3 \to \mathbf{R}$ where $T_1([x_1 x_2 x_3]) = \sqrt{x_1^2 + x_2^2 + x_3^2}$;

    (b) $T_2 : M_{m,n}(F) \to M_{n,m}(F)$ where $T_2(A) = A^T$;

    (c) $T_3 : M_n(F) \to F$ where $T_3(A) = \det(A)$.

2. If T is a linear transformation, prove that $T(-\mathbf{v}) = -T(\mathbf{v})$ for all vectors $\mathbf{v}$.

3. Let l be a fixed line in the xy-plane passing through the origin O. If P is any point in the plane, denote by P' the mirror image of P in the line l. Prove that the assignment $\mathbf{OP} \to \mathbf{OP'}$ determines a linear operator on $\mathbf{R}^2$. (This is called *reflection in the line l*).

4. A linear transformation $T : \mathbf{R}^4 \to \mathbf{R}^3$ is defined by

$$
T\left(\begin{pmatrix} x_1 \\ x_2 \\ x_3 \\ x_4 \end{pmatrix}\right) = \begin{pmatrix} x_1 & - x_2 & - x_3 & - x_4 \\ 2x_1 & + x_2 & - x_3 & \\ & x_2 & - x_3 & + x_4 \end{pmatrix}.
$$

Find the matrix that represents T with respect to the standard bases of $\mathbf{R}^4$ and $\mathbf{R}^3$.

5. A function $T : P_4(\mathbf{R}) \to P_4(\mathbf{R})$ is defined by the rule $T(f) = xf'' - 2xf' + f$. Show that T is a linear operator and find the matrix that represents T with respect to the standard basis of $P_4(\mathbf{R})$.

6. Find the matrix which represents the reflection in Exercise 3 with respect to the standard ordered basis of $\mathbf{R}^2$, given that the angle between the positive x-direction and the line l is ϕ.

7. Let $\mathcal{B}$ denote the standard basis of $\mathbf{R}^3$ and let $\mathcal{B}'$ be the basis consisting of

$$\begin{pmatrix} 2 \\ 0 \\ 0 \end{pmatrix}, \begin{pmatrix} -1 \\ 2 \\ 0 \end{pmatrix}, \begin{pmatrix} 1 \\ 1 \\ 1 \end{pmatrix}.$$

Find the matrices that represent the basis changes $\mathcal{B} \to \mathcal{B}'$ and $\mathcal{B}' \to \mathcal{B}$.

8. A linear transformation from $\mathbf{R}^3$ to $\mathbf{R}^2$ is defined by

$$T(\begin{pmatrix} x_1 \\ x_2 \\ x_3 \end{pmatrix}) = \begin{pmatrix} x_1 & - x_2 & - x_3 \\ -x_1 & & + x_3 \end{pmatrix}.$$

Let $\mathcal{B}$ and $\mathcal{C}$ be the ordered bases

$$\{\begin{pmatrix} 2 \\ 0 \\ 0 \end{pmatrix}, \begin{pmatrix} -1 \\ 2 \\ 0 \end{pmatrix}, \begin{pmatrix} 1 \\ 1 \\ 1 \end{pmatrix}\} \text{ and } \{\begin{pmatrix} 0 \\ -1 \end{pmatrix}, \begin{pmatrix} 1 \\ 2 \end{pmatrix}\}$$

of $\mathbf{R}^3$ and $\mathbf{R}^2$ respectively. Find the matrix that represents T with respect to these bases.

9. Explain why the matrices $\begin{pmatrix} 3 & 4 \\ 1 & 2 \end{pmatrix}$ and $\begin{pmatrix} 2 & -1 \\ 4 & 3 \end{pmatrix}$ cannot be similar.

10. If B is similar to A, prove that A is similar to B.

11. If B is similar to A and C is similar to B, prove that C is similar to A.

12. If B is similar to A, then B^T is similar to A^T: prove or disprove.

6.3 Kernel, Image and Isomorphism

If $T : V \rightarrow W$ is a linear transformation between two vector spaces, there are two important subspaces associated with T, the image and the kernel. The first of these has already been defined; the *image* of T,

$$\mathrm{Im}(T),$$

is the set of all images $T(\mathbf{v})$ of vectors $\mathbf{v}$ in V: thus $\mathrm{Im}(T)$ is a subset of W.

On the other hand, the *kernel* of T

$$\mathrm{Ker}(T)$$

is defined to be the set of all vectors $\mathbf{v}$ in V such that $T(\mathbf{v}) = \mathbf{0}_W$. Thus $\mathrm{Ker}(T)$ is a subset of V. Notice that by 6.2.1 the zero vector of V must belong to $\mathrm{Ker}(T)$, while the zero vector of W belongs to $\mathrm{Im}(T)$.

The first thing to observe is that we are actually dealing with subspaces here, not just subsets.

Theorem 6.3.1
If T is a linear transformation from a vector space V to a vector space W, then $\mathrm{Ker}(T)$ is a subspace of V and $\mathrm{Im}(T)$ is a subspace of W.

Proof
We need to check that $\mathrm{Ker}(T)$ and $\mathrm{Im}(T)$ contain the relevant zero vector, and that they are closed with respect to addition and scalar multiplication. The first point is settled by the equation $T(\mathbf{0}_V) = \mathbf{0}_W$, which was proved in 6.2.1. Also, by definition of a linear transformation, we have $T(\mathbf{v}_1 + \mathbf{v}_2) = T(\mathbf{v}_1) + T(\mathbf{v}_2)$ and $T(c\mathbf{v}_1) = cT(\mathbf{v}_1)$ for all vectors $\mathbf{v}_1$, $\mathbf{v}_2$ of V and scalars c. Therefore, if $\mathbf{v}_1$ and $\mathbf{v}_2$ belong to $\mathrm{Ker}(T)$, then $T(\mathbf{v}_1 + \mathbf{v}_2) = \mathbf{0}_W$, and $T(c\mathbf{v}_1) = \mathbf{0}_W$, so that $\mathbf{v}_1 + \mathbf{v}_2$ and

$c\mathbf{v}_1$ belong to $\mathrm{Ker}(T)$; thus $\mathrm{Ker}(T)$ is a subspace. For similar reasons $\mathrm{Im}(T)$ is a subspace.

Let us look next at some examples which relate these new concepts to some more familiar ones.

Example 6.3.1
Consider the homogeneous linear differential equation for a function y of the real variable x:

$$y^{(n)} + a_{n-1}(x)y^{(n-1)} + \cdots + a_1(x)y' + a_0(x)y = 0,$$

with x in the interval $[a, b]$ and $a_i(x)$ in $D_\infty[a, b]$. There is an associated linear operator T on the vector space $D_\infty[a, b]$ defined by

$$T(f) = f^{(n)} + a_{n-1}(x)f^{(n-1)} + \cdots + a_1(x)f' + a_0(x)f.$$

Then $\mathrm{Ker}(T)$ is the solution space of the differential equation.

Example 6.3.2
Let A be an $m \times n$ matrix over a field F. We have seen that the rule $T(X) = AX$ defines a linear transformation

$$T : F^n \rightarrow F^m.$$

Identify $\mathrm{Ker}(T)$ and $\mathrm{Im}(T)$.

In the first place, the definition shows that $\mathrm{Ker}(T)$ *is the null space of the matrix A*. Next an arbitrary element of $\mathrm{Im}(T)$ is a linear combination of the images of the standard basis elements of $\mathbf{R}^n$; but the latter are simply the columns of the matrix A. Consequently, *the image of T coincides with the column space of the matrix A*.

Example 6.3.3
After the last example it is natural to enquire if there is an interpretation of the row space of a matrix A as an image

space. That this is the case may be seen from a related linear transformation.

Given an $m \times n$ matrix A, define a linear transformation T_1 from F_m to F_n by the rule $T_1(X) = XA$. In this case $\mathrm{Im}(T_1)$ is generated by the images of the elements of the standard basis of F_m, that is, by the rows of A. Hence *the image of T_1 equals the row space of A*.

It is now time to consider what the kernel and image tell us about a linear transformation.

Theorem 6.3.2

Let T be a linear transformation from a vector space V to a vector space W. Then

   (i) *T is injective if and only if $\mathrm{Ker}(T)$ is the zero subspace of V;*

   (ii) *T is surjective if and only if $\mathrm{Im}(T) = W$.*

Proof
(i) Assume that T is an injective function. If $\mathbf{v}$ is a vector in the kernel of T, then $T(\mathbf{v}) = \mathbf{0}_W = T(\mathbf{0}_V)$. Therefore $\mathbf{v} = \mathbf{0}_V$ by injectivity, and $\mathrm{Ker}(T) = \mathbf{0}_V$. Conversely, suppose that $\mathrm{Ker}(T) = \mathbf{0}_V$. If $\mathbf{v}_1$ and $\mathbf{v}_2$ are vectors in V with the property $T(\mathbf{v}_1) = T(\mathbf{v}_2)$, then $T(\mathbf{v}_1 - \mathbf{v}_2) = T(\mathbf{v}_1) - T(\mathbf{v}_2) = \mathbf{0}_W$. Hence the vector $\mathbf{v}_1 - \mathbf{v}_2$ belongs to $\mathrm{Ker}(T)$ and $\mathbf{v}_1 = \mathbf{v}_2$.
(ii) This is true by definition of surjectivity.

For finite-dimensional vector spaces there is a simple formula which links the dimensions of the kernel and image of a linear transformation.

Theorem 6.3.3

Let $T : V \to W$ be a linear transformation where V and W are finite-dimensional vector spaces. Then

$$\dim(\mathrm{Ker}(T)) + \dim(\mathrm{Im}(T)) = \dim(V).$$

Proof

Here we may assume that V is not the zero space; otherwise the statement is true for obvious reasons. By 5.1.4 it is possible to choose a basis $\mathbf{v}_1, \ldots, \mathbf{v}_n$ of V such that part of it is a basis of $\mathrm{Ker}(T)$, say $\mathbf{v}_1, \ldots, \mathbf{v}_r$; here of course

$$n = \dim(V) \geq r = \dim(\mathrm{Ker}(T)).$$

We claim that the vectors $T(\mathbf{v}_{r+1}), \ldots, T(\mathbf{v}_n)$ are linearly independent. For if $c_{r+1}T(\mathbf{v}_{r+1}) + \cdots + c_nT(\mathbf{v}_n) = \mathbf{0}_W$ for some scalars c_i, then $T(c_{r+1}\mathbf{v}_{r+1} + \cdots + c_n\mathbf{v}_n) = \mathbf{0}_W$, so that $c_{r+1}\mathbf{v}_{r+1} + \cdots + c_n\mathbf{v}_n$ belongs to $\mathrm{Ker}(T)$ and is therefore expressible as a linear combination of $\mathbf{v}_1, \ldots, \mathbf{v}_r$. But $\mathbf{v}_1, \ldots, \mathbf{v}_r, \ldots, \mathbf{v}_n$ are certainly linearly independent. Hence $c_{r+1}, \ldots, c_n$ are all zero and our claim is established.

On the other hand, the vectors $T(\mathbf{v}_{r+1}), \ldots, T(\mathbf{v}_n)$ by themselves generate $\mathrm{Im}(T)$ since $T(\mathbf{v}_1) = \cdots = T(\mathbf{v}_r) = \mathbf{0}_W$; hence $T(\mathbf{v}_{r+1}), \ldots, T(\mathbf{v}_n)$ form a basis of $\mathrm{Im}(T)$. It follows that

$$\dim(\mathrm{Im}(T)) = n - r = \dim(V) - \dim(\mathrm{Ker}(T)),$$

from which the formula follows.

The dimension formula is in fact a generalization of something that we already know. For suppose we apply the formula to the linear transformation $T : F^n \to F^m$ defined by $T(X) = AX$, where A is an $m \times n$ matrix. Making the interpretations of $\mathrm{Ker}(T)$ and $\mathrm{Im}(T)$ as the null space and column space of A, we deduce that the sum of the dimensions of the null space and column space of A equals n. This is essentially the content of 5.1.7 and 5.2.4.

Isomorphism

Because of 6.3.2 we can tell whether a linear transformation $T : V \to W$ is bijective. And in view of 6.1.3 this is the same as asking whether T has an inverse. A bijective linear transformation is called an *isomorphism*.

Theorem 6.3.4

A linear transformation $T : V \to W$ is an isomorphism if and only if $\operatorname{Ker}(T)$ is the zero subspace of V and $\operatorname{Im}(T)$ equals W. Moreover, if T is an isomorphism, then so is its inverse $T^{-1} : W \to V$.

Proof

The first statement follows from 6.3.2. As for the second statement, all that need be shown is that T^{-1} is actually a linear transformation: for by 6.1.5 it certainly has an inverse. This is achieved by a trick. Let $\mathbf{v}_1$ and $\mathbf{v}_2$ be any two vectors in V. Then certainly

$$T(T^{-1}(\mathbf{v}_1 + \mathbf{v}_2)) = \mathbf{v}_1 + \mathbf{v}_2,$$

while on the other hand,

$$T(T^{-1}(\mathbf{v}_1) + T^{-1}(\mathbf{v}_2)) = T(T^{-1}(\mathbf{v}_1)) + T(T^{-1}(\mathbf{v}_2))$$
$$= \mathbf{v}_1 + \mathbf{v}_2,$$

because T is known to be a linear transformation. Since T is an injective function, this can only mean that the vectors $T^{-1}(\mathbf{v}_1 + \mathbf{v}_2)$ and $T^{-1}(\mathbf{v}_1) + T^{-1}(\mathbf{v}_2)$ are equal; for they have the same image under T.

In a similar way it can be demonstrated that $T^{-1}(c\mathbf{v}_1)$ equals $cT^{-1}(\mathbf{v}_1)$ where c is any scalar: just check that both sides have the same image under T. Hence T^{-1} is a linear transformation.

Two vector spaces V and W are said to be *isomorphic* if there is an isomorphism from one to the other. Observe that isomorphic vector spaces are necessarily over the same field of scalars. The notation

$$V \simeq W$$

is often used to express the fact that vector spaces V and W are isomorphic.

How can one tell if two finite-dimensional vector spaces are isomorphic? The answer is that the dimensions tell us all.

Theorem 6.3.5
Let V and W be finite-dimensional vector spaces over a field F. Then V and W are isomorphic if and only if $\dim(V) = \dim(W)$.

Proof
Suppose first that $\dim(V) = \dim(W) = n$. If $n = 0$, then V and W are both zero spaces and hence are surely isomorphic. Let $n > 0$. Then V and W have bases, say $\{\mathbf{v}_1, \ldots, \mathbf{v}_n\}$ and $\{\mathbf{w}_1, \ldots, \mathbf{w}_n\}$ respectively. There is a natural candidate for an isomorphism from V to W, namely the linear transformation $T : V \to W$ defined by

$$T(c_1\mathbf{v}_1 + \cdots + c_n\mathbf{v}_n) = c_1\mathbf{w}_1 + \cdots + c_n\mathbf{w}_n.$$

It is straightforward to check that T is a linear transformation. Hence V and W are isomorphic.

Conversely, let V and W be isomorphic via an isomorphism $T : V \to W$. Suppose that $\{\mathbf{v}_1, \ldots, \mathbf{v}_n\}$ is a basis of V. In the first place, notice that the vectors $T(\mathbf{v}_1), \ldots, T(\mathbf{v}_n)$ are linearly independent; for if $c_1T(\mathbf{v}_1) + \cdots + c_nT(\mathbf{v}_n) = \mathbf{0}_W$, then $T(c_1\mathbf{v}_1 + \cdots + c_n\mathbf{v}_n) = \mathbf{0}_W$. This implies that $c_1\mathbf{v}_1 + \cdots + c_n\mathbf{v}_n$ belongs to $\mathrm{Ker}(T)$ and so must be zero. This in turn implies that $c_1 = \cdots = c_n = 0$ because $\mathbf{v}_1, \ldots, \mathbf{v}_n$ are linearly independent. It follows by 5.1.1 that $\dim(W) \geq n = \dim(V)$. In the same way it may be shown that $\dim(W) \leq \dim(V)$; hence $\dim(V) = \dim(W)$.

Corollary 6.3.6
Every n-dimensional vector space V over a field F is isomorphic with the vector space F^n.

For both V and F^n have dimension n. This result makes it possible for some purposes to work just with vector spaces of column vectors.

Isomorphism theorems

There are certain theorems, known as isomorphism theorems, which provide a link between linear transformations and quotient spaces (which were defined in 5.3). Such theorems occur frequently in algebra. The first theorem of this type is:

Theorem 6.3.7

If $T : V \to W$ is a linear transformation between vector spaces V and W, then

$$V/\mathrm{Ker}(T) \simeq \mathrm{Im}(T).$$

Proof

Write $K = \mathrm{Ker}\ (T)$. We define a function $S : V/K \to \mathrm{Im}(T)$ by the rule $S(\mathbf{v} + K) = T(\mathbf{v})$. The first thing to notice is that S is well-defined: indeed if $\mathbf{u} \in K$, then

$$T(\mathbf{v} + \mathbf{u}) = T(\mathbf{v}) + T(\mathbf{u}) = T(\mathbf{v}) + \mathbf{0} = T(\mathbf{v}),$$

since $T(\mathbf{u}) = \mathbf{0}$. Thus $S(\mathbf{v} + K)$ does not depend on the choice of representative $\mathbf{v}$ of the coset $\mathbf{v} + K$.

Next it is simple to verify that S is a linear transformation: for example,

$$
\begin{aligned}
S((\mathbf{v}_1 + K) + (\mathbf{v}_2 + K)) &= S((\mathbf{v}_1 + \mathbf{v}_2) + K) \\
&= T(\mathbf{v}_1 + \mathbf{v}_2) \\
&= T(\mathbf{v}_1) + T(\mathbf{v}_2),
\end{aligned}
$$

which equals $S(\mathbf{v}_1 + K) + S(\mathbf{v}_2 + K)$. In a similar way it can be shown that $S(c(\mathbf{v} + K)) = cS(\mathbf{v} + K)$.

Clearly the function S is surjective, so all we need do to complete the proof is show it is injective. If $S(\mathbf{v} + K) = \mathbf{0}$, then $T(\mathbf{v}) = \mathbf{0}$; thus $\mathbf{v} \in K$ and $\mathbf{v} + K = \mathbf{0}_{V/K}$. Hence, by 6.3.2, S is injective.

The last result provides an alternative proof of the dimension formula in 6.3.3. Let $T : V \to W$ be a linear transformation. Then $\dim(V/\mathrm{Ker}(T)) = \dim(\mathrm{Im}(T)$ by 6.3.7. From the formula for the dimension of a quotient space (see 5.3.7), we obtain

$$\dim(V) - \dim(\mathrm{Ker}(T)) = \dim(\mathrm{Im}(T)),$$

so that $\dim(\mathrm{Ker}(T)) + \dim(\mathrm{Im}(T)) = \dim(V)$.

There is second isomorphism theorem which provides valuable insight into the relation between the sum of two subspaces and certain associated quotient spaces.

Theorem 6.3.8
If U and W are subspaces of a vector space V, then

$$(U + W)/W \simeq U/(U \cap W).$$

Proof
We begin by defining a function $T : U \to (U + W)/W$ by the rule $T(\mathbf{u}) = \mathbf{u} + W$, where $\mathbf{u} \in U$. It is a simple matter to check that T is a linear transformation. Since $\mathbf{u} + W$ is a typical vector in $(U + W)/W$, we see that T is surjective.

Next we need to compute the kernel of T. Now $T(\mathbf{u}) = \mathbf{u} + W$ equals the zero vector of $(U + W)/W$, i.e., the coset W, precisely when $\mathbf{u} \in W$, which is just to say that $\mathbf{u} \in U \cap W$. Therefore $\mathrm{Ker}(T) = U \cap W$. It now follows directly from 6.3.7 that $U/(U \cap W) \simeq (U + W)/W$.

We illustrate the usefulness of this last result by using it to give another proof of the dimension formula of 5.3.2.

Corollary 6.3.9
If U and W are subspaces of a finite dimensional vector space V, then

$$\dim(U + W) + \dim(U \cap W) = \dim(U) + \dim(W).$$

Proof

Since isomorphic vector spaces have the same dimension, we have $\dim((U + W)/W) = \dim(U/(U \cap W))$. Now use the formula for the dimension of a quotient space in 5.3.7 to obtain

$$\dim(U + W) - \dim(W) = \dim(U) - \dim(U \cap W),$$

from which the result follows.

The algebra of linear operators on a vector space

We conclude the chapter by observing that the set of all linear operators on a vector space has certain formal properties which are very similar to properties that have already been seen to hold for matrices. This similarity can be expressed by saying that both systems form what is called an *algebra*.

Consider a vector space V with finite dimension n over a field F. Let T_1 and T_2 be two linear operators on V. Then we define their *sum* $T_1 + T_2$ by the rule

$$T_1 + T_2(\mathbf{v}) = T_1(\mathbf{v}) + T_2(\mathbf{v})$$

and also the *scalar multiple* cT_1, where c is an element of F, by

$$cT_1(\mathbf{v}) = c(T_1(\mathbf{v})).$$

It is quite routine to verify that $T_1 + T_2$ and cT_1 are also linear operators on V. For example, to show that $T_1 + T_2$ is a linear operator we compute

$$\begin{aligned} T_1 + T_2(\mathbf{v}_1 + \mathbf{v}_2) &= T_1(\mathbf{v}_1 + \mathbf{v}_2) + T_2(\mathbf{v}_1 + \mathbf{v}_2) \\ &= T_1(\mathbf{v}_1) + T_1(\mathbf{v}_2) + T_2(\mathbf{v}_1) + T_2(\mathbf{v}_2), \end{aligned}$$

from which it follows that

$$T_1 + T_2(\mathbf{v}_1 + \mathbf{v}_2) = (T_1 + T_2(\mathbf{v}_1)) + (T_1 + T_2(\mathbf{v}_2)).$$

It is equally easy to show that $T_1 + T_2(c\mathbf{v}) = c(T_1 + T_2(\mathbf{v}))$. Thus the set of all linear operators on V, which will henceforth be written

$$L(V),$$

admits natural operations of addition and scalar multiplication.

Now there is a further natural operation that can be performed on elements of $L(V)$, namely functional composition as defined in 6.1. Thus, if T_1 and T_2 are linear operators on V, then the composite $T_1 \circ T_2$, which will in future be written

$$T_1 T_2,$$

is defined by the rule

$$T_1 T_2(\mathbf{v}) = T_1(T_2(\mathbf{v})).$$

One has of course to check that $T_1 T_2$ is actually a linear transformation, but again this is quite routine. So one can also form products in the set $L(V)$.

To illustrate these definitions, we consider an explicit example where sums, scalar multiples and products can be computed.

Example 6.3.4
Let T_1 and T_2 be the linear operators on $D_\infty[a, b]$ defined by $T_1(f) = f' - f$ and $T_2(f) = xf'' - 2f'$. The linear operators $T_1 + T_2$, cT_1 and $T_1 T_2$ may be found directly from the definitions as follows:

$$T_1 + T_2(f) = T_1(f) + T_2(f) = f' - f + xf'' - 2f'$$
$$= -f - f' + xf''.$$

Also

$$cT_1(f) = cf' - cf$$

and

$$T_1 T_2(f) = T_1(T_2(f)) = T_1(xf'' - 2f')$$
$$= (xf'' - 2f')' - (xf'' - 2f'),$$

which reduces to $T_1 T_2(f) = 2f' - (x + 1)f'' + xf^{(3)}$, after evaluation of the derivatives.

At this point one can sit down and check that those properties of matrices listed in 1.2.1 which relate to sums, scalar multiples and products are also valid for linear operators. Thus there is a similarity between the set of linear operators $L(V)$ and $M_n(F)$, the set of $n \times n$ matrices over F where $n = \dim(V)$. This similarity should come as no surprise since the action of a linear operator can be represented by multiplication by a suitable matrix.

The relation between $L(V)$ and $M_n(F)$ can be formalized by defining a new type of algebraic structure. This involves the concept of a ring, which was was described in 1.3, and that of a vector space.

An *algebra* A over a field F is a set which is simultaneously a ring with identity and a vector space over F, with the same rule of addition and zero element, which satisfies the additional axiom

$$c(xy) = (cx)y = x(cy)$$

for all x and y in A and all c in the field F. Notice that this axiom holds for the vector space $M_n(F)$ because of property (j) in 1.2.1. Hence $M_n(F)$ is an algebra over F. Now the additional axiom is also valid in $L(V)$, that is,

$$c(T_1 T_2) = (cT_1)T_2 = T_1(cT_2).$$

This is true because each of the three linear operators mentioned sends the vector $\mathbf{v}$ to $c(T_1(T_2(\mathbf{v})))$. It follows that

$L(V)$, the set of all linear operators on a vector space V over a field F, is an algebra over F.

Suppose now that we pick and fix an ordered basis $\mathcal{B}$ for the finite-dimensional vector space V. Then, with respect to $\mathcal{B}$, a linear operator T on V is represented by an $n \times n$ matrix, which will be denoted by

$$M(T).$$

By 6.2.3 the matrix $M(T)$ has the property

$$[T(\mathbf{v})]_\mathcal{B} = M(T)[\mathbf{v}]_\mathcal{B}.$$

It follows from 6.2.3 that the assignment of the matrix $M(T)$ to a linear operator T determines a bijective function from $L(V)$ to $M_n(F)$. The essential properties of this function are summarized in the next result.

Theorem 6.3.10

Let T_1 and T_2 be linear operators on an n-dimensional vector space V and let $M(T_i)$ denote the matrix representing T_i with respect to a fixed ordered basis $\mathcal{B}$ of V. Then the following equations hold:

(i) $M(T_1 + T_2) = M(T_1) + M(T_2)$;
(ii) $M(cT) = cM(T)$;
(iii) $M(T_1 T_2) = M(T_1)M(T_2)$ *for all scalars c.*

It is as well to restate this technical result in words to make sure that the reader grasps what is being asserted. According to part (i) of the theorem, if we *add* linear operators T_1 and T_2, the resulting linear operator $T_1 + T_2$ is represented by a matrix which is the *sum* of the matrices that represent T_1 and T_2. Also (ii) asserts that the scalar multiple cT_1 is represented by a matrix which is just c times the matrix representing T_1.

More unexpectedly, when we *compose* the linear operations T_1 and T_2, the resulting linear operator T_1T_2 is represented by the *product* of the matrices representing T_1 and T_2.

In technical language, the function which sends T to
$M(T)$ is an *algebra isomorphism from* $L(V)$ *to* $M_n(F)$. The
main point here is that isomorphic algebras, like isomorphic
vector spaces, are to be regarded as similar objects, which
exhibit the same essential features, even although their un-
derlying sets may be quite different.

In conclusion, our vague feeling that the algebras $L(V)$
and $M_n(F)$ are somehow quite closely related is made precise
by the assertion that *the algebra of all linear operators on an
n-dimensional vector space over a field* F *is isomorphic with
the algebra of all $n \times n$ matrices over F.*

Example 6.3.5

Prove part (iii) of Theorem 6.3.10.

Let $\mathbf{v}$ be any vector of the vector space; then, using the
fundamental equation $[T(\mathbf{v})]_{\mathcal{B}} = M(T)[\mathbf{v}]_{\mathcal{B}}$, we obtain

$$[T_1T_2(\mathbf{v})]_{\mathcal{B}} = M(T_1)[T_2(\mathbf{v})]_{\mathcal{B}} = M(T_1)(M(T_2)[\mathbf{v}]_{\mathcal{B}})$$
$$= M(T_1)M(T_2)[\mathbf{v}]_{\mathcal{B}},$$

which shows that $M(T_1T_2) = M(T_1)M(T_2)$, as required.

Exercises 6.3

1. Find bases for the kernel and image of the following linear
transformations:

    (a) $T : \mathbf{R}^4 \to \mathbf{R}$ where T sends a column to the sum
of its entries;

    (b) $T : P_3(\mathbf{R}) \to P_3(\mathbf{R})$ where $T(f) = f'$;

    (c) $T : \mathbf{R}^2 \to \mathbf{R}^2$ where $T(\begin{pmatrix} x \\ y \end{pmatrix}) = \begin{pmatrix} 2x + 3y \\ 4x + 6y \end{pmatrix}$.

2. Show that every subspace U of a finite-dimensional vector space V is the kernel and the image of suitable linear operators on V. [Hint: assume that U is non-zero, choose a basis for U and extend it to a basis of V].

3. Sort the following vector spaces into batches, so that those within the same batch are isomorphic:
$$\mathbf{R}^6, \ \mathbf{R}_6, \ \mathbf{C}^6, \ P_6(\mathbf{C}), \ M_{2,3}(\mathbf{R}), \ C[0,1].$$

4. Show that a linear transformation $T : V \to W$ is injective if and only if it has the property of mapping linearly independent subsets of V to linearly independent subsets of W.

5. Show that a linear transformation $T : V \to W$ is surjective if and only if it has the property of mapping any set of generators of V to a set of generators of W.

6. A linear operator on a finite-dimensional vector space is an isomorphism if and only if some representing matrix is invertible: prove or disprove.

7. Prove that the composite of two linear transformations is a linear transformation.

8. Prove parts (i) and (ii) of Theorem 6.3.10.

9. Let $T : V \to W$ and $S : W \to U$ be isomorphisms of vector spaces; show that the function $ST : V \to U$ is also an isomorphism.

10. Let T be a linear operator on a finite-dimensional vector space V. Prove that the following statements about T are equivalent:
   (a) T is injective;
   (b) T is surjective;
   (c) T is an isomorphism.
Are these statements still equivalent if V is infinitely generated?

11. Show that similar matrices have the same rank. [Use the fact that similar matrices represent the same linear operator].

12. (The third isomorphism theorem). Let U and W be subspaces of a vector space V such that $W \subseteq U$. Prove that U/W is a subspace of V/W and that $(V/W)/(U/W) \simeq V/U$. [Hint: define a function $T : V/W \to V/U$ by the rule $T(\mathbf{v} + W) = \mathbf{v} + U$. Show that T is a well defined linear transformation and apply 6.3.7].

13. Explain how to define a power T^m of a linear operator T on a vector space V, where $m \geq 0$. Then show that powers of T commute.

Chapter Seven

ORTHOGONALITY IN VECTOR SPACES

The notion of two lines being perpendicular, or orthogonal, is very familiar from analytical geometry. In this chapter we show how to extend the elementary concept of orthogonality to abstract vector spaces over **R** or **C**. Orthogonality turns out to be a tool of extraordinary utility with many applications, one of the most useful being the well-known Method of Least Squares. We begin with **R**^{n}, showing how to define orthogonality in this vector space in a way which naturally generalizes our intuitive notion of perpendicularity in three-dimensional space.

7.1 Scalar Products in Euclidean Space

Let X and Y be two vectors in **R**^{n}, with entries $x_1, \ldots, x_n$ and $y_1, \ldots, y_n$ respectively. Then the *scalar product* of X and Y is defined to be the matrix product

$$X^T Y = (x_1 x_2 \ \ldots \ x_n) \begin{pmatrix} y_1 \\ y_2 \\ \vdots \\ y_n \end{pmatrix} = x_1 y_1 + x_2 y_2 + \cdots + x_n y_n.$$

This is a real number. Notice that $X^T Y = Y^T X$, so the scalar product is symmetric in X and Y. Of particular interest is the scalar product of X with itself

$$X^T X = x_1^2 + x_2^2 + \cdots + x_n^2.$$

Since this expression cannot be negative, it has a real square root, which is called the *length* of X. It is written

$$\|X\| = \sqrt{X^T X} = \sqrt{x_1^2 + x_2^2 + \cdots + x_n^2}.$$

Notice that $\|X\| \geq 0$, and $\|X\| = 0$ if and only if all the x_i are zero, that is, $X = 0$. So the only vector of length 0 is the zero vector. A vector whose length is 1 is called a *unit vector*.

At this point it is as well to specialize to $\mathbf{R}^3$ where geometrical intuition can be used. Recall that a 3-column vector X in $\mathbf{R}^3$, with entries x_1, x_2, x_3, is represented by a line segment in three-dimensional space with arbitrary initial point (a_1, a_2, a_3) and endpoint $(a_1 + x_1, a_2 + x_2, a_3 + x_3)$. Thus the length of the vector X is just the length of any representing line segment.

This suggests that we look for a geometrical interpretation of the scalar product of two vectors in $\mathbf{R}^3$.

Theorem 7.1.1
Let X and Y be vectors in $\mathbf{R}^3$. Then

$$X^T Y = \|X\| \, \|Y\| \cos \, \theta.$$

where θ is the angle in the interval $[0, \pi]$ between line segments representing X and Y drawn from the same initial point.

Proof
Consider the triangle rule for adding the vectors X and $Y - X$ in the triangle IAB, as shown in the diagram below.

The idea is then to apply the cosine rule to this triangle.

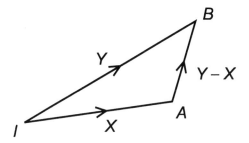

Thus we have

$$AB^2 = IA^2 + IB^2 - 2IA \cdot IB \cos \theta,$$

which becomes in vector form

$$\|Y - X\|^2 = \|X\|^2 + \|Y\|^2 - 2\|X\| \, \|Y\| \cos \theta.$$

As usual let the entries of X and Y be x_1, x_2, x_3 and y_1, y_2, y_3 respectively. Then

$$\|X\|^2 = x_1^2 + x_2^2 + x_3^2, \quad \|Y\|^2 = y_1^2 + y_2^2 + y_3^2$$

and

$$\|Y - X\|^2 = (y_1 - x_1)^2 + (y_2 - x_2)^2 + (y_3 - x_3)^2.$$

Now substitute these expressions in the equation for $\|Y-X\|^2$, and solve for the expression $\|X\| \, \|Y\| \cos \theta$. We obtain after some simplification the required result

$$\|X\| \, \|Y\| \cos \theta = x_1 y_1 + x_2 y_2 + x_3 y_3 = X^T Y.$$

The formula of 7.1.1 allows us to calculate quickly the angle θ between two non-zero vectors X and Y; for it yields the equation

$$\cos\ \theta = \frac{X^T Y}{\|X\|\ \|Y\|}.$$

Hence *the vectors X and Y are orthogonal if and only if*

$$X^T Y = 0.$$

There is another more or less immediate use for the formula of 7.1.1. Since $\cos\ \theta$ always lies between -1 and $+1$, we can derive a famous inequality.

Theorem 7.1.2 (*The Cauchy – Schwartz Inequality*)
If X and Y are any vectors in $\mathbf{R}^3$, then

$$|X^T Y| \leq \|X\|\ \|Y\|.$$

Projection of a vector on a line

Let X and Y be two vectors in $\mathbf{R}^3$ with Y non-zero.

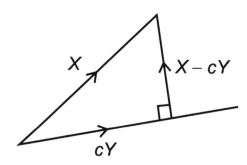

We wish to define the projection of X on Y. Now any vector parallel to Y will have the form cY for some scalar c. The idea is to try to choose c in such a way that the vector $X - cY$

is orthogonal to Y. For then cY will be the projection of X on Y, as one sees from the diagram.

The condition for $X - cY$ to be orthogonal to Y is

$$0 = (X - cY)^T Y = X^T Y - cY^T Y = X^T Y - c \, \|Y\|^2.$$

The correct value of c is therefore $X^T Y \, / \, \|Y\|^2$ and the *vector projection* of X on Y is

$$P = \frac{X^T Y}{\|Y\|^2} \, Y.$$

The *scalar projection* of X on Y is the length of P, that is,

$$\|P\| = \frac{|X^T Y|}{\|Y\|}.$$

We will see in 7.2 how to extend this concept to the projection of a vector on an arbitrary subspace.

Example 7.1.1
Consider the vectors

$$X = \begin{pmatrix} 1 \\ -1 \\ 2 \end{pmatrix}, \quad Y = \begin{pmatrix} 2 \\ 3 \\ 1 \end{pmatrix}$$

in $\mathbf{R}^3$. Here $\|X\| = \sqrt{6}$ $\|Y\| = \sqrt{14}$ and $X^T Y = 2 - 3 + 2 = 1$. The angle θ between X and Y is therefore given by

$$\cos \theta = \frac{1}{\sqrt{84}}$$

and θ is approximately $83.74°$. The vector projection of X on Y is $(1/14)Y$ and the scalar projection is $1/\sqrt{14}$.

The distance of a point from a plane

As an illustration of the usefulness of these ideas, we will find a formula for the shortest distance between the point (x_0, y_0, z_0) and the plane whose equation is

$$ax + by + cz = d.$$

First we need to recall a few basic facts about planes.

Suppose that (x_1, y_1, z_1) and (x_2, y_2, z_2) are two points on the given plane. Then $ax_1 + by_1 + cz_1 = d = ax_2 + by_2 + cz_2$, so that

$$a(x_1 - x_2) + b(y_1 - y_2) + c(z_1 - z_2) = 0.$$

Now this equation asserts that the vector

$$N \; = \; \begin{pmatrix} a \\ b \\ c \end{pmatrix}$$

is orthogonal to the vector with entries $x_1 - x_2$, $y_1 - y_2$, $z_1 - z_2$, and hence to every vector in the plane.

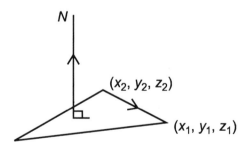

Thus N is a *normal vector* to the plane $ax + by + cz = d$, which is a familiar fact from the analytical geometry of three-dimensional space.

We are now in a position to calculate the shortest distance l from the point (x_0, y_0, z_0) to the plane. Let (x, y, z) be a point in the plane, and write

$$X = \begin{pmatrix} x \\ y \\ z \end{pmatrix} \quad \text{and} \quad Y = \begin{pmatrix} x_0 \\ y_0 \\ z_0 \end{pmatrix}.$$

Then l is simply the scalar projection of $X_0 - X$ on N, as may be seen from the diagram below:

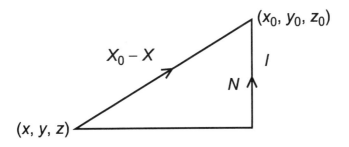

Therefore

$$l = \frac{|(X_0 - X)^T N|}{\|N\|}.$$

Now

$$
\begin{aligned}
(X_0 - X)^T N &= a(x_0 - x) + b(y_0 - y) + c(z_0 - z) \\
&= ax_0 + by_0 + cz_0 - d :
\end{aligned}
$$

for $ax + by + cz = d$ since the point (x, y, z) lies in the plane. Thus we arrive at the formula

$$l = \frac{|ax_0 + by_0 + cz_0 - d|}{\sqrt{a^2 + b^2 + c^2}}.$$

Vector products in $\mathbf{R}^3$

In addition to the scalar product, there is another well-known construction in $\mathbf{R}^3$ called the *vector product*. This is defined in the following manner.

Suppose that

$$X = \begin{pmatrix} x_1 \\ x_2 \\ x_3 \end{pmatrix} \text{ and } Y = \begin{pmatrix} y_1 \\ y_2 \\ y_3 \end{pmatrix}$$

are two vectors in $\mathbf{R}^3$. Then the *vector product* of X and Y

$$X \times Y$$

is defined to be the vector

$$\begin{pmatrix} x_2 y_3 - x_3 y_2 \\ x_3 y_1 - x_1 y_3 \\ x_1 y_2 - x_2 y_1 \end{pmatrix}.$$

Notice that each entry of this vector is a 2×2 determinant. Because of this, the vector product is best written as a 3×3 determinant. Following a commonly used notation, let us write $\mathbf{i}$, $\mathbf{j}$, $\mathbf{k}$ for the vectors of the standard basis of $\mathbf{R}^3$. Thus

$$\mathbf{i} = \begin{pmatrix} 1 \\ 0 \\ 0 \end{pmatrix}, \ \mathbf{j} = \begin{pmatrix} 0 \\ 1 \\ 0 \end{pmatrix} \text{ and } \mathbf{k} = \begin{pmatrix} 0 \\ 0 \\ 1 \end{pmatrix}.$$

Then the vector product $X \times Y$ can be expressed in the form

$$X \times Y = (x_2 y_3 - x_3 y_2)\mathbf{i} + (x_3 y_1 - x_1 y_3)\mathbf{j} + (x_1 y_2 - x_2 y_1)\mathbf{k}.$$

This expression is a row expansion of the 3×3 determinant

$$X \times Y = \begin{pmatrix} \mathbf{i} & \mathbf{j} & \mathbf{k} \\ x_1 & x_2 & x_3 \\ y_1 & y_2 & y_3 \end{pmatrix}.$$

Here the determinant is evaluated by expanding along row 1 in the usual manner.

Example 7.1.2

The vector product of $X = \begin{pmatrix} -1 \\ 3 \\ 2 \end{pmatrix}$ and $Y = \begin{pmatrix} 2 \\ -1 \\ 4 \end{pmatrix}$ is

$$X \times Y = \begin{pmatrix} \mathbf{i} & \mathbf{j} & \mathbf{k} \\ -1 & 3 & 2 \\ 2 & -1 & 4 \end{pmatrix},$$

which becomes on expansion

$$X \times Y = 14\mathbf{i} + 8\mathbf{j} - 5\mathbf{k} = \begin{pmatrix} 14 \\ 8 \\ -5 \end{pmatrix}.$$

The importance of the vector product $X \times Y$ arises from the fact that it is orthogonal to each of the vectors X and Y; thus it is represented by a line segment that is normal to the plane containing line segments corresponding to X and Y, in case these are not parallel. To see this we can simply form the scalar product of $X \times Y$ in turn with X and Y. For example,

$$X^T(X \times Y) = \begin{pmatrix} x_1 & x_2 & x_3 \\ x_1 & x_2 & x_3 \\ y_1 & y_2 & y_3 \end{pmatrix}.$$

Since rows 1 and 2 are identical, this is zero by a basic property of determinants (3.2.2).

In fact the vectors X, Y, $X \times Y$ form a *right-handed system* in the sense that their directions correspond to the thumb and first two index fingers of the right hand when held extended.

Theorem 7.1.3
If X and Y are vectors in $\mathbf{R}^3$, the vector $X \times Y$ is orthogonal to both X and Y, and the three vectors X, Y, $X \times Y$ form a right-handed system.

The length of the vector product, like the the scalar product, is a number with geometrical significance.

Theorem 7.1.4
If X and Y are vectors in $\mathbf{R}^3$ and θ is the angle in the interval $[0, \pi]$ between X and Y, then

$$\|X \times Y\| = \|X\|\,\|Y\|\sin\,\theta.$$

Proof
We compute the expression $\|X\|^2\|Y\|^2 - \|X \times Y\|^2$, by substituting $\|X\|^2 = x_1^2 + x_2^2 + x_3^2$, $\|Y\|^2 = y_1^2 + y_2^2 + y_3^2$ and

$$\|X \times Y\|^2 = (x_2 y_3 - x_3 y_2)^2 + (x_3 y_1 - x_1 y_3)^2 + (x_1 y_2 - x_2 y_1)^2.$$

After expansion and cancellation of some terms, we find that

$$\|X\|^2\|Y\|^2 - \|X \times Y\|^2 = (x_1 y_1 + x_2 y_2 + x_3 y_3)^2 = (X^T Y)^2.$$

Therefore, by 7.1.1,

$$\|X\|^2\|Y\|^2 - |X \times Y|^2 = \|X\|^2\|Y\|^2 \cos^2 \theta.$$

Consequently $\|X \times Y\|^2 = \|X\|^2\,\|Y\|^2 \sin^2 \theta$. Finally, take the square root of each side, noting that the positive sign is correct since $\sin\,\theta \geq 0$ in the interval $[0, \pi]$.

Theorem 7.1.4 provides another geometrical interpretation of the vector product $X \times Y$. For $\|X \times Y\|$ is simply the area of the parallelogram *IPRQ* formed by line segments representing the vectors X and Y. Indeed the area of this parallelogram equals

$$(IQ \sin \theta)IP = \|X\| \, \|Y\| \sin \theta = \|X \times Y\|.$$

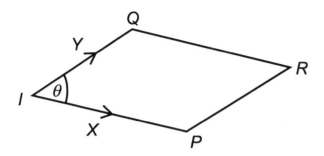

Orthogonality in $\mathbf{R}^n$

Having gained some insight from $\mathbf{R}^3$, we are now ready to define orthogonality in n-dimensional Euclidean space.

Let X and Y be two vectors in $\mathbf{R}^n$. Then X and Y are said to be *orthogonal* if

$$X^T Y = 0.$$

This a natural extension of orthogonality in $\mathbf{R}^3$. It follows from the definition that the zero vector is orthogonal to every vector in $\mathbf{R}^n$ and that no non-zero vector can be orthogonal to itself: indeed $X^T X = x_1^2 + x_2^2 + \cdots + x_n^2 > 0$ if $X \neq 0$.

It turns out that the inequality of 7.1.2 is valid for $\mathbf{R}^n$.

Theorem 7.1.5 (*Cauchy - Schwartz Inequality*)
If X and Y are vectors in $\mathbf{R}^n$, then

$$|X^T Y| \leq \|X\| \, \|Y\|.$$

We shall not prove 7.1.5 at this stage since a more general fact will be established in 7.2: see however Exercise 7.1.10.

Because of 7.1.5 it is meaningful to define the *angle between two non-zero vectors* X and Y in $\mathbf{R}^n$ to be the angle θ in the interval $[0, \pi]$ such that

$$\cos\ \theta = \frac{X^T Y}{\|X\|\ \|Y\|}.$$

An important consequence of 7.1.5 is

Theorem 7.1.6 (*The Triangle Inequality*)
If X and Y are vectors in $\mathbf{R}^n$, then

$$\|X + Y\| \le \|X\| + \|Y\|.$$

Proof
Let the entries of X and Y be $x_1, \ldots, x_n$ and $y_1, \ldots, y_n$ respectively. Then

$$\|X + Y\|^2 = (X + Y)^T(X + Y) = X^T X + X^T Y + Y^T X + Y Y^T$$

and, since $X^T Y = Y^T X$, this equals

$$\|X\|^2 + \|Y\|^2 + 2 X^T Y.$$

By the Cauchy-Schwartz Inequality $|X^T Y| \le \|X\|\ \|Y\|$, so it follows that

$$\|X + Y\|^2 \le \|X\|^2 + \|Y\|^2 + 2\|X\|\ \|Y\| = (\|X\| + \|Y\|)^2,$$

which yields the desired inequality.

When $n = 3$, the assertion of 7.1.6 is just the well-known fact that the sum of the lengths of two sides of a triangle is never less than the length of the third side, as can be seen from the triangle rule of addition for the vectors X and Y.

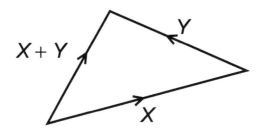

Complex matrices and orthogonality in $\mathbf{C}^n$

It is possible to define a notion of orthogonality in the complex vector space $\mathbf{C}^n$, a fact that will be important in Chapter Eight. However, a crucial change in the definition must be made. To see why a change is necessary, consider the complex vector $X = \begin{pmatrix} \sqrt{-1} \\ 1 \end{pmatrix}$. Then $X^T X = -1 + 1 = 0$. Since it does not seem reasonable to allow a non-zero vector to have length zero, we must alter the definition of a scalar product in order to exclude this phenomenon.

First it is necessary to introduce a new operation on complex matrices. Let A be an $m \times n$ matrix over the complex field $\mathbf{C}$. Define the *complex conjugate*

$$\bar{A}$$

of A to be the $m \times n$ matrix whose (i, j) entry is the complex conjugate of the (i, j) entry of A. Then define the *complex transpose* of A to be the transpose of the complex conjugate

$$A^* = (\bar{A})^T.$$

For example, if

$$A = \begin{pmatrix} 4 & -\sqrt{-1} & 3 \\ 1+\sqrt{-1} & -4 & 1-\sqrt{-1} \end{pmatrix},$$

then

$$A^* = \begin{pmatrix} 4 & 1-\sqrt{-1} \\ \sqrt{-1} & -4 \\ 3 & 1+\sqrt{-1} \end{pmatrix}.$$

Usually it is more appropriate to use the complex transpose when dealing with complex matrices. In many ways the complex transpose behaves like the transpose; for example, there is the following fact.

Theorem 7.1.7

If A and B are complex matrices, then $(AB)^* = B^*A^*$.

This follows at once from the equations $(\bar{A}\bar{B}) = (\bar{A})(\bar{B})$ and $(AB)^T = B^T A^T$.

Now let us use the complex transpose to define the *complex scalar product of vectors X and Y in $\mathbf{C}^n$*; this is to be

$$X^*Y = \bar{x}_1 y_1 + \cdots + \bar{x}_n y_n,$$

which is a complex number. Why is this definition any better than the previous one? The reason is that, if we define the *length* of the vector X in the natural way as

$$\|X\| = \sqrt{X^*X} = \sqrt{|x_1|^2 + \cdots + |x_n|^2},$$

then $\|X\|$ is always a non-negative real number, and it cannot equal 0 unless X is the zero vector. It is an important consequence of the definition that Y^*X equals the *complex conjugate* of X^*Y, so the complex scalar product is not symmetric in X and Y.

It remains to define orthogonality in $\mathbf{C}^n$. Two vectors X and Y in $\mathbf{C}^n$ are said to be *orthogonal* if

$$X^*Y = 0.$$

We now make the blanket assertion that all the results established for scalar products in $\mathbf{R}^n$ carry over to complex scalar products in $\mathbf{C}^n$. In particular the Cauchy-Schwartz and Triangle Inequalities are valid.

Exercises 7.1

1. Find the angle between the vectors $\begin{pmatrix} -2 \\ 4 \\ 3 \end{pmatrix}$ and $\begin{pmatrix} 1 \\ -2 \\ 3 \end{pmatrix}$.

2. Find the two unit vectors which are orthogonal to both of the vectors $\begin{pmatrix} -2 \\ 3 \\ -1 \end{pmatrix}$ and $\begin{pmatrix} 1 \\ 1 \\ 1 \end{pmatrix}$.

3. Compute the vector and scalar projections of $\begin{pmatrix} -2 \\ 3 \\ -1 \end{pmatrix}$ on $\begin{pmatrix} 1 \\ 2 \\ 3 \end{pmatrix}$.

4. Show that the planes $x - 3y + 4z = 12$ and $2x - 6y + 8z = 6$ are parallel and then find the shortest distance between them.

5. If $X = \begin{pmatrix} 2 \\ -1 \\ 3 \end{pmatrix}$ and $Y = \begin{pmatrix} 0 \\ 4 \\ 2 \end{pmatrix}$, find the vector product $X \times Y$. Hence compute the area of the parallelogram whose vertices have the following coordinates: $(1, 1, 1)$, $(3, 0, 4)$, $(1, 5, 3)$, $(3, 4, 6)$.

6. Establish the following properties of the vector product:
   (a) $X \times X = 0$; (b) $X \times (Y + Z) = X \times Y + X \times Z$;
   (c) $X \times Y = -Y \times X$; (d) $X \times (cY) = c(X \times Y) = (cX) \times Y$.

7. If X, Y, Z are vectors in $\mathbf{R}^3$, prove that

$$X^T(Y \times Z) = Y^T(Z \times X) = Z^T(X \times Y).$$

(This is called the *scalar triple product of X, Y, Z*). Then show that that the absolute value of this number equals the volume of the parallelopiped formed by line segments representing the vectors X, Y, Z drawn from the same initial point.

8. Use Exercise 7 to find the condition for the three vectors X, Y, Z to be represented by coplanar line segments.

9. Show that the set of all vectors in $\mathbf{R}^n$ which are orthogonal to a given vector X is a subpace of $\mathbf{R}^n$. What will its dimension be?

10. Prove the Cauchy-Schwartz Inequality for $\mathbf{R}^n$. [Hint: compute the expression $\|X\|^2\|Y\|^2 - |X^TY|^2$ and show that it is is non-negative].

11. Find the most general vector in $\mathbf{C}^3$ which is orthogonal to both of the vectors

$$\begin{pmatrix} -\sqrt{-1} \\ 2+\sqrt{-1} \\ 3 \end{pmatrix} \text{ and } \begin{pmatrix} 1 \\ 1 \\ \sqrt{-2} \end{pmatrix}.$$

12. Let A and B be complex matrices of appropriate sizes. Prove the following statements:

$$(a)(\bar{A})^T = (\overline{A^T}); \ (b)(A+B)^* = A^* + B^*; \ (c)(A^*)^* = A.$$

13. How should the vector projection of X on Y be defined in $\mathbf{C}^3$?

14. Show that the vector equation of the plane through the point (x_0, y_0, z_0) with normal vector N is

$$(X - X_0)^T N = 0$$

where X and X_0 are the vectors with entries x, y, z and x_0, y_0, z_0, respectively.

15. Prove the Cauchy-Schwartz Inequality for complex scalar products in $\mathbf{C}^n$.

16. Prove the Triangle Inequality for complex scalar products in $\mathbf{C}^n$.

17. Establish the following expression for the *vector triple product* in $\mathbf{R}^3$: $X \times (Y \times Z) = (X \cdot Z)Y - (X \cdot Y)Z$. [Hint: note that the vector on the right hand side is orthogonal to to both X and $Y \times Z$.]

7.2 Inner Product Spaces

We have seen how to introduce the notion of orthogonality in the vector spaces $\mathbf{R}^n$ and $\mathbf{C}^n$ for arbitrary n. But what about other vector spaces such as vector spaces of polynomials or continuous functions? It turns out that there is a general concept called an inner product which is a natural extension of the scalar products in $\mathbf{R}^n$ and $\mathbf{C}^n$. This allows the introduction of orthogonality in arbitrary real and complex vector spaces.

Let V be a real vector space, that is, a vector space over $\mathbf{R}$. An *inner product* on V is a rule which assigns to each pair of vectors $\mathbf{u}$ and $\mathbf{v}$ of V a real number $< \mathbf{u}, \mathbf{v} >$, their inner product, such that the following properties hold:

(i) $< \mathbf{v}, \mathbf{v} > \geq 0$ and $< \mathbf{v}, \mathbf{v} > = 0$ if and only if $\mathbf{v} = \mathbf{0}$;
(ii) $< \mathbf{u}, \mathbf{v} > = < \mathbf{v}, \mathbf{u} >$;
(iii) $< c\mathbf{u} + d\mathbf{v}, \mathbf{w} > = c < \mathbf{u}, \mathbf{w} > + d < \mathbf{v}, \mathbf{w} >$.

The understanding here is that these properties must hold for all vectors **u**, **v**, **w** and all real scalars c, d.

We now give some examples of inner products, the first one being the scalar product, which provided the original motivation.

Example 7.2.1

Define an inner product $< >$ on $\mathbf{R}^n$ by the rule

$$< X, Y > = X^T Y.$$

That this is an inner product follows from the laws of matrix algebra, and the fact that $X^T X$ is non-negative and equals 0 only if $X = 0$. This inner product will be referred to as the *standard inner product* on $\mathbf{R}^n$. It should be borne in mind that there are other possible inner products for this vector space; for example, an inner product on $\mathbf{R}^3$ is defined by

$$< X, Y > = 2x_1 y_1 + 3x_2 y_2 + 4x_3 y_3$$

where X and Y are the vectors with entries x_1, x_2, x_3 and y_1, y_2, y_3 respectively. The reader should verify that the axioms for an inner product hold in this case.

Example 7.2.2

Define an inner product $< >$ on the vector space $C[a, b]$ by the rule

$$< f, g > = \int_a^b f(x)g(x)dx.$$

This is very different type of inner product, which is important in the theory of orthogonal functions. Well-known properties of integrals show that the requirements for an inner product are satisfied. For example,

$$< f, f > = \int_a^b f(x)^2 dx \geq 0$$

since $f(x)^2 \geq 0$; also, if we think of the integral as the area under the curve $y = f(x)^2$, then it becomes clear that the integral cannot vanish unless $f(x)$ is identically equal to zero in $[a, b]$.

Example 7.2.3
Define an inner product on the vector space $P_n(\mathbf{R})$ of all real polynomials in x of degree less than n by the rule

$$< f, g > = \sum_{i=1}^{n} f(x_i)g(x_i)$$

where $x_1, \ldots, x_n$ are distinct real numbers.

Here it is not so clear why the first requirement for an inner product holds. Note that

$$< f, f > = \sum_{i=1}^{n} f(x_i)^2 \geq 0;$$

also the only way that this sum can vanish is if $f(x_1) = \ldots = f(x_n) = 0$. But f is a polynomial of degree at most $n-1$, so it cannot have n distinct roots unless it is the zero polynomial.

Orthogonality in inner product spaces

A *real inner product space* is a vector space V over $\mathbf{R}$ together with an inner product $< >$ on V. It will be convenient to speak of "the inner product space V", suppressing mention of the inner product where this is understood. Thus "the inner product space $\mathbf{R}^n$" refers to $\mathbf{R}^n$ with the scalar product as inner product: this is called the *Euclidean inner product space*.

Two vectors $\mathbf{u}$ and $\mathbf{v}$ of an inner product space V are said to be *orthogonal* if

$$< \mathbf{u}, \mathbf{v} > = 0.$$

It follows from the definition of an inner product that the zero vector is orthogonal to every vector and no non-zero vector can be orthogonal to itself.

Example 7.2.4
Show that the functions sin x, $m = 1, 2, \ldots$, are mutually orthogonal in the inner product space $C[0, \pi]$ where the inner product is given by the formula $< f, g > = \int_0^\pi f(x)g(x)dx$.

We have merely to compute the inner product of sin mx and sin nx :

$$< \sin\ mx,\ \sin\ nx > = \int_0^\pi \sin\ mx \sin\ nx\ dx.$$

Now, according to a well-known trigonometric identity,

$$\sin mx\ \sin\ nx = \frac{1}{2}(\cos(m - n)x - \cos(m + n)x).$$

Therefore, on evaluating the integrals, we obtain as the value of $< \sin\ mx,\ \sin\ nx >$

$$[\frac{1}{2(m - n)} \sin(m - n)x - \frac{1}{2(m + n)} \sin(m + n)x]_0^\pi = 0,$$

provided $m \neq n$. This is a very important set of orthogonal functions which plays a basic role in the theory of Fourier series.

If $\mathbf{v}$ is a vector in an inner product space V, then $< \mathbf{v}, \mathbf{v} > \geq 0$, so this number has a real square root. This allows us to define the *norm* of $\mathbf{v}$ to be the real number

$$\|\mathbf{v}\| = \sqrt{< \mathbf{v}, \mathbf{v} >}.$$

Thus $\|\mathbf{v}\| \geq 0$ and $\|\mathbf{v}\|$ equals zero if and only if $\mathbf{v} = \mathbf{0}$. A vector with norm 1 is called a *unit vector*. It is clear that norm is a generalization of length in Euclidean space.

Example 7.2.5
Find the norm of the function sin mx in the inner product space $C[0, \pi]$ of Example 7.2.4.

Once again we have to compute an integral:

$$\| \sin mx \|^2 = \int_0^\pi \sin^2 mx \ dx = \int_0^\pi \frac{1}{2}(1 - \cos 2mx)dx = \pi/2.$$

Hence $\| \sin mx \| = \sqrt{(\pi/2)}$. It follows that the functions

$$\sqrt{\frac{2}{\pi}} \sin mx, \ m = 1, 2, \ldots,$$

form a set of mutually orthogonal unit vectors. Such sets are called orthonormal and will be studied in 7.3.

There is an important inequality relating inner product and norm which has already been encountered for Euclidean spaces.

Theorem 7.2.1 (*The Cauchy - Schwartz Inequality*)
Let **u** *and* **v** *be vectors in an inner product space. Then*

$$| < \mathbf{u}, \mathbf{v} > | \leq \| \mathbf{u} \| \ \| \mathbf{v} \|.$$

Proof
We can assume that $\mathbf{v} \neq \mathbf{0}$ or else the result is obvious. Let t denote an arbitrary real number. Then, using the defining properties of the inner product, we find that $< \mathbf{u} - t\mathbf{v}, \mathbf{u} - t\mathbf{v} >$ equals

$$< \mathbf{u}, \mathbf{u} > - < \mathbf{u}, \mathbf{v} > t - < \mathbf{v}, \mathbf{u} > t + < \mathbf{v}, \mathbf{v} > t^2,$$

which reduces to

$$\| \mathbf{u} \|^2 - 2 < \mathbf{u}, \mathbf{v} > t + \| \mathbf{v} \|^2 t^2 \geq 0.$$

For brevity write $a = \|\mathbf{v}\|^2$, $b = <\mathbf{u}, \mathbf{v}>$ and $c = \|\mathbf{u}\|^2$. Thus

$$at^2 - 2bt + c = <\mathbf{u} - t\mathbf{v}, \mathbf{u} - t\mathbf{v}> \geq 0.$$

To see what this implies, complete the square in the usual manner;

$$at^2 - 2bt + c = a\left(\left(t - \frac{b}{a}\right)^2 + \left(\frac{c}{a} - \frac{b^2}{a^2}\right)\right).$$

Since $a > 0$ and the expression on the left hand side of the equation is non-negative for all values of t, it follows that $c/a \geq b^2/a^2$, that is, $b^2 \leq ac$. On substituting the values of a, b and c, and taking the square root, we obtain the desired inequality.

Example 7.2.6

If 7.2.1 is applied to the vector space $C[a, b]$ with the inner product specified in Example 7.2.2, we obtain the inequality

$$\left| \int_a^b f(x)g(x)dx \right| \leq \left(\int_a^b f(x)^2 dx \right)^{1/2} \left(\int_a^b g(x)^2 dx \right)^{1/2}.$$

Normed linear spaces

The next step in our series of generalizations is to extend the notion of length of a vector in Euclidean space. Let V denote a real vector space. By a *norm* on V is meant a rule which assigns to each vector $\mathbf{v}$ a real number $\|\mathbf{v}\|$, its *norm*, such that the following properties hold:

(i) $\|\mathbf{v}\| \geq 0$ and $\|\mathbf{v}\| = 0$ if and only if $v = \mathbf{0}$;
(ii) $\|c\mathbf{v}\| = |c| \, \|\mathbf{v}\|$;
(iii) $\|\mathbf{u} + \mathbf{v}\| \leq \|\mathbf{u}\| + \|\mathbf{v}\|$. (The Triangle Inequality).

These are to hold for all vectors $\mathbf{u}$ and $\mathbf{v}$ in V and all scalars c. A vector space together with a norm is called a *normed linear space*.

We already know an example of a normed linear space; for the length function on $\mathbf{R}^n$ is a norm. To see why this is so, we need to remember that the Triangle Inequality was established for the length function in 7.1.6.

The reader will have noticed that the term "norm" has already been used in the context of an inner product space. Let us show that these two usages are consistent.

Theorem 7.2.2

Let V be an inner product space and define

$$\|\mathbf{v}\| = \sqrt{<\mathbf{v}, \mathbf{v}>}.$$

Then $\|\ \|$ is a norm on V and V is a normed linear space.

Proof

We need to check the three axioms for a norm. In the first place, $\|\mathbf{v}\| = \sqrt{<\mathbf{v}, \mathbf{v}>} \geq 0$, and this cannot vanish unless $\mathbf{v} = \mathbf{0}$, by the definition of an inner product. Next, if c is a scalar, then

$$\|c\mathbf{v}\| = \sqrt{<c\mathbf{v}, c\mathbf{v}>} = \sqrt{(c^2 <\mathbf{v}, \mathbf{v}>)} = |c|\,\|\mathbf{v}\|.$$

Finally, the Triangle Inequality must be established. By the defining properties of the inner product:

$$\|\mathbf{u} + \mathbf{v}\|^2 = <\mathbf{u} + \mathbf{v}, \mathbf{u} + \mathbf{v}> = \|\mathbf{u}\|^2 + 2 <\mathbf{u}, \mathbf{v}> + \|\mathbf{v}\|^2,$$

which, by 7.2.1, cannot exceed

$$\|\mathbf{u}\|^2 + 2\|\mathbf{u}\|\,\|\mathbf{v}\| + \|\mathbf{v}\|^2 = (\|\mathbf{u}\| + \|\mathbf{v}\|)^2.$$

On taking square roots, we derive the required inequality.

Theorem 7.2.2 enables us to give many examples of normed linear spaces.

Example 7.2.7

The Euclidean space $\mathbf{R}^n$ is a normed linear space if length is taken as the norm. Thus

$$\|X\| = \sqrt{X^T X} = \sqrt{x_1^2 + x_2^2 + \cdots + x_n^2}.$$

Example 7.2.8

The vector space $C[a, b]$ becomes a normed linear space if $\|f\|$ is defined to be

$$(\int_a^b f(x)^2 dx)^{1/2}.$$

Example 7.2.9 (*Matrix norms*)

A different type of normed linear space arises if we consider the vector space of all real $m \times n$ matrices and introduce a norm on it as follows. If $A = [a_{ij}]_{m,n}$, define $\|A\|$ to be

$$(\sum_{i=1}^m \sum_{j=1}^n a_{ij}^2)^{1/2}.$$

On the face of it this is a reasonable measure of the "size" of the matrix. But of course one has to show that this is really a norm. A neat way to do this is as follows: put $\hat{A}$ equal to the mn-column vector whose entries are the elements of A listed by rows. The key point to note is that $\|A\|$ is just the length of the vector $\hat{A}$ in $\mathbf{R}^{mn}$. It follows at once that $\| \ \|$ is a norm since we know that length is a norm.

Inner products on complex vector spaces

So far inner products have only been defined on real vector spaces. Now it has already been seen that there is a reasonable concept of orthogonality in the complex vector space $\mathbf{C}^n$, although it differs from orthogonality in $\mathbf{R}^n$ in that a different scalar product must be used. This suggests that if an inner product is to be defined on an arbitrary complex vector space, there will have to be a change in the definition of the inner product.

Let V be a vector space over $\mathbf{C}$. An *inner product* on V is a rule that assigns to each pair of vectors $\mathbf{u}$ and $\mathbf{v}$ in V a complex number $< \mathbf{u}, \mathbf{v} >$ such that the following rules hold:

(i) $< \mathbf{v}, \mathbf{v} > \geq 0$ and $< \mathbf{v}, \mathbf{v} > = 0$ if and only if $\mathbf{v} = \mathbf{0}$;
(ii) $< \mathbf{u}, \mathbf{v} > = \overline{< \mathbf{v}, \mathbf{u} >}$;
(iii) $< c\mathbf{u} + d\mathbf{v}, \mathbf{w} > = \bar{c} < \mathbf{u}, \mathbf{w} > + \bar{d} < \mathbf{v}, \mathbf{w} >$.

These are to hold for all vectors $\mathbf{u}, \mathbf{v}, \mathbf{w}$ and all complex scalars c, d. Observe that property (ii) implies that $< \mathbf{v}, \mathbf{v} >$ is real: for this complex number equals its complex conjugate. A complex vector space which is equipped with an inner product is called a *complex inner product space*.

Our prime example of a complex inner product space is $\mathbf{C}^n$ with the complex scalar product $< X, Y > = X^*Y$. To see that this is a complex inner product, we need to note that $X^*Y = \overline{Y^*X}$ and

$$< cX + dY, Z > = (cX + dY)^*Z = \bar{c}X^*Z + \bar{d}Y^*Z,$$

which is just $\bar{c} < X, Z > + \bar{d} < Y, Z >$.

Provided that the changes implied by the altered conditions (ii) and (iii) are made, the concepts and results already established for real inner product spaces can be extended to complex inner product spaces. In addition, results stated for real inner product spaces in the remainder of this section hold for complex inner product spaces, again with the appropriate changes.

Orthogonal complements

We return to the study of orthogonality in real inner product spaces. We wish to introduce the important notion of the orthogonal complement of a subspace. Here what we have in mind as our model is the simple situation in three-dimensional space where the orthogonal complement of a plane is the set of line segments perpendicular to it.

Let S be a subspace of a real inner product space V. The *orthogonal complement* of S is defined to be the set of all vectors in V that are orthogonal to every vector in S: it is denoted by the symbol

$$S^\perp.$$

Example 7.2.10

Let S be the subspace of $\mathbf{R}^3$ consisting of all vectors of the form

$$\begin{pmatrix} a \\ b \\ 0 \end{pmatrix}$$

where a and b are real numbers. Thus elements of S correspond to line segments in the xy-plane. Equally clearly $S^\perp$ is the set of all vectors of the form

$$\begin{pmatrix} 0 \\ 0 \\ c \end{pmatrix}.$$

These correspond to line segments along the z-axis, hardly a surprising conclusion.

The most fundamental property of an orthogonal complement is that it is a subspace.

Theorem 7.2.3

Let S be a subspace of a real inner product space V. Then
(a) $S^\perp$ *is a subspace of V;*
(b) $S \cap S^\perp = 0$;
(c) *if S is finitely generated, a vector $\mathbf{v}$ belongs to $S^\perp$ if and only if it is orthogonal to every vector in some set of generators of S.*

Proof

To show that $S^\perp$ is a subspace we need to verify that it contains the zero vector and is closed with respect to addition and scalar multiplication. The first statement is true since the zero vector is orthogonal to every vector. As for the remaining ones, take two vectors $\mathbf{v}$ and $\mathbf{w}$ in $S^\perp$, let $\mathbf{s}$ be an arbitrary vector in S and let c be a scalar. Then

$$< c\mathbf{v}, \mathbf{s} > = c < \mathbf{v}, \mathbf{s} > = 0,$$

and

$$< \mathbf{v} + \mathbf{w}, \mathbf{s} > = < \mathbf{v}, \mathbf{s} > + < \mathbf{w}, \mathbf{s} > = 0.$$

Hence $c\mathbf{v}$ and $\mathbf{v} + \mathbf{w}$ belong to $S^\perp$.

Now suppose that $\mathbf{v}$ belongs to the intersection $S \cap S^\perp$. Then $\mathbf{v}$ is orthogonal to itself, which can only mean that $\mathbf{v} = \mathbf{0}$.

Finally, assume that $\mathbf{v}_1, \ldots, \mathbf{v}_m$ are generators of S and that $\mathbf{v}$ is orthogonal to each $\mathbf{v}_i$. A general vector of S has the form $\sum_{i=1}^{m} c_i \mathbf{v}_i$ for some scalars c_i. Then

$$< \mathbf{v}, \sum_{i=1}^{m} c_i \mathbf{v}_i > = \sum_{i=1}^{m} c_i < \mathbf{v}, \mathbf{v}_i > = 0.$$

Hence $\mathbf{v}$ is orthogonal to every vector in S and so it belongs to $S^\perp$. The converse is obvious.

Example 7.2.11

In the inner product space $P_3(\mathbf{R})$ with

$$< f, g > = \int_0^1 f(x)g(x)dx,$$

find the orthogonal complement of the subspace S generated by 1 and x.

Let $f = a_0 + a_1 x + a_2 x^2$ be an element of $P_3(\mathbf{R})$. By 7.2.3, a polynomial f belongs to $S^\perp$ if and only if it is orthogonal to 1 and x; the conditions for this are

$$< f, 1 > = \int_0^1 f(x)dx = a_0 + \frac{1}{2}a_1 + \frac{1}{3}a_2 = 0$$

and

$$< f, x >= \int_0^1 xf(x)dx = \frac{1}{2}a_0 + \frac{1}{3}a_1 + \frac{1}{4}a_2 = 0.$$

Solving these equations, we find that $a_0 = t/6$, $a_1 = -t$ and $a_2 = t$, where t is arbitrary. Hence $f = t(x^2 - x + \frac{1}{6})$ is the most general element of $S^\perp$. It follows that $S^\perp$ is the 1-dimensional subspace generated by the polynomial $x^2 - x + \frac{1}{6}$.

Notice in the last example that $\dim(S) + \dim(S^\perp) = 3$, the dimension of $P_3(\mathbf{R})$. This is no coincidence, as the following fundamental theorem shows.

Theorem 7.2.4

Let S be a subspace of a finite-dimensional real inner product space V; then

$$V = S \oplus S^\perp \quad \text{and} \quad \dim(V) = \dim(S) + \dim(S^\perp).$$

Proof
According to the definition in 5.3, we must prove that $V = S + S^\perp$ and $S \cap S^\perp = 0$. The second statement is true by 7.2.3, but the first one requires proof.

Certainly, if $S = 0$, then $S^\perp = V$ and the result is clear. Having disposed of this case, we may assume that S is non-zero and choose a basis $\mathbf{v}_1, \ldots, \mathbf{v}_m$ for S. Extend this basis of

S to a basis of V, say $\mathbf{v}_1, \ldots, \mathbf{v}_m, \mathbf{v}_{m+1}, \ldots, \mathbf{v}_n$: this possible by 5.1.4. If $\mathbf{v}$ is an arbitrary vector of V, we can write

$$\mathbf{v} = \sum_{j=1}^{n} c_j \mathbf{v}_j.$$

By 7.2.3 the vector $\mathbf{v}$ belongs to $S^{\perp}$ if and only if it is orthogonal to each of the vectors $\mathbf{v}_1, \ldots, \mathbf{v}_m$; the conditions for this are

$$< \mathbf{v}_i, \mathbf{v} > = \sum_{j=1}^{n} < \mathbf{v}_i, \mathbf{v}_j > c_j = 0, \text{ for } i = 1, 2, \ldots, m.$$

Now the above equations constitute a linear system of m equations in the n unknowns $c_1, c_2, \ldots, c_n$. Therefore the dimension of $S^{\perp}$ equals the dimension of the solution space of the linear system, which we know from 5.1.7 to be $n - r$ where r is the rank of the $m \times n$ coefficient matrix $A = [< \mathbf{v}_i, \mathbf{v}_j >]$. Obviously $r \leq m$; we shall show that in fact $r = m$. If this is false, then the m rows of A must be linearly dependent and there exist scalars $d, d_2, \ldots, d_m$, not all of them zero, such that

$$0 = \sum_{i=1}^{m} d_i < \mathbf{v}_i, \mathbf{v}_j > = < \sum_{i=1}^{m} d_i \mathbf{v}_i, \mathbf{v}_j >$$

for $j = 1, \ldots, n$. But a vector which is orthogonal to every vector in a basis of V must be zero. Hence $\sum_{i=1}^{m} d_i \mathbf{v}_i = 0$, which can only mean that $d_1 = d_2 = \cdots = d_m = 0$ since $\mathbf{v}_1, \ldots, \mathbf{v}_m$ are linearly independent. By this contradiction $r = m$.

We conclude that $\dim(S^{\perp}) = n - m = n - \dim(S)$, which implies that $\dim(S) + \dim(S^{\perp}) = n = \dim(V)$. It follows from 5.3.2 that

$$\dim(S + S^{\perp}) = \dim(S) + \dim(S^{\perp}) = \dim(V).$$

Hence $V = S + S^\perp$, as required.

An important consequence of the theorem is

Corollary 7.2.5
If S is a subspace of a finite-dimensional real inner product space V, then

$$(S^\perp)^\perp = S.$$

Proof
Every vector in S is certainly orthogonal to every vector in $S^\perp$; thus S is a subspace of $(S^\perp)^\perp$. On the other hand, a computation with dimensions using 7.2.4 yields

$$\begin{aligned}
\dim((S^\perp)^\perp) &= \dim(V) - \dim(S^\perp) \\
&= \dim(V) - (\dim(V) - \dim(S)) \\
&= \dim(S)
\end{aligned}$$

Therefore $S = (S^\perp)^\perp$.

Projection on a subspace
The direct decomposition of an inner product space into a subspace and its orthogonal complement afforded by 7.2.4 leads to wide generalization of the elementary notion of projection of one vector on another, as described in 7.1. This generalized projection will prove invaluable during the discussion of least squares in 7.4.

Let V be a finite-dimensional real inner product space, let S be a subspace and let $\mathbf{v}$ an element of V. Since $V = S \oplus S^\perp$, there is a unique expression for $\mathbf{v}$ of the form

$$\mathbf{v} = \mathbf{s} + \mathbf{s}^\perp$$

where $\mathbf{s}$ and $\mathbf{s}^\perp$ belong to S and $S^\perp$ respectively. Call $\mathbf{s}$ the *projection of $\mathbf{v}$ on the subspace S*. Of course, $\mathbf{s}^\perp$ is the projection of $\mathbf{v}$ on the subspace $S^\perp$. For example, if V is $\mathbf{R}^3$, and

S is the subspace generated by a given vector $\mathbf{u}$, then $\mathbf{s}$ is the projection of $\mathbf{v}$ on $\mathbf{u}$ in the sense of 7.1.

Example 7.2.12
Find the projection of the vector X on the column space of the matrix A where

$$X = \begin{pmatrix} 1 \\ 1 \\ 1 \end{pmatrix} \quad \text{and } A = \begin{pmatrix} 1 & 3 \\ 2 & -1 \\ 1 & 4 \end{pmatrix}.$$

Let S denote the column space of A. Now the columns of A are linearly independent, so they form a basis of S. We have to find a vector Y in S such that $X - Y$ is orthogonal to both columns of A; for then $X - Y$ will belong to $S^{\perp}$ and Y will be the projection of X on S. Now Y must have the form

$$Y = x \begin{pmatrix} 1 \\ 2 \\ 1 \end{pmatrix} + y \begin{pmatrix} 3 \\ -1 \\ 4 \end{pmatrix} = \begin{pmatrix} x + 3y \\ 2x - y \\ x + 4y \end{pmatrix}$$

for some scalars x and y. Then if A_1 and A_2 are the columns of A, the conditions for $X - Y$ to belong to $S^{\perp}$ are

$$< X - Y, A_1 > = (1 - x - 3y) + 2(1 - 2x + y) + (1 - x - 4y) = 0$$

and

$$< X - Y, A_2 > = 3(1 - x - 3y) - (1 - 2x + y) + 4(1 - x - 4y) = 0.$$

These equations yield $x = 74/131$ and $y = 16/131$. The projection of X on the subspace S is therefore

$$Y = \frac{1}{131} \begin{pmatrix} 122 \\ 132 \\ 138 \end{pmatrix}.$$

Orthogonality and the fundamental subspaces of a matrix

We saw in Chapter Four that there are three natural subspaces associated with a matrix A, namely the null space, the row space and the column space. There are of course corresponding subspaces associated with the transpose A^T, so in all six subspaces may be formed. However there is very little difference between the row space of A and the column space of A^T; indeed, if we transpose the vectors in the row space of A, we get the vectors of the column space of A^T. Similarly the vectors in the row space of A^T arise by transposing vectors in the column space of A. Thus there are essentially four interesting subspaces associated with A, namely, the null and column spaces of A and of A^T. These subspaces are connected by the orthogonality relations indicated in the next result.

Theorem 7.2.6

Let A be a real matrix. Then the following statements hold:
     (i) *null space of A = (column space of A^T)$^{\perp}$;*
     (ii) *null space of A^T = (column space of A)$^{\perp}$;*
     (iii) *column space of A = (null space of A^T)$^{\perp}$;*
     (iv) *column space of A^T = (null space of A)$^{\perp}$.*

Proof

To establish (i) observe that a column vector X belongs to the null space of A if and only if it is orthogonal to every column of A^T, that is, X is in (column space of A^T)$^{\perp}$. To deduce (ii) simply replace A by A^T in (i). Equations (iii) and (iv) follow on taking the orthogonal complement of each side of (ii) and (i) respectively, if we remember that $S = (S^{\perp})^{\perp}$ by 7.2.5.

Exercises 7.2

1. Which of the following are inner product spaces?
  (a) $\mathbf{R}^n$ where $< X, Y > = -X^T Y$;
  (b) $\mathbf{R}^n$ where $< X, Y > = 2X^T Y$;
  (c) $C[0,1]$ where $< f, g > = \int_0^1 (f(x) + g(x))dx$.

2. Consider the inner product space $C[0, \pi]$ where $< f, g > = \int_0^\pi f(x)g(x)dx$; show that the functions $1/\sqrt{\pi}$, $\sqrt{2/\pi}\cos mx$, $m = 1, 2, \ldots$, form a set of mutually orthogonal unit vectors.

3. Let w be a fixed, positive valued function in the vector space $C[a,b]$. Show that if $< f, g >$ is defined to be

$$\int_a^b f(x)w(x)g(x)dx,$$

then $< \ >$ is an inner product on $C[a,b]$. [Here w is called a *weight function*].

4. Which of the following are normed linear spaces?
  (a) $\mathbf{R}^3$ where $\|X\| = x_1^2 + x_2^2 + x_3^2$;
  (b) $\mathbf{R}^3$ where $\|X\| = \sqrt{x_1^2 + x_2^2 - x_3^2}$;
  (c) $\mathbf{R}^3$ where $\|X\| = $ the maximum of $|x_1|, |x_2|, |x_3|$.

5. Let V be a finite-dimensional real inner product space with an ordered basis $\mathbf{v}_1, \ldots, \mathbf{v}_n$. Define a_{ij} to be $< \mathbf{v}_i, \mathbf{v}_j >$. If $A = [a_{ij}]$ and $\mathbf{u}$ and $\mathbf{w}$ are any vectors of V, show that $< \mathbf{u}, \mathbf{w} > = [\mathbf{u}]^T A[\mathbf{w}]$ where $[\,\mathbf{u}]$ is the coordinate vector of $\mathbf{u}$ with respect to the given ordered basis.

6. Prove that the matrix A in Exercise 5 has the following properties:
  (a) $X^T AX \geq 0$ for all X;
  (b) $X^T AX = 0$ only if $X = 0$;
  (c) A is symmetric.
Deduce that A must be non-singular.

7. Let A be a real $n \times n$ matrix with properties (a), (b) and (c) of Exercise 6. Prove that $< X, Y > = X^T A Y$ defines an inner product on $\mathbf{R}^n$. Deduce that $\|X\| = \sqrt{X^T A X}$ defines a norm on $\mathbf{R}^n$.

8. Let S be the subspace of the inner product space $P_3(\mathbf{R})$ generated by the polynomials $1 - x^2$ and $2 - x + x^2$, where $< f, g > = \int_0^1 f(x) g(x) dx$. Find a basis for the orthogonal complement of S.

9. Find the projection of the vector with entries $1, -2, 3$ on the column space of the matrix $\begin{pmatrix} 1 & 0 \\ 2 & -4 \\ 3 & 5 \end{pmatrix}$.

10. Prove the following statements about subspaces S and T of a finite dimensional real inner product space:
  (a) $(S + T)^\perp = S^\perp \cap T^\perp$;
  (b) $S^\perp = T^\perp$ always implies that $S = T$;
  (c) $(S \cap T)^\perp = S^\perp + T^\perp$.

11. If S is a subspace of a finite dimensional real inner product space V, prove that $S^\perp \simeq V/S$.

7.3 Orthonormal Sets and the Gram-Schmidt Process

Let V be an inner product space. A set of vectors in V is called *orthogonal* if every pair of distinct vectors in the set is orthogonal. If in addition each vector in the set is a unit vector, that is, has norm is 1, then the set is called *orthonormal.*

Example 7.3.1

In the Euclidean space $\mathbf{R}^3$ the vectors

$$\begin{pmatrix} 1 \\ 1 \\ 2 \end{pmatrix}, \begin{pmatrix} 1 \\ 1 \\ -1 \end{pmatrix}, \begin{pmatrix} -1 \\ 1 \\ 0 \end{pmatrix}$$

form an orthogonal set since the scalar product of any two of them vanishes. To obtain an orthonormal set, simply multiply each vector by the reciprocal of its length:

$$\frac{1}{\sqrt{6}} \begin{pmatrix} 1 \\ 1 \\ 2 \end{pmatrix}, \quad \frac{1}{\sqrt{3}} \begin{pmatrix} 1 \\ 1 \\ -1 \end{pmatrix}, \quad \frac{1}{\sqrt{2}} \begin{pmatrix} -1 \\ 1 \\ 0 \end{pmatrix}.$$

Example 7.3.2

The standard basis of $\mathbf{R}^n$ consisting of the columns of the identity matrix 1_n is an orthonormal set.

Example 7.3.3

The functions

$$\sqrt{2/\pi} \sin mx, \ m = 1, 2, \ldots$$

form an orthonormal subset of the inner product space $C[0, \pi]$. For we observed in Examples 7.2.4 and 7.2.5 that these vectors are mutually orthogonal and have norm 1.

A basic property of orthogonal subsets is that they are always linearly independent.

Theorem 7.3.1

Let V be a real inner product space; then any orthogonal subset of V consisting of non-zero vectors is linearly independent.

Proof

Suppose that the subset $\{\mathbf{v}_1, \ldots, \mathbf{v}_n\}$ is orthogonal, so that $<\mathbf{v}_i, \mathbf{v}_j> = 0$ if $i \neq j$. Assume that there is a linear relation of the form $c_1\mathbf{v}_1 + \cdots + c_n\mathbf{v}_n = \mathbf{0}$. Then, on taking the inner product of both sides with $\mathbf{v}_j$, we get

$$0 = \sum_{i=1}^{n} <c_i\mathbf{v}_i, \mathbf{v}_j> = \sum_{i=1}^{n} c_i <\mathbf{v}_i, \mathbf{v}_j> = c_j <\mathbf{v}_j, \mathbf{v}_j>$$
$$= c_j\|\mathbf{v}_j\|^2$$

since $< \mathbf{v}_i, \mathbf{v}_j > \, = 0$ if $i \neq j$. Now $\|\mathbf{v}_j\| \neq 0$ since $\mathbf{v}_j$ is not the zero vector; therefore $c_j = 0$ for all j. It follows that the $\mathbf{v}_j$ are linearly independent.

This result raises the possibility of an orthonormal basis, and indeed we have already seen in Example 7.3.2 that the standard basis of $\mathbf{R}^n$ is orthonormal. While at present there are no grounds for believing that such a basis always exists, it is instructive to record at this stage some useful properties of orthonormal bases.

Theorem 7.3.2
Suppose that $\{\mathbf{v}_1, \dots, \mathbf{v}_n\}$ is an orthonormal basis of a real inner product space V. If $\mathbf{v}$ is an arbitrary vector of V, then

$$\mathbf{v} = \sum_{i=1}^{n} < \mathbf{v}, \mathbf{v}_i > \mathbf{v}_i \ \text{ and } \ \|\mathbf{v}\|^2 = \sum_{i=1}^{n} < \mathbf{v}, \mathbf{v}_i >^2 .$$

Proof
Let $\mathbf{v} = \sum_{i=1}^{n} c_i \mathbf{v}_i$ be the expression for $\mathbf{v}$ in terms of the given basis. Forming the inner product of both sides with $\mathbf{v}_j$, we obtain

$$< \mathbf{v}, \mathbf{v}_j > \, = \, < \sum_{i=1}^{n} c_i \mathbf{v}_i, \mathbf{v}_j > \, = \sum_{i=1}^{n} c_i < \mathbf{v}_i, \mathbf{v}_j > \, = c_j$$

since $< \mathbf{v}_i, \mathbf{v}_j > \, = 0$ if $i \neq j$ and $< \mathbf{v}_j, \mathbf{v}_j > \, = 1$. Finally,

$$\|\mathbf{v}\|^2 = \, < \mathbf{v}, \mathbf{v} > \, = \, < \sum_{i=1}^{n} c_i \mathbf{v}_i, \ \sum_{j=1}^{n} c_j \mathbf{v}_j >$$

$$= \sum_{i=1}^{n} \sum_{j=1}^{n} c_i c_j < \mathbf{v}_i, \mathbf{v}_j >,$$

which reduces to $\sum_{j=1}^{n} c_j^2$.

Another useful feature of orthonormal bases is that they greatly simplify the procedure for calculating projections.

Theorem 7.3.3
Let V be an inner product space and let S be a subspace and
v *a vector of V. Assume that $\{s_1, \ldots, s_m\}$ is an orthonormal basis of S. Then the projection of* **v** *on S is*

$$\sum_{i=1}^{m} <\mathbf{v}, \mathbf{s}_i> \mathbf{s}_i.$$

Proof
Put $\mathbf{p} = \sum_{i=1}^{m} <\mathbf{v}, \mathbf{s}_i> \mathbf{s}_i$, a vector which quite clearly belongs to S. Now $<\mathbf{p}, \mathbf{s}_j> = <\mathbf{v}, \mathbf{s}_j>$, so

$$\begin{aligned}
<\mathbf{v} - \mathbf{p}, \mathbf{s}_j> &= <\mathbf{v}, \mathbf{s}_j> - <\mathbf{p}, \mathbf{s}_j> \\
&= <\mathbf{v}, \mathbf{s}_j> - <\mathbf{v}, \mathbf{s}_j> \\
&= 0.
\end{aligned}$$

Hence $\mathbf{v} - \mathbf{p}$ is orthogonal to each basis element of S, which shows that $\mathbf{v} - \mathbf{p}$ belongs to $S^{\perp}$. Since $\mathbf{v} = \mathbf{p} + (\mathbf{v} - \mathbf{p})$, and the expression for **v** as the sum of an element of S and an element of $S^{\perp}$ is unique, it follows that **p** is the projection of **v** on S.

Example 7.3.4
The vectors

$$X_1 = \frac{1}{\sqrt{6}} \begin{pmatrix} 1 \\ -2 \\ 1 \end{pmatrix}, \quad X_2 = \frac{1}{\sqrt{5}} \begin{pmatrix} 2 \\ 1 \\ 0 \end{pmatrix}$$

form an orthonormal basis of a subspace S of $\mathbf{R}^3$; find the projection on S of the column vector X with entries $1, -1, 1$.

Apply 7.3.3 with $s_1 = X_1$ and $s_2 = X_2$; we find that the projection of X on S is

$$P = <X, X_1> X_1 + <X, X_2> X_2$$

$$= \frac{4}{\sqrt{6}} X_1 + \frac{1}{\sqrt{5}} X_2 = \frac{1}{15} \begin{pmatrix} 16 \\ -17 \\ 10 \end{pmatrix}.$$

Having seen that orthonormal bases are potentially useful, let us now address the problem of finding such bases.

Gram-Schmidt orthogonalization

Suppose that V is a finite-dimensional real inner product space with a given basis $\{u_1, \ldots, u_n\}$; we shall describe a method of constructing an orthonormal basis of V which is known as the *Gram-Schmidt process*.

The orthonormal basis of V is constructed one element at a time. The first step is to get a unit vector;

$$v_1 = \frac{1}{\|u_1\|} u_1.$$

Notice that u_1 and v_1 generate the same subspace; let us call it S_1. Then v_1 clearly forms an orthonormal basis of S_1. Next let

$$p_1 = <u_2, v_1> v_1.$$

By 7.3.3 this is the projection of u_2 on S_1. Thus $u_2 - p_1$ belongs to $S_1^{\perp}$ and $u_2 - p_1$ is orthogonal to v_1. Notice that $u_2 - p_1 \neq 0$ since u_1 and u_2 are linearly independent. The second vector in the orthonormal basis is taken to be

$$v_2 = \frac{1}{\|u_2 - p_1\|} (u_2 - p_1).$$

By definition of v_1 and v_2, these vectors generate the same subspace as u_1, u_2, say S_2. Also v_1 and v_2 form an orthonormal basis of S_2.

The next step is to define

$$\mathbf{p}_2 = <\mathbf{u}_3, \mathbf{v}_1> \mathbf{v}_1 + <\mathbf{u}_3, \mathbf{v}_2> \mathbf{v}_2,$$

which by 7.3.3 is the projection of $\mathbf{u}_3$ on S_2. Then $\mathbf{u}_3 - \mathbf{p}_2$ belongs to $S_2^{\perp}$ and so it is orthogonal to $\mathbf{v}_1$ and $\mathbf{v}_2$. Again one must observe that $\mathbf{u}_3 - \mathbf{p}_2 \neq \mathbf{0}$, the reason being that $\mathbf{u}_1$, $\mathbf{u}_2$, $\mathbf{u}_3$ are linearly independent. Now define the third vector of the orthonormal basis to be

$$\mathbf{v}_3 = \frac{1}{\|\mathbf{u}_3 - \mathbf{p}_2\|}(\mathbf{u}_3 - \mathbf{p}_2).$$

Then $\mathbf{v}_1$, $\mathbf{v}_2$, $\mathbf{v}_3$ form an orthonormal basis of the subspace S_3 generated by $\mathbf{u}_1$, $\mathbf{u}_2$, $\mathbf{u}_3$.

The procedure is repeated n times until we have constructed n vectors $\mathbf{v}_1, \ldots, \mathbf{v}_n$; these will form an orthonormal basis of V.

Our conclusions are summarised in the following fundamental theorem.

Theorem 7.3.4 (*The Gram - Schmidt Process*)
Let $\{\mathbf{u}_1, \ldots, \mathbf{u}_n\}$ be a basis of a finite-dimensional real inner product space V. Define recursively vectors $\mathbf{v}_1, \ldots, \mathbf{v}_n$ by the rules

$$\mathbf{v}_1 = \frac{1}{\|\mathbf{u}_1\|}\mathbf{u}_1 \quad and \quad \mathbf{v}_{i+1} = \frac{1}{\|\mathbf{u}_{i+1} - \mathbf{p}_i\|}(\mathbf{u}_{i+1} - \mathbf{p}_i),$$

where

$$\mathbf{p}_i = <\mathbf{u}_{i+1}, \mathbf{v}_1> \mathbf{v}_1 + \cdots + <\mathbf{u}_{i+1}, \mathbf{v}_i> \mathbf{v}_i$$

is the projection of $\mathbf{u}_{i+1}$ on the subspace $S_i = <\mathbf{v}_1, \ldots, \mathbf{v}_i>$. Then $\mathbf{v}_1, \ldots, \mathbf{v}_n$ form an orthonormal basis of V.

The Gram-Schmidt process furnishes a practical method for constructing orthonormal bases, although the calculations can become tedious if done by hand.

Example 7.3.5

Find an orthonormal basis for the column space S of the matrix

$$\begin{pmatrix} 1 & 1 & 2 \\ 1 & 2 & 3 \\ 1 & 2 & 1 \\ 1 & 1 & 6 \end{pmatrix}.$$

In the first place the columns X_1, X_2, X_3 of the matrix are linearly independent and so constitute a basis of S. We shall apply the Gram-Schmidt process to this basis to produce an orthonormal basis $\{Y_1, Y_2, Y_3\}$ of S, following the steps in the procedure.

$$Y_1 = \frac{1}{\|X_1\|} X_1 = \frac{1}{2}\begin{pmatrix} 1 \\ 1 \\ 1 \\ 1 \end{pmatrix}.$$

Now compute the projection of X_2 on $S_1 =< Y_1 >$;

$$P_1 = < X_2, Y_1 > Y_1 = 3Y_1 = \frac{3}{2}\begin{pmatrix} 1 \\ 1 \\ 1 \\ 1 \end{pmatrix}.$$

The next vector in the orthonormal basis is

$$Y_2 = \frac{1}{\|X_2 - P_1\|}(X_2 - P_1) = \frac{1}{2}\begin{pmatrix} -1 \\ 1 \\ 1 \\ -1 \end{pmatrix}.$$

The projection of X_3 on $S_2 =< Y_1, Y_2 >$ is

$$P_2 = < X_3, Y_1 > Y_1 + < X_3, Y_2 > Y_2 = 6Y_1 - 2Y_2 = \begin{pmatrix} 4 \\ 2 \\ 2 \\ 4 \end{pmatrix}.$$

The final vector in the orthonormal basis of S is therefore

$$Y_3 = \frac{1}{\|X_3 - P_2\|}(X_3 - P_2) = \frac{1}{\sqrt{10}}\begin{pmatrix} -2 \\ 1 \\ -1 \\ 2 \end{pmatrix}.$$

Example 7.3.6
Find an orthonormal basis of the inner product space $P_3(\mathbf{R})$ where $< f, g >$ is defined to be $\int_{-1}^{+1} f(x)g(x)dx$.

We begin with the standard basis $\{1, x, x^2\}$ of $P_3(\mathbf{R})$ and then use the Gram-Schmidt process to construct an orthonormal basis $\{f_1, f_2, f_3\}$. Since $\|1\| = \sqrt{(\int_{-1}^{+1} x)} = \sqrt{2}$, the first member of the basis is

$$f_1 = \frac{1}{\|1\|}1 = \frac{1}{\sqrt{2}}.$$

Next $< x, f_1 > = \int_{-1}^{+1}(x/\sqrt{2})dx = 0$, so $p_1 = < x, f_1 > f_1 = 0$. Hence

$$f_2 = \frac{1}{\|x\|}x = \sqrt{\frac{3}{2}}\,x$$

since $\|x\| = \sqrt{(\int_{-1}^{+1} x^2 dx)} = \sqrt{\frac{2}{3}}$.

Continuing the procedure, we find that $< x^2, f_1 > = \sqrt{2}/3$ and $< x^2, f_2 > = 0$. Hence $p_2 = < x^2, f_1 > f_1 + < x^2, f_2 > f_2 = 1/3$, and so the final vector of the orthonormal basis is

$$f_3 = \frac{1}{\|x^2 - \frac{1}{3}\|}(x^2 - \frac{1}{3}) = \frac{3\sqrt{5}}{2\sqrt{2}}(x^2 - \frac{1}{3}).$$

Consequently the polynomials

$$\frac{1}{\sqrt{2}}, \quad \sqrt{\frac{3}{2}}\,x \quad \text{and} \quad \frac{3\sqrt{5}(x^2 - 1/3)}{2\sqrt{2}}$$

form an orthonormal basis of $P_3(\mathbf{R})$.

QR-factorization

In addition to being a practical tool for computing orthonormal bases, the Gram-Schmidt procedure has important theoretical implications. For example, it leads to a valuable way of factorizing an arbitrary real matrix. This is generally referred to as *QR-factorization* from the standard notation for the factors Q and R.

Theorem 7.3.5
Let A be a real $m \times n$ real matrix with rank n. Then A can be written as a product QR where Q is a real $m \times n$ matrix whose columns form an orthonormal set and R is a real $n \times n$ upper triangular matrix with positive entries on its principal diagonal.

Proof
Let V denote the column space of the matrix A. Then V is a subspace of the Euclidean inner product space $\mathbf{R}^m$. Since A has rank n, the n columns $X_1, \ldots, X_n$ of A are linearly independent, and thus form a basis of V. Hence the Gram-Schmidt process can be applied to this basis to produce an orthonormal basis of V, say $Y_1, \ldots, Y_n$.

Now we see from the way that the Y_i in the Gram-Schmidt procedure are defined that these vectors have the form

$$
\begin{cases}
Y_1 = b_{11}X_1 \\
Y_2 = b_{12}X_1 + b_{22}X_2 \\
\quad \cdot \qquad\qquad \cdot \qquad\quad \cdot \qquad\quad \cdot \\
Y_n = b_{1n}X_1 + b_{1n}X_2 + \cdots + b_{nn}X_n
\end{cases}
$$

for certain real numbers b_{ij} with b_{ii} positive. Solving the equations for $X_1, \ldots, X_n$ by back-substitution, we get a linear system of the same general form:

$$
\begin{cases}
X_1 = r_{11}Y_1 \\
X_2 = r_{12}Y_1 + r_{22}Y_2 \\
\quad . \qquad\quad . \qquad\quad . \qquad\quad . \\
X_n = r_{1n}Y_1 + r_{2n}Y_2 + \cdots + r_{nn}Y_n
\end{cases}
$$

for certain real numbers r_{ij}, with r_{ii} positive again. These equations can be written in matrix form

$$
A = [X_1\ X_2\ \ldots\ X_n]
$$

$$
= [Y_1\ Y_2\ \ldots\ Y_n]
\begin{pmatrix}
r_{11} & r_{12} & \cdots & r_{1n} \\
0 & r_{22} & \cdots & r_{2n} \\
. & . & \cdots & . \\
0 & 0 & \cdots & r_{nn}
\end{pmatrix}.
$$

The columns of the $m \times n$ matrix $Q = [Y_1\ Y_2\ \ldots\ Y_n]$ form an orthonormal set since they constitute an orthonormal basis of $\mathbf{R}^m$, while the matrix $R = [r_{ij}]_{n,n}$ is plainly upper triangular.

The most important case of this theorem is when A is a non-singular square matrix. Then the matrix Q is $n \times n$, and its columns form a orthonormal set; equivalently it has the property

$$
Q^T Q = I_n,
$$

which, by 3.3.4, is just to say that $Q^{-1} = Q^T$.

A square matrix A such that

$$
A^T = A^{-1}
$$

is called an *orthogonal matrix*. We shall see in Chapter 9 that orthogonal matrices play an important role in the study of canonical forms of matrices.

It is instructive to determine to investigate the possible forms of an orthogonal 2×2 matrix.

Example 7.3.7

Find all real orthogonal 2×2 matrices.

Suppose that the real matrix

$$A = \begin{pmatrix} a & b \\ c & d \end{pmatrix}$$

is orthogonal; thus $A^T A = I_2$. Equating the entries of the matrix $A^T A$ to those of I_2, we obtain the equations

$$a^2 + c^2 = 1 = b^2 + d^2 \quad \text{and} \quad ab + cd = 0.$$

Now the first equation asserts that the point (a, c) lies on the circle $x^2 + y^2 = 1$. Hence there is an angle θ in the interval $[0, 2\pi]$ such that $a = \cos \theta$ and $c = \sin \theta$. Similarly there is an angle ϕ in this interval such that $b = \cos \phi$ and $d = \sin \phi$.

Now we still have to satisfy the third equation $ab + cd = 0$, which requires that

$$\cos \theta \cos \phi + \sin \theta \sin \phi = 0$$

that is, $\cos(\phi - \theta) = 0$. Hence $\phi - \theta = \pm \pi/2$ or $\pm 3\pi/2$. We need to solve for b and d in each case. If $\phi = \theta + \pi/2$ or $\phi = \theta - 3\pi/2$, we find that $b = -\sin \theta$ and $d = \cos \theta$. If, on the other hand, $\phi = \theta - \pi/2$ or $\phi = \theta + 3\pi/2$, it follows that $b = \sin \theta$ and $d = -\cos \theta$.

We conclude that A has of one of the forms

$$\begin{pmatrix} \cos \theta & -\sin \theta \\ \sin \theta & \cos \theta \end{pmatrix}, \quad \begin{pmatrix} \cos \theta & \sin \theta \\ \sin \theta & -\cos \theta \end{pmatrix}$$

with θ in the interval $[0, 2\pi]$. Conversely, it is easy to verify that such matrices are orthogonal. Thus the real orthogonal 2×2 matrices are exactly the matrices of the above types.

We remark that these matrices have already appeared in other contexts. The first matrix represents an anticlockwise rotation in $\mathbf{R}^2$ through angle θ: see Example 6.2.6. The second matrix corresponds to a reflection in $\mathbf{R}^2$ in the line through the origin making angle $\theta/2$ with the positive x-direction; see Exercises 6.2.3 and 6.2.6. Thus a connection has been established between 2×2 real orthogonal matrices on the one hand, and rotations and reflections in 2-dimensional Euclidean space on the other.

It is worthwhile restating the QR-factorization principle in the important case where the matrix A is invertible.

Theorem 7.3.6
Every invertible real matrix A can be written as a product QR where Q is a real orthogonal matrix and R is a real upper triangular matrix with positive entries on its principal diagonal.

Example 7.3.8
Write the following matrix in the QR-factorized form:

$$A = \begin{pmatrix} 1 & 1 & 2 \\ 1 & 2 & 3 \\ 1 & 1 & 1 \end{pmatrix}.$$

The method is to apply the Gram-Schmidt process to the columns X_1, X_2, X_3 of A, which are linearly independent and so form a basis for the column space of A. This yields an orthonormal basis $\{Y_1, Y_2, Y_3\}$ where

$$Y_1 = \frac{1}{\sqrt{3}} \begin{pmatrix} 1 \\ 1 \\ 1 \end{pmatrix} = \frac{1}{\sqrt{6}} X_1,$$

$$Y_2 = \frac{1}{\sqrt{6}} \begin{pmatrix} -1 \\ 2 \\ -1 \end{pmatrix} = 2\frac{\sqrt{6}}{3} X_1 + \frac{\sqrt{6}}{2} X_2$$

and

$$Y_3 = \frac{1}{\sqrt{2}} \begin{pmatrix} 1 \\ 0 \\ -1 \end{pmatrix} = -3\frac{\sqrt{2}}{2} X_2 + \sqrt{2}X_3.$$

Solving back, we obtain the equations

$$\begin{cases} X_1 = \sqrt{3}Y_1 \\ X_2 = 4\sqrt{3}/3 \ Y_1 + \sqrt{6}/3 \ Y_2 \\ X_3 = 2\sqrt{3}Y_1 + \sqrt{6}/2 \ Y_2 + \sqrt{2}/2 \ Y_3 \end{cases}$$

Therefore $A = QR$ where

$$Q = \begin{pmatrix} 1/\sqrt{3} & -1/\sqrt{6} & 1/\sqrt{2} \\ 1/\sqrt{3} & 2/\sqrt{6} & 0 \\ 1/\sqrt{3} & -1/\sqrt{6} & -1/\sqrt{2} \end{pmatrix}$$

and

$$R = \begin{pmatrix} \sqrt{3} & 4/\sqrt{3} & 2\sqrt{3} \\ 0 & \sqrt{6}/3 & \sqrt{6}/2 \\ 0 & 0 & \sqrt{2}/2 \end{pmatrix}.$$

Unitary matrices

We point out, without going through the details, that there is a version of the Gram-Schmidt procedure applicable to complex inner product spaces. In this the formulas of 7.3.4 are carried over with minor changes, to reflect the properties of complex inner products.

There is also a QR-factorization theorem. In this an important change must be made; the matrix Q which is produced by the Gram-Schmidt process has the property that its columns are orthogonal with respect to the *complex* inner product on $\mathbf{C}^m$. In the case where Q is square this is equivalent to the equation

$$Q^*Q = I_n$$

or

$$Q^{-1} = Q^*.$$

Recall here that $Q^* = (\bar{Q})^T$. A complex matrix Q with the above property is said to be *unitary*. Thus unitary matrices are the complex analogs of real orthogonal matrices. For example, the matrix

$$\begin{pmatrix} \cos\theta & i\sin\theta \\ i\sin\theta & \cos\theta \end{pmatrix}.$$

is unitary for all real values of θ; here of course $i = \sqrt{-1}$.

Exercises 7.3

1. Show that the following vectors constitute an orthogonal basis of $\mathbf{R}^3$:

$$\begin{pmatrix} 1 \\ 0 \\ -1 \end{pmatrix}, \begin{pmatrix} 2 \\ 1 \\ 2 \end{pmatrix}, \begin{pmatrix} 1 \\ -4 \\ 1 \end{pmatrix}.$$

2. Modify the basis in Exercise 1 to obtain an orthonormal basis.

3. Find an orthonormal basis for the column space of the matrix

$$\begin{pmatrix} 0 & 1 & 1 \\ 1 & -2 & 1 \\ 1 & 2 & 0 \end{pmatrix}.$$

4. Find an orthonormal basis for the subspace of $P_3(\mathbf{R})$ generated by the polynomials $1 - 6x$ and $1 - 6x^2$ where $< f, g > = \int_0^1 f(x)g(x)dx$.

5. Find the projection of the vector $\begin{pmatrix} 3 \\ 4 \\ -2 \end{pmatrix}$ on the subspace

of $\mathbf{R}^3$ which has the orthonormal basis consisting of

$$\frac{1}{3}\begin{pmatrix} 2 \\ 1 \\ 2 \end{pmatrix} \text{ and } \sqrt{18}\begin{pmatrix} 1 \\ -4 \\ 1 \end{pmatrix}.$$

6. Express the matrix of Exercise 3 in QR-factorized form.

7. Show that a non-singular complex matrix can be expressed as the product of a unitary matrix and an upper triangular matrix whose diagonal elements are real and positive.

8. Find a factorization of the type described in the previous exercise for the matrix

$$\begin{pmatrix} -i & i \\ 1+i & 2 \end{pmatrix},$$

where $i = \sqrt{-1}$.

9. If A and B are orthogonal matrices, show that A^{-1} and AB are also orthogonal. Deduce that the set of all real orthogonal $n \times n$ matrices is a group with respect to matrix multiplication in the sense of 1.3.

10. If $A = QR = Q'R'$ are two QR-factorizations of the real non-singular square matrix A, what can you say about the relationship between the Q and Q', and R and R'?

11. Let L be a linear operator on the Euclidean inner product space $\mathbf{R}^n$. Call L *orthogonal* if it preserves lengths, that is, if $\|L(X)\| = \|X\|$ for all vectors X in $\mathbf{R}^n$.

(a) Give some natural examples of orthogonal linear operators.

(b) Show that L is orthogonal if and only if it preserves inner products, that is, $< L(X), L(Y) > = < X, Y >$ for all X and Y.

12. Let L be a linear operator on the Euclidean space $\mathbf{R}^n$. Prove that L is orthogonal if and only if $L(X) = AX$ where A is an orthogonal matrix.

13. Deduce from Exercise 12 and Example 7.3.7 that a linear operator on $\mathbf{R}^2$ is orthogonal if and only if it is either a rotation or a reflection.

7.4 The Method of Least Squares

A well known application of linear algebra is a method of fitting a function to experimental data called the Method of Least Squares. In order to illustrate the practical problem involved, let us consider an experiment involving two measurable variables x and y where it is suspected that y is, approximately at least, a linear function of x.

Assume that we have some supporting data in the form of observed values of the variables and x and y, which can be thought of as a set of points in the xy-plane

$$(a_1, b_1), \ldots, (a_m, b_m).$$

This means that when $x = a_i$, it was observed that $y = b_i$. Now if there really were a linear relation, and if the data were free from errors, all of these points would lie on a straight line, whose equation could then be determined, and the linear relation would be known. But in practice it is highly unlikely that this will be the case. What is needed is a way of finding the straight line which "bests fits" the given data. The equation of this best-fitting line will furnish a linear relation which is an approximation to y.

It remains to explain what is meant by the best-fitting straight line. It is here that the "least squares" arise.

Consider the linear relation $y = cx + d$; this is the equation of a straight line in the xy-plane. The conditions for the line to pass through the m data points are

$$\begin{cases} ca_1 + d &= b_1 \\ ca_2 + d &= b_2 \\ \quad . & \quad . \\ ca_m + d &= b_m \end{cases}$$

Now in all probability these equations will be inconsistent. However, we can ask for real numbers c and d which come as close to satisfying the equations of the linear system as possible, in the sense that they minimize the "total error". A good measure of this total error is the expression

$$(ca_1 + d - b_1)^2 + \cdots + (ca_m + d - b_m)^2.$$

This is the sum of the squares of the vertical deviations of the line from the data points in the diagram above. Here the squares are inserted to take care of any negative signs that might appear.

It should be clear the line-fitting problem is just a particular instance of a general problem about inconsistent linear systems. Suppose that we have a linear system of m equations in n unknowns $x_1, \ldots, x_n$

$$AX = B.$$

Since the system may be inconsistent, the problem of interest is to find a vector X which minimizes the length of the vector $AX - B$, or what is equivalent and also a good deal more convenient, its square,

$$E = \|AX - B\|^2.$$

In our original example, where a straight line was to be fitted to the data, the matrix A has two columns $[a_1 a_2 \ldots a_m]$ and $[1\ 1\ \ldots\ 1]$, while $X = \begin{pmatrix} c \\ d \end{pmatrix}$ and B is the column $[b_1 b_2 \ldots b_m]^T$. Then E is the sum of the squares of the quantities $c a_i + d - b_i$.

A vector X which minimizes E is called a *least squares solution* of the linear system $AX = B$. A least squares solution will be an actual solution of the system if and and only if the system is consistent.

The normal system

Once again consider a linear system $AX = B$ and write $E = \|AX - B\|^2$. We will show how to minimize E. Put $A = [a_{ij}]_{m,n}$ and let the entries of X and B be $x_1, \ldots, x_n$ and $b_1, \ldots, b_m$ respectively. The ith entry of $AX - B$ is clearly $(\sum_{j=1}^{n} a_{ij} x_j) - b_i$. Hence

$$E = \|AX - B\|^2 = \sum_{i=1}^{m} \left((\sum_{j=1}^{n} a_{ij} x_j) - b_i \right)^2,$$

which is a quadratic function of $x_1, \ldots, x_n$.

At this juncture it is necessary to recall from calculus the procedure for finding the absolute minima of a function of several variables. First one finds the critical points of the function E, by forming its partial derivatives and setting them equal to zero:

$$\frac{\partial E}{\partial x_k} = \sum_{i=1}^{m} 2 \left((\sum_{j=1}^{n} a_{ij} x_j) - b_i \right) a_{ik} = 0.$$

Hence

$$\sum_{i=1}^{m} \sum_{j=1}^{n} a_{ik} a_{ij} x_j = \sum_{i=1}^{m} a_{ik} b_i,$$

for $k = 1, 2, \ldots, n$. This is a new linear system of equations in $x_1, \ldots, x_n$ whose matrix form is

$$(A^T A)X = A^T B.$$

It is called the *normal system* of the linear system $AX = B$. The solutions of the normal system are the critical points of E.

Now E surely has an absolute minimum – after all it is a continuous function with non-negative values. Since the function E is unbounded when $|x_j|$ is large, its absolute minima must occur at critical points. Therefore we can state:

Theorem 7.4.1
Every least squares solution of the linear system $AX = B$ is a solution of the normal system $(A^T A)X = A^T B$.

At this point potential difficulties appear: what if the normal system is inconsistent? If this were to happen, we would have made no progress whatsoever. And even if the normal system is consistent, need all its solutions be least squares solutions?

To help answer these questions, we establish a simple result about matrices.

Lemma 7.4.2
Let A be a real $m \times n$ matrix. Then $A^T A$ is a symmetric $n \times n$ matrix whose null space equals the null space of A and whose column space equals the column space of A^T.

Proof
In the first place $(A^T A)^T = A^T (A^T)^T = A^T A$, so $A^T A$ is certainly symmetric. Let S be the column space of A. Then by 7.2.6 the null space of A^T equals $S^\perp$.

Let X be any n-column vector. Then X belongs to the null space of $A^T A$ if and only if $A^T (AX) = 0$; this amounts to saying that AX belongs to the null space of A^T or, what

is the same thing, to $S^{\perp}$. But AX also belongs to S; for it is
a linear combination of the columns of A. Now $S \cap S^{\perp}$ is the
zero space by 7.2.3. Hence $AX = 0$ and X belongs to the null
space of A. On the other hand, it is obvious that if X belongs
to the null space of A, then it must belong to the null space
of $A^T A$. Hence the null space of $A^T A$ equals the null space of
A.

Finally, by 7.2.6 and the last paragraph we can assert
that the column space of $A^T A$ equals

$$(\text{null space of } A^T A)^{\perp} = (\text{null space of } A)^{\perp}.$$

This equals the column space of A^T, as claimed.

We come now to the fundamental theorem on the Method
of Least Squares.

Theorem 7.4.3
Let $AX = B$ be a linear system of m equations in n unknowns.

(a) *The normal system $(A^T A)X = A^T B$ is always con-*
sistent and its solutions are exactly the least squares solutions
of the linear system $AX = B$;

(b) *if A has rank n, then $A^T A$ is invertible and there is*
a unique least squares solution of the normal system, namely
$X = (A^T A)^{-1} A^T B.$

Proof
By 7.4.2 the column space of $A^T A$ equals the column space of
A^T. Therefore the column space of the matrix

$$[A^T A \mid A^T B]$$

equals the column space of $A^T A$; for the extra column $A^T B$
is a linear combination of the columns of A^T and thus belongs
to the column space of $A^T A$. It follows that the coefficient
matrix and the augmented matrix of the normal system have

the same rank. By 5.2.5 this is just the condition for the normal system to be consistent.

The next point to establish is that every solution of the normal system is a least squares solution of $AX = B$. Suppose that X_1 and X_2 are two solutions of the normal system. Then $A^T A(X_1 - X_2) = A^T B - A^T B = 0$, so that $Y = X_1 - X_2$ belongs to the null space of $A^T A$. By 7.4.2 the latter equals the null space of A. Thus $AY = 0$. Since $X_1 = Y + X_2$, we have

$$AX_1 - B = A(Y + X_2) - B = AX_2 - B.$$

This means that $E = \|AX - B\|^2$ has the same value for $X = X_1$ and $X = X_2$. Thus all solutions of the normal system give the same value of E. Since by 7.4.1 every least squares solution is a solution of the normal system, it follows that the solutions of the normal system constitute the set of all least squares solutions, as claimed.

Finally, suppose that A has rank n. Then the matrix $A^T A$ also has rank n since by 7.4.2 the column space of $A^T A$ equals the column space of A^T, which has dimension n. Since $A^T A$ is $n \times n$, it is invertible by 5.2.4 and 2.3.5. Hence the equation $A^T AX = A^T B$ leads to the unique solution

$$X = (A^T A)^{-1} A^T B,$$

which completes the proof On the other hand, if the rank of A is less than n, there will be infinitely many least squares solutions. We shall see later how to select one that is in some sense optimal.

Example 7.4.1

Find the least squares solution of the following linear system:

$$\begin{cases} x_1 & + x_2 & + x_3 & = 4 \\ -x_1 & + x_2 & + x_3 & = 0 \\ & - x_2 & + x_3 & = 1 \\ x_1 & & + x_3 & = 2 \end{cases}$$

Here

$$A = \begin{pmatrix} 1 & 1 & 1 \\ -1 & 1 & 1 \\ 0 & -1 & 1 \\ 1 & 0 & 1 \end{pmatrix} \text{ and } B = \begin{pmatrix} 4 \\ 0 \\ 1 \\ 2 \end{pmatrix},$$

so A has has rank 3. Since the augmented matrix has rank 4, the linear system is inconsistent. We know from 7.4.3 that there is a unique least squares solution in this case. To find it, first compute

$$A^T A = \begin{pmatrix} 3 & 0 & 1 \\ 0 & 3 & 1 \\ 1 & 1 & 4 \end{pmatrix} \text{ and } (A^T A)^{-1} = \frac{1}{30} \begin{pmatrix} 11 & 1 & -3 \\ 1 & 11 & -3 \\ -3 & -3 & 9 \end{pmatrix}.$$

Hence the least squares solution is

$$X = (A^T A)^{-1} A^T B = \frac{1}{5} \begin{pmatrix} 8 \\ 3 \\ 6 \end{pmatrix},$$

that is, $x_1 = 8/5$, $x_2 = 3/5$, $x_3 = 6/5$.

Example 7.4.2
A certain experiment yields the following data:

x	-1	0	1	2
y	0	1	3	9

It is suspected that y is a quadratic function of x. Use the Method of Least Squares to find the quadratic function that best fits the data.

Suppose that the function is $y = a + bx + cx^2$. We need to find a least squares solution of the linear system

$$\begin{cases} a & - & b & + & c & = 0 \\ a & & & & & = 1 \\ a & + & b & + & c & = 3 \\ a & & + 2b & + 4c & & = 9 \end{cases}$$

Again the linear system is inconsistent. Here

$$A = \begin{pmatrix} 1 & -1 & 1 \\ 1 & 0 & 0 \\ 1 & 1 & 1 \\ 1 & 2 & 4 \end{pmatrix} \text{ and } B = \begin{pmatrix} 0 \\ 1 \\ 3 \\ 9 \end{pmatrix}$$

and A has rank 3. We find that

$$A^T A = \begin{pmatrix} 4 & 2 & 6 \\ 2 & 6 & 8 \\ 6 & 8 & 18 \end{pmatrix}$$

and

$$(A^T A)^{-1} = \frac{1}{80} \begin{pmatrix} 44 & 12 & -20 \\ 12 & 36 & -20 \\ -20 & -20 & 20 \end{pmatrix}.$$

The unique least squares solution is therefore

$$X = (A^T A)^{-1} A^T B = \frac{1}{20} \begin{pmatrix} 11 \\ 33 \\ 25 \end{pmatrix},$$

that is, $a = 11/20$, $b = 33/20$, $c = 5/4$. Hence the quadratic function that best fits the data is

$$y = \frac{11}{20} + \frac{33}{20}x + \frac{5}{4}x^2.$$

Least squares and QR-factorization

Consider once again the least squares problem for the linear system $AX = B$ where A is $m \times n$ with rank n; we have seen that in this case there is a unique least squares solution $X = (A^T A)^{-1} A^T B$. This expression assumes a simpler form when A is replaced by its QR-factorization. Let this be

$A = QR$, as in 7.3.5. Thus Q is an $m \times n$ matrix with orthonormal columns and R is an $n \times n$ upper triangular matrix with positive diagonal elements. Since the columns of Q form an orthonormal set, $Q^T Q = I_n$. Hence

$$A^T A = R^T Q^T Q R = R^T R.$$

Thus $X = (R^T R)^{-1} R^T Q^T B$, which reduces to

$$X = R^{-1} Q^T B,$$

a considerable simplification of the original formula. However Q and R must already be known before this formula can be used.

Example 7.4.3
Consider the least squares problem

$$\begin{cases} x_1 & + & x_2 & + & 2x_3 & = 1 \\ x_1 & + & 2x_2 & + & 3x_3 & = 2 \\ x_1 & + & 2x_2 & + & x_3 & = 1 \end{cases}$$

Here

$$A = \begin{pmatrix} 1 & 1 & 2 \\ 1 & 2 & 3 \\ 1 & 2 & 1 \end{pmatrix} \quad \text{and} \quad B = \begin{pmatrix} 1 \\ 1 \\ 1 \end{pmatrix}.$$

It was shown in Example 7.3.8 that $A = QR$ where

$$Q = \begin{pmatrix} 1/\sqrt{3} & -1/\sqrt{6} & 1/\sqrt{2} \\ 1/\sqrt{3} & 2/\sqrt{6} & 0 \\ 1/\sqrt{3} & -1/\sqrt{6} & -1/\sqrt{2} \end{pmatrix}$$

and

$$R = \begin{pmatrix} \sqrt{3} & 4/\sqrt{3} & 2\sqrt{3} \\ 0 & \sqrt{6}/3 & \sqrt{6}/2 \\ 0 & 0 & \sqrt{2}/2 \end{pmatrix}.$$

Hence the least squares solution is

$$X = R^{-1}Q^T B = \begin{pmatrix} 0 \\ 1 \\ 0 \end{pmatrix},$$

that is, $x_1 = 1$, $x_2 = 0$, $x_3 = 0$.

Geometry of the least squares process

There is a suggestive geometric interpretation of the least squares process in terms of projections. Consider the least squares problem for the linear system $AX = B$ where A has m rows. Let S denote the column space of the coefficient matrix A. The least squares solutions are the solutions of the normal system $A^T AX = A^T B$, or equivalently

$$A^T(B - AX) = 0.$$

The last equation asserts that $B - AX$ belongs to the null space of A^T, which by 7.2.6 is equal to $S^\perp$. Our condition can therefore be reformulated as follows: X is a least squares solution of $AX = B$ if and only if $B - AX$ belongs to $S^\perp$.

Now $B = AX + (B - AX)$ and AX belongs to S. Recall from 7.2.4 that B is uniquely expressible as the sum of its projections on the subspaces S and $S^\perp$; we conclude that $B - AX$ belongs to $S^\perp$ precisely when AX is the projection of B on S. In short we have a discovered a geometric description of the least squares solutions.

Theorem 7.4.4

Let $AX = B$ be an arbitrary linear system and let S denote the column space of A. Then a column vector X is a least squares solution of the linear system if and only if AX is the projection of B on S.

Notice that the projection AX is uniquely determined by the linear system $AX = B$. However there is a unique least squares solution X if and only if X is uniquely determined by AX, that is, if $AX = A\hat{X}$ implies that $X = \hat{X}$. Hence X is unique if and only if the null space of A is zero, that is, if the rank of A is n. Therefore we can state

Corollary 7.4.5

There is a unique least squares solution of the linear system $AX = B$ if and only if the rank of A equals the number of columns of A.

Optimal least squares solutions

Returning to the general least squares problem for the linear system $AX = B$ with n unknowns, we would like to be able to say something about the least squares solutions in the case where the rank of A is less than n. In this case there will be many least square solutions; what we have in mind is to find a sensible way of picking one of them. Now a natural way to do this would be to select a least squares solution with minimal length. Accordingly we define an *optimal least squares solution* of $AX = B$ to be a least squares solution X whose length $\|X\|$ is as small as possible.

There is a simple method of finding an optimal least squares solution. Let U denote the null space of A; then U equals (column space of $A^T)^\perp$, by 7.2.6. Suppose X is a least squares solution of the system $AX = B$. Now there is a unique expression $X = X_0 + X_1$ where X_0 belongs to U and X_1 belongs to $U^\perp$; this is by 7.2.4. Then $AX = AX_0 + AX_1 = AX_1$; for $AX_0 = 0$ since X_0 belongs to the null space of A. Thus $AX - B = AX_1 - B$, so that X_1 is also a least squares solution of $AX = B$. Now we compute

$$\|X\|^2 = \|X_0 + X_1\|^2 = (X_0 + X_1)^T(X_0 + X_1) = X_0^T X_0 + X_1^T X_1.$$

For $X_0^T X_1 = 0 = X_1^T X_0$ since X_0 and X_1 belong to U and $U^\perp$ respectively. Therefore

$$\|X\|^2 = \|X_0\|^2 + \|X_1\|^2 \geq \|X_1\|^2.$$

Now, if X is an optimal solution, then $\|X\| = \|X_1\|$, so that $\|X_0\| = 0$ and hence $X_0 = 0$. Thus $X = X_1$ belongs to $U^\perp$. It follows that each optimal least squares solution must belong to $U^\perp$, the column space of A^T.

Finally, we show that there is a *unique* least squares solution in $U^\perp$. Suppose that X and $\hat{X}$ are two least squares solutions in $U^\perp$. Then from 7.4.4 we see that AX and $A\hat{X}$ are both equal to the projection of B on the column space of A. Thus $A(X - \hat{X}) = 0$ and $X - \hat{X}$ belongs to U, the null space of A. But X and $\hat{X}$ also belong to $U^\perp$, whence so does $X - \hat{X}$. Since $U \cap U^\perp = 0$, it follows that $X - \hat{X} = 0$ and $X = \hat{X}$. Hence X is the unique optimal least squares solution and it belongs to $U^\perp$. Combining these conclusions with 7.4.4, we obtain:

Theorem 7.4.6

A linear system $AX = B$ has a unique optimal least squares solution, namely the unique vector X in the column space of A^T such that AX is the projection of B on the column space of A^T.

The proof of 7.4.6 has the useful feature that it tells us how to find the optimal least squares solution of a linear system $AX = B$. First find any least squares solution, and then compute its projection on the column space of A^T.

Example 7.4.4

Find the optimal least squares solution of the linear system

$$\begin{cases} x_1 & - x_2 & + x_3 & = 1 \\ x_1 & + x_2 & - 2x_3 & = 2 \\ 2x_1 & & - x_3 & = 4 \end{cases}$$

The first step is to identify the normal system $(A^T A)X = A^T B$;

$$\begin{cases} 6x_1 & & - 3x_3 & = 11 \\ & 2x_2 & - 3x_3 & = 1 \\ -3x_1 & - 3x_2 & + 6x_3 & = -7 \end{cases}$$

Any solution of this will do; for example, we can take the solution vector

$$X = \begin{pmatrix} 7/3 \\ 2 \\ 1 \end{pmatrix}.$$

To obtain an optimal least squares solution, find the projection of X on the column space of A^T; the first two columns of A^T form a basis of this space. Proceeding as in Example 7.2.12, we find the optimal solution to be

$$\frac{1}{42} \begin{pmatrix} 67 \\ -9 \\ -20 \end{pmatrix},$$

so that $x_1 = 67/42$, $x_2 = -3/14$, $x_3 = -10/21$ is the optimal least squares solution of the linear system.

Least squares in inner product spaces

In 7.4.4 we obtained a geometrical interpretation of the least squares process in $\mathbf{R}^n$ in terms of projections on subspaces. This raises the question of least squares processes in an arbitrary finite-dimensional real inner product space V.

First we must formulate the least squares problem in V. This consists in approximating a vector $\mathbf{v}$ in V by a vector in a subspace S of V. A natural way to do this is to choose $\mathbf{x}$ in S so that $\|\mathbf{x} - \mathbf{v}\|^2$ is as small as possible. This is a direct generalization of the least squares problem in $\mathbf{R}^n$. For, if we are given the linear system $AX = B$ and we take S to be the column space of A, $\mathbf{v}$ to be B and $\mathbf{x}$ to be the vector AX of S, then the least squares problem is to minimize $\|AX - B\|^2$.

It turns out that the solution of this general least squares problem is the projection of $\mathbf{v}$ on S, just as in the special case of $\mathbf{R}^n$.

Theorem 7.4.7

Let V be a finite-dimensional, real inner product space, and let $\mathbf{v}$ be an element and S a subspace of V. Denote the projection of $\mathbf{v}$ on S by $\mathbf{p}$. Then, if $\mathbf{x}$ is any vector in S other than $\mathbf{p}$, the inequality $\|\mathbf{x} - \mathbf{v}\| > \|\mathbf{p} - \mathbf{v}\|$ holds.

Thus $\mathbf{p}$ is the vector in S which most closely approximates $\mathbf{v}$ in the sense that it makes $\|\mathbf{p} - \mathbf{v}\|$ as small as possible.

Proof

Since $\mathbf{x}$ and $\mathbf{p}$ both belong to S, so does $\mathbf{x} - \mathbf{p}$. Also $\mathbf{p} - \mathbf{v}$ belongs to $S^\perp$ since $\mathbf{p}$ is the projection of $\mathbf{v}$ on S. Hence $< \mathbf{p} - \mathbf{v}, \mathbf{x} - \mathbf{p} > = 0$. It follows that

$$
\begin{aligned}
\|\mathbf{x} - \mathbf{v}\|^2 &= \; < (\mathbf{x} - \mathbf{p}) + (\mathbf{p} - \mathbf{v}), (\mathbf{x} - \mathbf{p}) + (\mathbf{p} - \mathbf{v}) > \\
&= \; < \mathbf{x} - \mathbf{p}, \mathbf{x} - \mathbf{p} > + < \mathbf{p} - \mathbf{v}, \mathbf{p} - \mathbf{v} > \\
&= \|\mathbf{x} - \mathbf{p}\|^2 + \|\mathbf{p} - \mathbf{v}\|^2 > \|\mathbf{p} - \mathbf{v}\|^2
\end{aligned}
$$

since $\mathbf{x} - \mathbf{p} \neq \mathbf{0}$. Hence $\|\mathbf{x} - \mathbf{v}\| > \|\mathbf{p} - \mathbf{v}\|$.

In applying 7.4.7 it is advantageous to have at hand an orthonormal basis $\{\mathbf{v}_1, \ldots, \mathbf{v}_m\}$ of S. For the task of computing $\mathbf{p}$, the projection of $\mathbf{v}$ on S, is then much easier since the formula of 7.3.3 is available:

$$
\mathbf{p} = \sum_{i=1}^{m} < \mathbf{v}, \mathbf{s}_i > \mathbf{s}_i \, .
$$

Example 7.4.5

Use least squares to find a quadratic polynomial that approximates the function e^x in the interval $[-1, 1]$.

Here it is assumed that we are working in the inner product space $C[-1, 1]$ where $< f, g > = \int_{-1}^{1} f(x)g(x)dx$. Let S denote the subspace consisting of all quadratic polynomials in x. An orthonormal basis for S was found in Example 7.3.6:

$$f_1 = \frac{1}{\sqrt{2}}, \quad f_2 = \sqrt{\frac{3}{2}}x, \quad f_3 = \frac{3\sqrt{5}}{2\sqrt{2}}(x^2 - \frac{1}{3}).$$

By 7.4.7 the least squares approximation to e^x in S is simply the projection of e^x on S; this is given by the formula

$$p = < e^x, f_1 > f_1 + < e^x, f_2 > f_2 + < e^x, f_3 > f_3.$$

Evaluating the integrals by integration by parts, we obtain

$$< e^x, f_1 > = \frac{(e - e^{-1})}{\sqrt{2}}, \quad < e^x, f_2 > = \sqrt{6}e^{-1}$$

and

$$< e^x, f_3 > = \sqrt{\frac{5}{2}}(e - 7e^{-1}).$$

The desired approximation to e^x is therefore

$$p = \frac{1}{2}(e - e^{-1}) + 3e^{-1}x + \frac{15}{4}(e - 7e^{-1})(x^2 - \frac{1}{3}).$$

Alternatively one can calculate the projection by using the standard basis $1, x, x^2$.

Exercises 7.4

1. Find least squares solutions of the following linear systems:

$$\text{(a)} \begin{cases} x_1 & + x_2 & & = 0 \\ & x_2 & + x_3 & = 0 \\ x_1 & - x_2 & - x_3 & = 3 \\ x_1 & & + x_3 & = 0 \end{cases}$$

$$(b) \begin{cases} x_1 & + x_2 & - 2x_3 & = 3 \\ 2x_1 & - x_2 & + 3x_3 & = 4 \\ x_1 & & + x_3 & = 1 \\ x_1 & + x_2 & + x_3 & = 1 \end{cases}$$

2. The following data were collected for the mean annual temperature t and rainfall r in a certain region; use the Method of Least Squares to find a linear approximation for r in terms of t (a calculator is necessary):

t	24	27	22	24
r	47	30	35	38

3. In a tropical rain forest the following data was collected for the numbers x and y (per square kilometer) of a prey species and a predator species over a number of years. Use least squares to find a quadratic function of x that approximates y (a calculator is necessary):

x	2	3	4	5
y	1	2	2	1

4. Find the optimal least squares solution of the linear system

$$\begin{cases} x_1 & & + 2x_3 & = 1 \\ & x_2 & + 3x_3 & = 0 \\ -x_1 & + x_2 & + x_3 & = 0 \\ & - x_2 & - 3x_3 & = 1 \end{cases}$$

5. Find a least squares approximation to the function e^{-x} by a linear function in the interval $[1, 2]$. [Use the inner product $< f, g > = \int_1^2 f(x)g(x)dx$].

6. Find a least squares approximation for the function $\sin x$ as a quadratic function of x in the interval $[0, \pi]$. [Here the inner product $< f, g > = \int_0^\pi f(x)g(x)dx$ is to be used].

Chapter Eight

EIGENVECTORS AND EIGENVALUES

An eigenvector of an $n \times n$ matrix A is a non-zero n-column vector X such that $AX = cX$ for some scalar c, which is called an eigenvalue of A. Thus the effect of left multiplication of an eigenvector by A is merely to multiply it by a scalar, and when $n \leq 3$, a parallel vector is obtained. Similarly, if T is a linear operator on a vector space V, an eigenvector of T is a non-zero vector $\mathbf{v}$ of V such that $T(\mathbf{v}) = c\mathbf{v}$ for some scalar c called an *eigenvalue*. For example, if T is a rotation in $\mathbf{R}^3$, the eigenvectors of T are the non-zero vectors parallel to the axis of rotation and the eigenvalues are all equal to 1.

A large amount of information about a matrix or linear operator is carried by its eigenvectors and eigenvalues. In addition, the theory of eigenvectors and eigenvalues has important applications to systems of linear recurrence relations, Markov processes and systems of linear differential equations. We shall describe the basic theory in the first section and then we give applications in the following two sections of the chapter.

8.1 Basic Theory of Eigenvectors and Eigenvalues

We begin with the fundamental definition. Let A be an $n \times n$ matrix over a field of scalars F. An *eigenvector* of A is a non-zero n-column vector X over F such that

$$AX = cX$$

for some scalar c in F; the scalar c is then referred to as the *eigenvalue* of A associated with the eigenvector X.

In order to clarify the definition and illustrate the technique for finding eigenvectors and eigenvalues, an example will be worked out in detail.

Example 8.1.1
Consider the real 2×2 matrix

$$A = \begin{pmatrix} 2 & -1 \\ 2 & 4 \end{pmatrix}.$$

The condition for the vector

$$X = \begin{pmatrix} x_1 \\ x_2 \end{pmatrix}$$

to be an eigenvector of A is that $AX = cX$ for some scalar c. This is equivalent to $(A - cI_2)X = 0$, which simply asserts that X is a solution of the linear system

$$\begin{pmatrix} 2 - c & -1 \\ 2 & 4 - c \end{pmatrix} \begin{pmatrix} x_1 \\ x_2 \end{pmatrix} = \begin{pmatrix} 0 \\ 0 \end{pmatrix}.$$

Now by 3.3.2 this linear system will have a non-trivial solution x_1, x_2 if and only if the determinant of the coefficient matrix vanishes,

$$\begin{vmatrix} 2 - c & -1 \\ 2 & 4 - c \end{vmatrix} = 0,$$

that is, $c^2 - 6c + 10 = 0$. The roots of this quadratic equation are $c_1 = 3 + \sqrt{-1}$ and $c_2 = 3 - \sqrt{-1}$, so these are the eigenvalues of A.

The eigenvectors for each eigenvalue are found by solving the linear systems $(A - c_1 I_2)X = 0$ and $(A - c_2 I_2)X = 0$. For example, in the case of c_1 we have to solve

$$\begin{cases} (-1 - \sqrt{-1})x_1 - & x_2 = 0 \\ 2x_1 + (1 - \sqrt{-1})x_2 = 0 \end{cases}$$

The general solution of this system is $x_1 = \frac{d}{2}(-1 + \sqrt{-1})$ and $x_2 = d$, where d is an arbitrary scalar. Thus the eigenvectors of A associated with the eigenvalue c_1 are the non-zero vectors of the form

$$d \begin{pmatrix} (-1 + \sqrt{-1})/2 \\ 1 \end{pmatrix}.$$

Notice that these, together with the zero vector, form a 1-dimensional subspace of $\mathbf{C}^2$. In a similar manner the eigenvectors for the eigenvalue $3 - \sqrt{-1}$ are found to be the vectors of the form

$$d \begin{pmatrix} -(1 + \sqrt{-1})/2 \\ 1 \end{pmatrix}$$

where $d \neq 0$. Again these form with the zero vector a subspace of $\mathbf{C}^2$.

It should be clear to the reader that the method used in this example is in fact a general procedure for finding eigenvectors and eigenvalues. This will now be described in detail.

The characteristic equation of a matrix

Let A be an $n \times n$ matrix over a field of scalars F, and let X be a non-zero n-column vector over F. The condition for X to be an eigenvector of A is $AX = cX$, or

$$(A - cI_n)X = 0,$$

where c is the corresponding eigenvalue. Hence the eigenvectors associated with c, together with the zero vector, form the null space of the matrix $A - cI_n$. This subspace is often referred to as the *eigenspace* of the eigenvalue c.

Now $(A - cI_n)X = 0$ is a linear system of n equations in n unknowns. By 3.3.2 the condition for there to be a non-trivial solution of the system is that the coefficient matrix have zero determinant,

$$\det(A - cI_n) = 0.$$

Conversely, if the scalar c satisfies this equation, there will be a non-zero solution of the system and c will be an eigenvalue. These considerations already make it clear that the determinant

$$\det(A - xI_n) = \begin{vmatrix} a_{11} - x & a_{12} & \cdots & a_{1n} \\ a_{21} & a_{22} - x & \cdots & a_{2n} \\ \cdot & \cdot & \cdots & \cdot \\ a_{n1} & a_{n2} & \cdots & a_{nn} - x \end{vmatrix}$$

must play an important role. This is a polynomial of degree n in x which is called the *characteristic polynomial* of A. The equation obtained by setting the characteristic polynomial equal to zero is the *characteristic equation*. Thus the eigenvalues of A are the roots of the characteristic equation (or characteristic polynomial) which lie in the field F.

At this point it is necessary to point out that A may well have no eigenvalues in F. For example, the characteristic polynomial of the real matrix

$$\begin{pmatrix} 0 & 1 \\ -1 & 0 \end{pmatrix}$$

is $x^2 + 1$, which has no real roots, so the matrix has no eigenvalues in **R**.

However, if A is a complex $n \times n$ matrix, its characteristic equation will have n complex roots, some of which may be equal. The reason for this is a well-known result known as *The Fundamental Theorem of Algebra*; it asserts that every polynomial f of positive degree n with complex coefficients can be expressed as a product of n linear factors; thus the equation $f(x) = 0$ has exactly n roots in **C**. Because of this we can be sure that complex matrices always have all their eigenvalues and eigenvectors in **C**. It is this case that principally concerns us here.

Let us sum up our conclusions about the eigenvalues of complex matrices so far.

Theorem 8.1.1

Let A be an $n \times n$ complex matrix.

(i) *The eigenvalues of A are precisely the n roots of the characteristic polynomial* $\det(A - xI_n)$;

(ii) *the eigenvectors of A associated with an eigenvalue c are the non-zero vectors in the null space of the matrix $A - cI_n$.*

Thus in Example 8.1.1 the characteristic polynomial of the matrix is

$$\begin{vmatrix} 2 - x & -1 \\ 2 & 4 - x \end{vmatrix} = x^2 - 6x + 10.$$

The eigenvalues are the roots of the characteristic equation $x^2 - 6x + 10 = 0$, that is, $c_1 = 3 + \sqrt{-1}$ and $c_2 = 3 - \sqrt{-1}$; the eigenspaces of c_1 and c_2 are generated by the vectors

$$\begin{pmatrix} (-1 + \sqrt{-1})/2 \\ 1 \end{pmatrix} \text{ and } \begin{pmatrix} -(1 + \sqrt{-1})/2 \\ 1 \end{pmatrix}$$

respectively.

Example 8.1.2

Find the eigenvalues of the upper triangular matrix

$$\begin{pmatrix} a_{11} - x & a_{12} & a_{13} & \cdots & a_{1n} \\ 0 & a_{22} - x & a_{23} & \cdots & a_{2n} \\ \cdot & \cdot & \cdot & \cdots & \cdot \\ 0 & 0 & 0 & \cdots & a_{nn} - x \end{pmatrix}.$$

The characteristic polynomial of this matrix is

$$\begin{vmatrix} a_{11} - x & a_{12} & a_{13} & \cdots & a_{1n} \\ 0 & a_{22} - x & a_{23} & \cdots & a_{2n} \\ \cdot & \cdot & \cdot & \cdots & \cdot \\ 0 & 0 & 0 & \cdots & a_{nn} - x \end{vmatrix},$$

which, by 3.1.5, equals $(a_{11} - x)(a_{22} - x) \ldots (a_{nn} - x)$. The eigenvalues of the matrix are therefore just the diagonal entries $a_{11}, a_{22}, \ldots, a_{nn}$.

Example 8.1.3

Consider the 3×3 matrix

$$A = \begin{pmatrix} 2 & -1 & -1 \\ -1 & 2 & -1 \\ -1 & -1 & 0 \end{pmatrix}.$$

The characteristic polynomial of this matrix is

$$\begin{vmatrix} 2-x & -1 & -1 \\ -1 & 2-x & -1 \\ -1 & -1 & -x \end{vmatrix} = -x^3 + 4x^2 - x - 6.$$

Fortunately one can guess a root of this cubic polynomial, namely $x = -1$. Dividing the polynomial by $x + 1$ using long division, we obtain the quotient $-x^2 + 5x - 6 = -(x-2)(x-3)$. Hence the characteristic polynomial can be factorized completely as $-(x+1)(x-2)(x-3)$, and the eigenvalues of A are $-1, 2$ and 3.

To find the corresponding eigenvectors, we have to solve the three linear systems $(A + I_3)X = 0$, $(A - 2I_3)X = 0$ and $(A - 3I_3)X = 0$. On solving these, we find that the respective eigenvectors are the non-zero scalar multiples of the vectors

$$\begin{pmatrix} 1 \\ 1 \\ 2 \end{pmatrix}, \quad \begin{pmatrix} 1 \\ 1 \\ -1 \end{pmatrix}, \quad \begin{pmatrix} 1 \\ -1 \\ 0 \end{pmatrix}.$$

The eigenspaces are generated by these three vectors and so each has dimension 1.

Properties of the characteristic polynomial

Now let us see what can be said in general about the characteristic polynomial of an $n \times n$ matrix A. Let $p(x)$ denote this polynomial; thus

$$p(x) = \begin{vmatrix} a_{11} - x & a_{12} & \cdots & a_{1n} \\ a_{21} & a_{22} - x & \cdots & a_{2n} \\ \cdot & \cdot & \cdots & \cdot \\ a_{n1} & a_{n2} & \cdots & a_{nn} - x \end{vmatrix}.$$

At this point we need to recall the definition of a determinant as an alternating sum of terms, each term being a product of entries, one from each row and column. The term of $p(x)$ with highest degree in x arises from the product

$$(a_{11} - x) \cdots (a_{nn} - x)$$

and is clearly $(-x)^n$. The terms of degree $n - 1$ are also easy to locate since they arise from the same product. Thus the coefficient of x^{n-1} is

$$(-1)^{n-1}(a_{11} + \cdots + a_{nn})$$

and the sum of the diagonal entries of A is seen to have significance; it is given a special name, the *trace* of A,

$$\text{tr}(A) = a_{11} + a_{22} + \cdots + a_{nn}.$$

The term in $p(x)$ of degree $n - 1$ is therefore $\text{tr}(A)(-x)^{n-1}$.

The constant term in $p(x)$ may be found by simply putting $x = 0$ in $p(x) = \det(A - xI_n)$, thereby leaving $\det(A)$. Our knowledge of $p(x)$ so far is summarized in the formula

$$p(x) = (-x)^n + \text{tr}(A)(-x)^{n-1} + \cdots + \det(A).$$

The other coefficients in the characteristic polynomial are not so easy to describe, but they are in fact expressible as subdeterminants of $\det(A)$. For example, take the case of x^{n-2}. Now terms in x^{n-2} arise in two ways: from the product $(a_{11} - x) \cdots (a_{nn} - x)$ or from products like

$$-a_{12}a_{21}(a_{33} - x) \cdots (a_{nn} - x).$$

So a typical contribution to the coefficient of x^{n-2} is

$$(-1)^{n-2}(a_{11}a_{22} - a_{12}a_{21}) = (-1)^{n-2} \begin{vmatrix} a_{11} & a_{12} \\ a_{21} & a_{22} \end{vmatrix}.$$

From this it is clear that the term of degree $n - 2$ in $p(x)$ is just $(-x)^{n-2}$ times the sum of all the 2×2 determinants of the form

$$\begin{vmatrix} a_{ii} & a_{ij} \\ a_{ji} & a_{jj} \end{vmatrix}$$

where $i < j$.

In general one can prove by similar considerations that the following is true.

Theorem 8.1.2

The characteristic polynomial of the $n \times n$ matrix A equals

$$\sum_{i=0}^{n} d_i(-x)^{n-i}$$

where d_i is the sum of all the $i \times i$ subdeterminants of $\det(A)$ whose principal diagonals are part of the principal diagonal of A.

Now assume that the matrix A has complex entries. Let $c_1, c_2, \ldots, c_n$ be the eigenvalues of A. These are the n roots of the characteristic polynomial $p(x)$. Therefore, allowing for the fact that the term of $p(x)$ with highest degree has coefficient $(-1)^n$, one has

$$p(x) = (c_1 - x)(c_2 - x) \cdots (c_n - x).$$

The constant term in this product is evidently just $c_1 c_2 \ldots c_n$, while the term in x^{n-1} has coefficient $(-1)^{n-1}(c_1 + \cdots + c_n)$. On the other hand, we previously found these to be $\det(A)$ and $(-1)^{n-1}\text{tr}(A)$ respectively. Thus we arrive at two important relations between the eigenvalues and the entries of A.

Corollary 8.1.3
If A is any complex square matrix, the product of the eigenvalues equals the determinant of A and the sum of the eigenvalues equals the trace of A

Recall from Chapter Six that matrices A and B are said to be similar if there is an invertible matrix S such that $B = SAS^{-1}$. The next result indicates that similar matrices have much in common, and really deserve their name.

Theorem 8.1.4
Similar matrices have the same characteristic polynomial and hence they have the same eigenvalues, trace and determinant.

Proof
The characteristic polynomial of $B = SAS^{-1}$ is

$$\begin{aligned}
\det(SAS^{-1} - xI) &= \det(S(A - xI)S^{-1}) \\
&= \det(S)\det(A - xI)\det(S)^{-1} \\
&= \det(A - xI).
\end{aligned}$$

Here we have used two fundamental properties of determinants established in Chapter Three, namely 3.3.3 and 3.3.5. The statements about trace and determinant now follow from 8.1.3.

On the other hand, one cannot expect similar matrices to have the same eigenvectors. Indeed the condition for X to be an eigenvector of SAS^{-1} with eigenvalue c is $(SAS^{-1})X = cX$, which is equivalent to $A(S^{-1}X) = c(S^{-1}X)$. Thus X is an eigenvector of SAS^{-1} if and only if $S^{-1}X$ is an eigenvector of A.

Eigenvectors and eigenvalues of linear transformations

Because of the close relationship between square matrices and linear operators on finite-dimensional vector spaces observed in Chapter Six, it is not surprising that one can also define eigenvectors and eigenvalues for a linear operator.

Let $T : V \rightarrow V$ be a linear operator on a vector space V over a field of scalars F. An *eigenvector* of T is a non-zero vector $\mathbf{v}$ of V such that $T(\mathbf{v}) = c\mathbf{v}$ for some scalar c in F: here c is the *eigenvalue* of T associated with the eigenvector $\mathbf{v}$.

Suppose now that V is a finite-dimensional vector space over F with dimension n. Choose an ordered basis for V, say $\mathcal{B}$. Then with respect to this ordered basis T is represented by an $n \times n$ matrix over F, say A; this means that

$$[T(\mathbf{v})]_{\mathcal{B}} = A[\mathbf{v}]_{\mathcal{B}}.$$

Here $[\mathbf{u}]_{\mathcal{B}}$ is the coordinate column vector of a vector $\mathbf{u}$ in V with respect to basis $\mathcal{B}$. The condition $T(\mathbf{v}) = c\mathbf{v}$ for $\mathbf{v}$ to be an eigenvector of T with associated eigenvalue c, becomes $A[\mathbf{v}]_{\mathcal{B}} = c[\mathbf{v}]_{\mathcal{B}}$, which is just the condition for $[\mathbf{v}]_{\mathcal{B}}$ to be an eigenvector of the representing matrix A; also the eigenvalues of T and A are the same.

If the ordered basis of V is changed, the effect is to replace A by a similar matrix. Of course any such matrix will have the same eigenvalues as T; thus we have another proof of the fact that similar matrices have the same eigenvalues.

These observations permit us to carry over to linear operators concepts such as characteristic polynomial and trace, which were introduced for matrices.

Example 8.1.4

Consider the linear transformation $T : D_\infty[a, b] \to D_\infty[a, b]$ where $T(f) = f'$, the derivative of the function f. The condition for f to be an eigenvector of T is $f' = cf$ for some constant c. The general solution of this simple differential equation is $f = de^{cx}$ where d is a constant. Thus the eigenvalues of T are all real numbers c, while the eigenvectors are the exponential functions de^{cx} with $d \neq 0$.

Diagonalizable matrices

We wish now to consider the question: when is a square matrix similar to a diagonal matrix? In the first place, why is this an interesting question? The essential reason is that diagonal matrices behave so much more simply than arbitrary matrices. For example, when a diagonal matrix is raised to the nth power, the effect is merely to raise each element on the diagonal to the nth power, whereas there is no simple expression for the nth power of an arbitrary matrix. Suppose that we want to compute A^n where A is similar to a diagonal matrix D, with say $A = SDS^{-1}$. It is easily seen that $A^n = SD^nS^{-1}$. Thus it is possible to calculate A^n quite simply if we have explicit knowledge of S and D. It will emerge in 8.2 and 8.3 that this provides the basis for effective methods of solving systems of linear recurrences and linear differential equations.

Now for the important definition. Let A be a square matrix over a field F. Then A is said to be *diagonalizable* over F if it is similar to a diagonal matrix D over F, that is, there is an invertible matrix S over F such that $A = SDS^{-1}$ or equivalently, $D = S^{-1}AS$. One also says that S *diagonalizes* A. A diagonalizable matrix need not be diagonal: the reader should give an example to demonstrate this. It is an important observation that if A is diagonalizable and its eigenvalues are $c_1, \ldots, c_n$, then A must be similar to the diagonal matrix with $c_1, \ldots, c_n$ on the principal diagonal. This is because similar matrices have the same eigenvalues and the eigenvalues of a diagonal matrix are just the entries on the principal diagonal – see Example 8.1.2.

What we are aiming for is a criterion which will tell us exactly which matrices are diagonalizable. A key step in the search for this criterion comes next.

Theorem 8.1.5
Let A be an $n \times n$ matrix over a field F and let $c_1, \ldots, c_r$ be distinct eigenvalues of A with associated eigenvectors $X_1, \ldots, X_r$. Then $\{X_1, \ldots, X_r\}$ is a linearly independent subset of F^n.

Proof
Assume the theorem is false; then there is a positive integer i such that $\{X_1, \ldots, X_i\}$ is linearly independent, but the addition of the next vector X_{i+1} produces a linearly *dependent* set $\{X_1, \ldots, X_{i+1}\}$. So there are scalars $d_1, \ldots, d_{i+1}$, not all of them zero, such that

$$d_1 X_1 + \cdots + d_{i+1} X_{i+1} = 0.$$

Premultiply both sides of this equation by A and use the equations $AX_j = c_j X_j$ to get

$$c_1 d_1 X_1 + \cdots + c_{i+1} d_{i+1} X_{i+1} = 0.$$

On subtracting c_{i+1} times the first equation from the second, we arrive at the relation

$$(c_1 - c_{i+1}) d_1 X_1 + \cdots + (c_i - c_{i+1}) d_i X_i = 0.$$

Since $X_1, \ldots, X_i$ are linearly independent, all the coefficients $(c_j - c_{i+1})d_j$ must vanish. But $c_1, \ldots, c_{i+1}$ are all different, so we can conclude that $d_j = 0$ for $j = 1, \ldots, i$; hence $d_{i+1}X_{i+1} = 0$ and so $d_{i+1} = 0$, in contradiction to the original assumption. Therefore the statement of the theorem must be correct.

The criterion for diagonalizability can now be established.

Theorem 8.1.6
Let A be an $n \times n$ matrix over a field F. Then A is diagonalizable if and only if A has n linearly independent eigenvectors in F^n.

Proof
First of all suppose that A has n linearly independent eigenvectors in F^n, say $X_1, \ldots, X_n$, and that the associated eigenvalues are $c_1, \ldots, c_n$. Define S to be the $n \times n$ matrix whose columns are the eigenvectors; thus

$$S = (X_1 \ldots X_n).$$

The first thing to notice is that S is invertible; for by 8.1.5 its columns are linearly independent. Forming the product of A and S in partitioned form, we find that

$$AS = (AX_1 \ldots AX_n) = (c_1 X_1 \quad \cdots \quad c_n X_n),$$

which equals

$$(X_1 \ \ldots \ X_n) \begin{pmatrix} c_1 & 0 & 0 & \cdots & 0 \\ 0 & c_2 & 0 & \cdots & 0 \\ . & . & . & \cdots & . \\ 0 & 0 & . & \cdots & c_n \end{pmatrix} = SD,$$

where D is the diagonal matrix with entries $c_1, \ldots, c_n$. Therefore $S^{-1}AS = D$ and A is diagonalizable.

Conversely, assume that A is diagonalizable and that $S^{-1}AS = D$ is a diagonal matrix with entries $c_1, \ldots, c_n$. Then $AS = SD$. This implies that if X_i is the ith column of S, then AX_i equals the ith column of SD, which is c_iX_i. Hence $X_1, \ldots, X_n$ are eigenvectors of A associated with eigenvalues $c_1, \ldots, c_n$. Since $X_1, \ldots, X_n$ are columns of the invertible matrix S, they must be linearly independent. Consequently A has n linearly independent eigenvectors.

Corollary 8.1.7

An $n \times n$ complex matrix which has n distinct eigenvalues is diagonalizable.

This follows at once from 8.1.5 and 8.1.6. On the other hand, it is easy to think of matrices which are not diagonalizable: for example, there is the matrix

$$A = \begin{pmatrix} 1 & 1 \\ 0 & 1 \end{pmatrix}.$$

Indeed if A were diagonalizable, it would be similar to the identity matrix I_2 since both its eigenvalues equal 1, and $S^{-1}AS = I_2$ for some S; but the last equation implies that $A = SI_2S^{-1} = I_2$, which is not true.

An interesting feature of the proof of 8.1.6 is that it provides us with a method of finding a matrix S which diagonalizes A. One has simply to find a set of linearly independent eigenvectors of A; if there are enough of them, they can be taken to form the columns of the matrix S.

Example 8.1.5

Find a matrix which diagonalizes $A = \begin{pmatrix} 2 & -1 \\ 2 & 4 \end{pmatrix}$.

In Example 8.1.1 we found the eigenvalues of A to be $3 + \sqrt{-1}$ and $3 - \sqrt{-1}$; hence A is diagonalizable by 8.1.7. We also found eigenvectors for A; these form a matrix

$$S = \begin{pmatrix} (-1 + \sqrt{-1})/2 & -(1 + \sqrt{-1})/2 \\ 1 & 1 \end{pmatrix}.$$

Then by the preceding theory we may be sure that

$$S^{-1}AS = \begin{pmatrix} 3 + \sqrt{-1} & 0 \\ 0 & 3 - \sqrt{-1} \end{pmatrix}.$$

Triangularizable matrices

It has been seen that not every complex square matrix is diagonalizable. Compensating for this failure is the fact such a matrix is always similar to an upper triangular matrix; this is a result with many applications.

Let A be a square matrix over a field F. Then A is said to be *triangularizable over F* if there is an invertible matrix S over F such that $S^{-1}AS = T$ is upper triangular. It will also be convenient to say that S *triangularizes A*. Note that the diagonal entries of the triangular matrix T will necessarily be the eigenvalues of A. This is because of Example 8.1.2 and the fact that similar matrices have the same eigenvalues. Thus a necessary condition for A to be triangularizable is that it have n eigenvalues in the field F. When $F = \mathbf{C}$, this condition is always satisfied, and this is the case in which we are interested.

Theorem 8.1.8
Every complex square matrix is triangularizable.

Proof
Let A denote an $n \times n$ complex matrix. We show by induction on n that A is triangularizable. Of course, if $n = 1$, then A is already upper triangular: let $n > 1$. We shall use induction on n and assume that the result is true for square matrices with $n - 1$ rows.

We know that A has at least one eigenvalue c in $\mathbf{C}$, with associated eigenvector X say. Since $X \neq 0$, it is possible to adjoin vectors to X to produce a basis of $\mathbf{C}^n$, say $X = X_1, X_2, \ldots, X_n$; here we have used 5.1.4. Next, recall that left multiplication of the vectors of $\mathbf{C}^n$ by A gives rise to linear operator T on $\mathbf{C}^n$. With respect to the basis $\{X_1, \ldots, X_n\}$, the linear operator T will be represented by a matrix with the special form

$$B_1 = \begin{pmatrix} c & A_2 \\ 0 & A_1 \end{pmatrix}$$

where A_1 and A_2 are certain complex matrices, A_1 having $n - 1$ rows and columns. The reason for the special form is that $T(X_1) = AX_1 = cX_1$ since X_1 is an eigenvalue of A. Notice that the matrices A and B_1 are similar since they represent the same linear operator T; suppose that in fact $B_1 = S_1^{-1}AS_1$ where S_1 is an invertible $n \times n$ matrix.

Now by induction hypothesis there is an invertible matrix S_2 with $n - 1$ rows and columns such that $B_2 = S_2^{-1}A_1S_2$ is upper triangular. Write

$$S = S_1 \begin{pmatrix} 1 & 0 \\ 0 & S_2 \end{pmatrix}.$$

This is a product of invertible matrices, so it is invertible. An easy matrix computation shows that $S^{-1}AS$ equals

$$\begin{pmatrix} 1 & 0 \\ 0 & S_2^{-1} \end{pmatrix} (S_1^{-1}AS_1) \begin{pmatrix} 1 & 0 \\ 0 & S_2 \end{pmatrix},$$

which equals

$$\begin{pmatrix} 1 & 0 \\ 0 & S_2^{-1} \end{pmatrix} B_1 \begin{pmatrix} 1 & 0 \\ 0 & S_2 \end{pmatrix}.$$

Replace B_1 by $\begin{pmatrix} c & A_2 \\ 0 & A_1 \end{pmatrix}$ and multiply the matrices together to get

$$S^{-1}AS = \begin{pmatrix} c & A_2 S_2 \\ 0 & S_2^{-1} A_1 S_2 \end{pmatrix} = \begin{pmatrix} c & A_2 S_2 \\ 0 & B_2 \end{pmatrix}.$$

This matrix is clearly upper triangular, so the theorem is proved.

The proof of the theorem provides a method for triangularizing a matrix.

Example 8.1.6

Triangularize the matrix $A = \begin{pmatrix} 1 & 1 \\ -1 & 3 \end{pmatrix}$.

The characteristic polynomial of A is $x^2 - 4x + 4$, so both eigenvalues equal 2. Solving $(A - 2I_2)X = 0$, we find that all the eigenvectors of A are scalar multiples of $X_1 = \begin{pmatrix} 1 \\ 1 \end{pmatrix}$. Hence A is not diagonalizable by 8.1.6.

Let T be the linear operator on $\mathbf{C}^2$ arising from left multiplication by A. Adjoin a vector to X_2 to X_1 to get a basis $\mathcal{B}_2 = \{X_1, X_2\}$ of $\mathbf{C}^2$, say $X_2 = \begin{pmatrix} 0 \\ 1 \end{pmatrix}$. Denote by $\mathcal{B}_1$ the standard basis of $\mathbf{C}^2$. Then the change of basis $\mathcal{B}_1 \to \mathcal{B}_2$ is described by the matrix $S_1 = \begin{pmatrix} 1 & 0 \\ -1 & 1 \end{pmatrix}$. Therefore by 6.2.6 the matrix A which represents T with respect to the basis $\mathcal{B}_2$ is

$$S_1 A S_1^{-1} = \begin{pmatrix} 2 & 1 \\ 0 & 2 \end{pmatrix}$$

Hence $S = S_1^{-1} = \begin{pmatrix} 1 & 0 \\ 1 & 1 \end{pmatrix}$ triangularizes A.

Exercises 8.1

1. Find all the eigenvectors and eigenvalues of the following matrices:

$$
\begin{pmatrix} 1 & 5 \\ 3 & 3 \end{pmatrix}; \quad
\begin{pmatrix} 1 & 2 & -1 \\ 1 & 0 & 1 \\ 4 & -4 & 5 \end{pmatrix}; \quad
\begin{pmatrix} 1 & 0 & 0 & 0 \\ 2 & 2 & 0 & 0 \\ 1 & 0 & 3 & 0 \\ 0 & 1 & -1 & 4 \end{pmatrix}.
$$

2. Prove that $\mathrm{tr}(A+B) = \mathrm{tr}(A) + \mathrm{tr}(B)$ and $\mathrm{tr}(cA) = c\,\mathrm{tr}(A)$ where A and B are $n \times n$ matrices and c is a scalar.

3. If A and B are $n \times n$ matrices, show that AB and BA have the same eigenvalues. [Hint: let c be an eigenvalue of AB and prove that it is an eigenvalue of BA].

4. Suppose that A is a square matrix with real entries and real eigenvalues. Prove that every eigenvalue of A has an associated *real* eigenvector.

5. If A is a real matrix with distinct eigenvalues, then A is diagonalizable over $\mathbf{R}$: true or false?

6. Let $p(x)$ be the polynomial

$$(-1)^n (x^n + a_{n-1}x^{n-1} + a_{n-2}x^{n-2} + \cdots + a_0).$$

Show that $p(x)$ is the characteristic polynomial of the following matrix (which is called the *companion matrix* of $p(x)$):

$$
\begin{pmatrix}
0 & 0 & \cdots & 0 & -a_0 \\
1 & 0 & \cdots & 0 & -a_1 \\
0 & 1 & \cdots & 0 & -a_2 \\
\cdot & \cdot & \cdots & \cdot & \cdot \\
0 & 0 & \cdots & 1 & -a_{n-1}
\end{pmatrix}.
$$

7. Find matrices which diagonalize the following:

(a) $\begin{pmatrix} 1 & 5 \\ 3 & 3 \end{pmatrix}$; (b) $\begin{pmatrix} 1 & 2 & -1 \\ 1 & 0 & 1 \\ 4 & -4 & 5 \end{pmatrix}$.

8. For which values of a and b is the matrix $\begin{pmatrix} 0 & a \\ b & 0 \end{pmatrix}$ diagonalizable over $\mathbf{C}$?

9. Prove that a complex 2×2 matrix is *not* diagonalizable if and only if it is similar to a matrix of the form $\begin{pmatrix} a & b \\ 0 & a \end{pmatrix}$ where $b \neq 0$.

10. Let A be a diagonalizable matrix and assume that S is a matrix which diagonalizes A. Prove that a matrix T diagonalizes A if and only if it is of the form $T = CS$ where C is a matrix such that $AC = CA$.

11. If A is an invertible matrix with eigenvalues $c_1, \ldots, c_n$, show that the eigenvalues of A^{-1} are $c_1^{-1}, \ldots, c_n^{-1}$.

12. Let $T : V \to V$ be a linear operator on a complex n-dimensional vector space V. Prove that there is a basis $\{\mathbf{v}_1, \ldots, \mathbf{v}_n\}$ of V such that $T(\mathbf{v}_i)$ is a linear combination of $\mathbf{v}_i, \ldots, \mathbf{v}_n$ for $i = 1, \ldots, n$.

13. Let $T : P_n(\mathbf{R}) \to P_n(\mathbf{R})$ be the linear operator corresponding to differentiation. Show that all the eigenvalues of T are zero. What are the eigenvectors?

14. Let $c_1, \ldots, c_n$ be the eigenvalues of a complex matrix A. Prove that the eigenvalues of A^m are $c_1^m, \ldots, c_n^m$ where m is any positive integer. [Hint: A is triangularizable].

15. Prove that a square matrix and its transpose have the same eigenvalues.

8.2 Applications to Systems of Linear Recurrences

A *recurrence relation* is an equation involving a function y of a non-negative integral variable n, the value of y at n being written y_n. The equation relates the values of the function at certain consecutive integers, typically $y_{n+1}, y_n, \ldots, y_{n-r}$. In addition there may be some initial conditions to be satisfied, which specify certain values of y_i. If the equation is linear in y, the recurrence relation is said to be *linear*. The problem is to solve the recurrence, that is, to find the most general function which satisfies the equation and the initial conditions. Linear recurrence relations, and more generally systems of linear recurrence relations, occur in many real-life problems. We shall see that the theory of eigenvalues provides an effective means for solving such problems.

To understand how systems of linear relations can arise we consider a predator-prey problem.

Example 8.2.1
In a population of rabbits and weasels it is observed that each year the number of rabbits is equal to four times the number of rabbits less twice the number of weasels in the previous year. The number of weasels in any year equals the sum of the numbers of rabbits and weasels in the previous year. If the initial numbers of rabbits and weasels were 100 and 10 respectively, find the numbers of each species after n years.

Let r_n and w_n denote the respective numbers of rabbits and weasels after n years. The information given in the statement of the problem translates into the equations

$$\begin{cases} r_{n+1} = 4r_n - 2w_n \\ w_{n+1} = r_n + w_n \end{cases}$$

together with the initial conditions $r_0 = 100$, $w_0 = 10$. Thus we have to solve a system of two linear recurrence relations for r_n and w_n, subject to two initial conditions.

At first sight it may not seem clear how eigenvalues enter into this problem. However, let us put the system of linear recurrences in matrix form by writing

$$X_n = \begin{pmatrix} r_n \\ w_n \end{pmatrix} \text{ and } A = \begin{pmatrix} 4 & -2 \\ 1 & 1 \end{pmatrix}.$$

Then the two recurrences are equivalent to the single matrix equation

$$X_{n+1} = AX_n,$$

while the initial conditions assert that

$$X_0 = \begin{pmatrix} 100 \\ 10 \end{pmatrix}.$$

These equations enable us to calculate successive vectors X_n; thus $X_1 = AX_0$, $X_2 = A^2 X_0$, and in general

$$X_n = A^n X_0.$$

In principle this equation provides the solution of our problem. However the equation is difficult to use since it involves calculating powers of A; these soon become very complicated and there is no obvious formula for A^n.

The key observation is that powers of a *diagonal* matrix are easy to compute; one simply forms the appropriate power of each diagonal element. Fortunately the matrix A is diagonalizable since it has distinct eigenvalues 2 and 3. Corresponding eigenvectors are found to be $\begin{pmatrix} 1 \\ 1 \end{pmatrix}$ and $\begin{pmatrix} 2 \\ 1 \end{pmatrix}$; therefore the matrix $S = \begin{pmatrix} 1 & 2 \\ 1 & 1 \end{pmatrix}$ diagonalizes A, and

$$D = S^{-1}AS = \begin{pmatrix} 2 & 0 \\ 0 & 3 \end{pmatrix}.$$

It is now easy to find X_n; for $A^n = (SDS^{-1})^n = SD^nS^{-1}$. Therefore

$$X_n = A^n X_0 = SD^n S^{-1} X_0$$

$$= \begin{pmatrix} 1 & 2 \\ 1 & 1 \end{pmatrix} \begin{pmatrix} 2^n & 0 \\ 0 & 3^n \end{pmatrix} \begin{pmatrix} -1 & 2 \\ 1 & -1 \end{pmatrix} \begin{pmatrix} 100 \\ 10 \end{pmatrix},$$

which leads to

$$X_n = \begin{pmatrix} 180 \cdot 3^n - 80 \cdot 2^n \\ 90 \cdot 3^n - 80 \cdot 2^n \end{pmatrix}.$$

The solution to the problem can now be read off:

$$r_n = 180 \cdot 3^n - 80 \cdot 2^n \text{ and } w_n = 90 \cdot 3^n - 80 \cdot 2^n.$$

Let us consider for a moment the implications of these equations. Notice that r_n and w_n both increase without limit as $n \to \infty$ since 3^n is the dominant term; however

$$\lim_{n \to \infty} \left(\frac{r_n}{w_n} \right) = 2.$$

The conclusion is that, while both populations explode, in the long run there will be twice as many rabbits as weasels.

Having seen that eigenvalues provide a satisfactory solution to the rabbit-weasel problem, we proceed to consider systems of linear recurrences in general.

Systems of first order linear recurrence relations

A *system of first order (homogeneous) linear recurrence relations* in functions $y_n^{(1)}, \ldots, y_n^{(m)}$ of an integral variable n is a set of equations of the form

$$\begin{cases} y_{n+1}^{(1)} = a_{11}y_n^{(1)} + \cdots + a_{1m}y_n^{(m)} \\ y_{n+1}^{(2)} = a_{21}y_n^{(1)} + \cdots + a_{2m}y_n^{(m)} \\ \quad \vdots \qquad\qquad \vdots \qquad\qquad \vdots \\ y_{n+1}^{(m)} = a_{m1}y_n^{(1)} + \cdots + a_{mm}y_n^{(m)} \end{cases}$$

We shall only consider the case where the coefficients a_{ij} are constants. One objective might be to find all the functions $y_n^{(1)}, \ldots, y_n^{(m)}$ which satisfy the equations of the system, i.e., the general solution. Alternatively, one might want to find a solution which satisfies certain given conditions,

$$y_0^{(1)} = b_1, \ y_0^{(2)} = b_2, \ \ldots, \ y_0^{(m)} = b_m$$

where $b_1, \ldots, b_m$ are constants. Clearly the rabbit and weasel problem is of this type.

The method adopted in Example 8.2.1 can be applied with advantage to the general case. First convert the given system of recurrences to matrix form by introducing the matrix $A = [a_{ij}]_{m,m}$, the *coefficient matrix*, and defining

$$Y_n = \begin{pmatrix} y_n^{(1)} \\ y_n^{(2)} \\ \vdots \\ y_n^{(m)} \end{pmatrix} \quad \text{and } B = \begin{pmatrix} b_1 \\ b_2 \\ \vdots \\ b_m \end{pmatrix}.$$

Then the system of recurrences becomes simply

$$Y_{n+1} = AY_n,$$

with the initial condition $Y_0 = B$. The general solution of this is

$$Y_n = A^n B_0.$$

Now assume that A is diagonalizable: suppose that in fact $D = S^{-1}AS$ is diagonal with diagonal entries $d_1, \ldots, d_m$. Then $A = SDS^{-1}$ and $A^n = SD^nS^{-1}$, so that

$$Y_n = SD^nS^{-1}B.$$

Here of course D^n is the diagonal matrix with entries

$d_1{}^n, d_2{}^n \ldots, d_m{}^n$. Since we know how to find S and D, all we need do is compute the product Y_n, and read off its entries to obtain the functions $y_n{}^{(1)}, \ldots, y_n{}^{(m)}$.

At this point the reader may ask: what if A is not diagonalizable? A complete discussion of this case would take us too far afield. However one possible approach is to exploit the fact that the coefficient matrix A is certainly triangularizable by 8.1.8. Thus we can find S such that $S^{-1}AS = T$ is upper triangular. Now write $U_n = S^{-1}Y_n$, so that $Y_n = SU_n$. Then the recurrence $Y_{n+1} = AY_n$ becomes $SU_{n+1} = ASU_n$, or $U_{n+1} = (S^{-1}AS)U_n = TU_n$. In principle this "triangular" system of recurrence relations can be solved by a process of back substitution: first solve the last recurrence for $u_n^{(m)}$, then substitute for $u_n^{(m)}$ in the second last recurrence and solve for $u_n^{(m-1)}$, and so on. What makes the procedure effective is the fact that powers of a triangular matrix are easier to compute than those of an arbitrary matrix.

Example 8.2.2
Consider the system of linear recurrences

$$\begin{cases} y_{n+1} &= y_n + z_n \\ z_{n+1} &= -y_n + 3z_n \end{cases}$$

The coefficient matrix $A = \begin{pmatrix} 1 & 1 \\ -1 & 3 \end{pmatrix}$ is not diagonalizable, but it was triangularized in Example 8.1.6; there it was found that

$$T = S^{-1}AS = \begin{pmatrix} 2 & 1 \\ 0 & 2 \end{pmatrix} \text{ where } S = \begin{pmatrix} 1 & 0 \\ 1 & 1 \end{pmatrix}.$$

Put $U_n = S^{-1}Y_n$; here the entries of U_n and Y_n are written u_n, v_n and y_n, z_n respectively. The recurrence relation $Y_{n+1} = AY_n$ becomes $U_{n+1} = TU_n$. This system of linear recurrences is in triangular form:

$$\begin{cases} u_{n+1} & = 2u_n & + v_n \\ v_{n+1} & = & 2v_n \end{cases}$$

The second recurrence has the obvious solution $v_n = d_2 2^n$ with d_2 constant. Substitute for v_n in the first equation to get $u_{n+1} = 2u_n + d_2 2^n$. This recurrence can be solved in a simple-minded fashion by calculating successively $u_1, u_2, \ldots$ and looking for the pattern. It turns out that $u_n = d_1 2^n + d_2 n\, 2^{n-1}$ where d_1 is another constant. Finally, y_n and z_n can be found from the equation $Y_n = SU_n$; the general solution is therefore

$$\begin{cases} y_n = d_1 2^n + d_2 n 2^{n-1} \\ z_n = d_1 2^n + d_2 (n+2) 2^{n-1} \end{cases}$$

Higher order recurrence relations

A system of recurrence relations for $y_n^{(1)}, \ldots, y_n^{(m)}$ which expresses each $y_{n+1}^{(i)}$ in terms of the $y_j^{(k)}$ for $j = n-r+1, \ldots, n$, is said to be of *order r*. When $r \geq 2$, such a system can be converted into a first order system by introducing more unknowns. The method works well even for a single recurrence relation, as the next example shows.

Example 8.2.3 (*The Fibonacci sequence*)
The sequence of integers 0, 1, 1, 2, 3, 5,. .. is generated by adding pairs of consecutive terms to get the next term. Thus, if the terms are written $y_0, y_1, y_2, \ldots$, then y_n satisfies

$$y_{n+1} = y_n + y_{n-1}, \quad n \geq 1,$$

which is a second order recurrence relation.

To convert this into a first order system we introduce the new function $z_n = y_{n-1}, (n \geq 1)$. This results in an equivalent system of first order recurrences

$$\begin{cases} y_{n+1} = y_n + z_n \\ z_{n+1} = y_n \end{cases}$$

with initial conditions $y_0 = 0$ and $z_0 = 1$. The coefficient matrix $A = \begin{pmatrix} 1 & 1 \\ 1 & 0 \end{pmatrix}$ has eigenvalues $(1 + \sqrt{5})/2$ and $(1 - \sqrt{5})/2$, so it is diagonalizable. Diagonalizing A as in Example 8.1.5, we find that

$$D = S^{-1}AS = \begin{pmatrix} (1 + \sqrt{5})/2 & 0 \\ 0 & (1 - \sqrt{5})/2 \end{pmatrix}$$

where

$$S = \begin{pmatrix} (1 + \sqrt{5})/2 & (1 - \sqrt{5})/2 \\ 1 & 1 \end{pmatrix}.$$

Then $Y_n = A^n Y_0 = (SDS^{-1})^n Y_0 = SD^n S^{-1} Y_0$. This yields the rather unexpected formula

$$y_n = \frac{1}{\sqrt{5}} \left\{ \left(\frac{1 + \sqrt{5}}{2} \right)^n - \left(\frac{1 - \sqrt{5}}{2} \right)^n \right\}$$

for the $(n + 1)$th Fibonacci number.

Markov processes

In order to motivate the concept of a Markov process, we consider a problem about population movement.

Example 8.2.4

Each year 10% of the population of California leave the state for some other part of the United States, while 20% of the U.S. population outside California enter the state. Assuming a constant total population of the country, what will the ultimate population distribution be?

Let y_n and z_n be the numbers of people inside and outside California after n years; then the information given translates into the system of linear recurrences

$$\begin{cases} y_{n+1} = .9y_n + .2z_n \\ z_{n+1} = .1y_n + .8z_n \end{cases}$$

Writing

$$X_n = \begin{pmatrix} y_n \\ z_n \end{pmatrix} \text{ and } A = \begin{pmatrix} .9 & .2 \\ .1 & .8 \end{pmatrix},$$

we have $X_{n+1} = AX_n$. The matrix A has eigenvalues 1 and .7, so we could proceed to solve for y_n and z_n in the usual way. However this is unnecessary in the present example since it is only the ultimate behavior of y_n and z_n that is of interest.

Assuming that the limits exist, we see that the real object of interest is the vector

$$X_\infty = \lim_{n \to \infty} X_n = \begin{pmatrix} \lim_{n \to \infty} y_n \\ \lim_{n \to \infty} z_n \end{pmatrix}.$$

Taking the limit as $n \to \infty$ of both sides of the equation $X_{n+1} = AX_n$, we obtain $X = AX$; hence X is an eigenvector of A associated with the eigenvalue 1. An eigenvector is quickly found to be $\begin{pmatrix} 2 \\ 1 \end{pmatrix}$. Thus X_∞ must be a scalar multiple of this vector. Now the sum of the entries of X_∞ equals the total U.S. population, p say, and it follows that

$$X_\infty = \frac{p}{3} \begin{pmatrix} 2 \\ 1 \end{pmatrix}.$$

So the (alarming) conclusion is that ultimately two thirds of the U.S. population will be in California and one third elsewhere. This can be confirmed by explicitly calculating y_n and z_n and taking the limit as $n \to \infty$.

The preceding problem is an example of what is known as a Markov process. For an understanding of this concept some knowledge of elementary probability is necessary. A *Markov process* is a system which has a finite set of states $S_1, \ldots, S_n$. At any instant the system is in a definite state and over a fixed period of time it changes to another state. The probability that the system changes from state S_j to state S_i over one time period is assumed to be a constant p_{ij}. The matrix

$$P = [p_{ij}]_{n,n}$$

is called the *transition matrix* of the system. In Example 8.2.4 there are two states: a person is either in or not in California. The transition matrix is the matrix A.

Clearly all the entries of P lie in the interval $[0, 1]$; more importantly P has the property that *the sum of the entries in any column equals* 1. Indeed $\sum_{i=1}^{n} p_{ij} = 1$ since it is certain that the system will change from state S_j to *some* state S_i. This property guarantees that 1 *is an eigenvalue of* P; indeed $\det(P - I) = 0$ because the sum of the entries in any column of the matrix $P - I$ is equal to zero, so its determinant is zero.

Suppose that we are interested in the behavior of the system over two time periods. For this we need to know the probability of going from state S_j to state S_i over two periods. Now the probability of the system going from S_j to S_i via S_k is $p_{ik}p_{kj}$, so the probability of going from state S_j to S_i over two periods is

$$\sum_{k=1}^{n} p_{ik}p_{kj}.$$

But this is immediately recognizable as the (i, j) entry of P^2; therefore the transition matrix for the system over two time periods is P^2. More generally the transition matrix for the system over k time periods is seen to be P^k by similar considerations.

The interesting problem for a Markov process is to determine the ultimate behavior of the system over a long period of time, that is to say, $\lim_{k \to \infty}(P^k)$. For the (i, j) entry of this matrix is the probability that the system will go from state S_i to state S_j in the long run.

The first question to be addressed is whether this limit always exists. In general the answer is negative, as a very simple example shows: if $P = \begin{pmatrix} 0 & 1 \\ 1 & 0 \end{pmatrix}$, then P^k equals either $\begin{pmatrix} 1 & 0 \\ 0 & 1 \end{pmatrix}$ or $\begin{pmatrix} 0 & 1 \\ 1 & 0 \end{pmatrix}$, according to whether k is even or odd; so the limit does not exist in this case. Nevertheless it turns out that under some mild assumptions about the matrix the limit does exist. Let us call a transition matrix P *regular* if some positive power of P has all its entries positive. For example, the matrix $\begin{pmatrix} 0 & .5 \\ 1 & .5 \end{pmatrix}$ is regular; indeed all powers after the first have positive entries. But, as we have seen, the matrix $\begin{pmatrix} 0 & 1 \\ 1 & 0 \end{pmatrix}$ is not regular. A Markov system is said to be *regular* if its transition matrix is regular.

The fundamental theorem about Markov processes can now be stated. A proof may be found in [15], for example.

Theorem 8.2.1
Let P be the transition matrix of a regular Markov system. Then $\lim_{k \to \infty}(P^k)$ exists and has the form $(X \ X \ \ldots \ X)$ where X is the unique eigenvector of P associated with the eigenvalue 1 which has entry sum equal to 1.

Our second example of a Markov process is the library book problem from Chapter One (see Exercise 1.2.12).

Example 8.2.5
A certain library owns 10,000 books. Each month 20% of the books in the library are lent out and 80% of the books lent out

are returned, while 10% remain lent out and 10% are reported lost. Finally, 25% of books listed as lost the previous month are found and returned to the library. How many books will be in the library, lent out, and lost in the long run?

Here there are three states that a book may be in: $S_1 =$ in the library: $S_2 =$ lent out: $S_3 =$ lost. The transition matrix for this Markov process is

$$P = \begin{pmatrix} .8 & .8 & .25 \\ .2 & .1 & 0 \\ 0 & .1 & .75 \end{pmatrix}.$$

Clearly P^2 has positive entries, so P is regular. Of course P has the eigenvalue 1; the corresponding eigenvector with entry sum equal to 1 is found to be

$$X = \frac{1}{59} \begin{pmatrix} 45 \\ 10 \\ 4 \end{pmatrix}.$$

So the probabilities that a book is in states S_1, S_2, S_3 after a long period of time are $45/59$, $10/59$, $4/59$ respectively. Therefore the expected numbers of books in the library, lent out, and lost, *in the long run*, are obtained by multiplying these probabilities by the total number of books, 10,000. These numbers are therefore 7627, 1695, 678 respectively.

Exercises 8.2

1. Solve the following systems of linear recurrences with the specified initial conditions:

(a) $\begin{cases} y_{n+1} = & -12x_n \\ x_{n+1} = y_n + 7z_n \end{cases}$ where $y_0 = 0$, $z_0 = 1$;

(b) $\begin{cases} y_{n+1} = 2y_n + 10z_n \\ z_{n+1} = 2y_n + 3z_n \end{cases}$ where $y_0 = 0$, $z_0 = 1$.

2. In a certain nature reserve there are two competing animal species A and B. It is observed that the number of species A equals three times the number of A last year less twice the number of species B last year. Also the number of species B is twice the number of B last year less the number of species A last year. Write down a system of linear recurrence relations for a_n and b_n, the numbers of each species after n years, and solve the system. What are the long term prospects for each species?

3. A pair of newborn rabbits begins to breed at age one month, and each successive month produces one pair of offspring (one of each sex). Initially there were two pairs of rabbits. If r_n is the total number of pairs of rabbits at the beginning of the nth month, show that r_n satisfies $r_{n+1} = r_n + r_{n-1}$ and $r_1 = 2 = r_2$. Solve this second order recurrence relation for r_n.

4. A tower n feet high is to be built from red, white and blue blocks. Each red block is 1 foot high, while the white and blue blocks are 2 feet high. If u_n denotes the number of different designs for the tower, show that the recurrence relation $u_{n+1} = u_n + 2u_{n-1}$ must hold. By solving this recurrence, find a formula for u_n.

5. Solve the system of recurrence relations $y_{n+1} = 3y_n - 2z_n$, $z_{n+1} = 2y_n - z_n$, with the initial conditions $y_0 = 1$, $z_0 = 0$.

6. Solve the second order system $y_{n+1} = y_{n-1}$, $z_{n+1} = y_n + 4z_n$, with the initial conditions $y_0 = 0$, $y_1 = 1 = z_1$.

7. In a certain city 90% of employed persons retain their jobs at the end of each year, while 60% of the unemployed find a job during the year. Assuming that the total employable population remains constant, find the unemployment rate in the long run.

8. A certain species of bird nests in three locations A, B and C. It is observed that each year half of the birds at A and half of the birds at B move their nests to C, while the others stay in the same nesting place. The birds nesting at C are evenly split between A and B. Find the ultimate distribution of birds among the three nesting sites, assuming that the total bird population remains constant.

9. There are three political parties in a certain city, conservatives, liberals and socialists. The probabilities that someone who voted conservative last time will vote liberal or socialist at the next election are .3 and .2 respectively. The probabilities of a liberal voting conservative or socialist are .2 and .1. Finally, the probabilities of a socialist voting conservative or liberal are .1 and .2. What percentages of the electorate will vote for the three parties in the long run, assuming that everyone votes and the number of voters remains constant?

8.3   Applications to Systems of Linear Differential Equations

In this section we show how the theory of eigenvalues developed in 8.1 can be applied to solve systems of linear differential equations. Since there is a close analogy between linear recurrence relations and linear differential equations, the reader will soon notice a similarity between the methods used here and in 8.2.

For simplicity we consider initially a system of *first order linear (homogeneous) differential equations* for functions $y_1, \ldots, y_n$ of x. This has the general form

$$
\begin{cases}
y_1' = a_{11}y_1 + \cdots + a_{1n}y_n \\
y_2' = a_{21}y_1 + \cdots + a_{2n}y_n \\
\quad \cdot \qquad \cdot \qquad \cdots \qquad \cdot \\
y_n' = a_{n1}y_1 + \cdots + a_{nn}y_n
\end{cases}
$$

Here the a_{ij} are assumed to be constants. The object is to find the most general functions $y_1, \ldots, y_n$, differentiable in some interval $[a, b]$, which satisfy the equations of the system. Alternatively one may wish to find functions which satisfy in addition a set of initial conditions of the form

$$y_1(x_0) = b_1, \ y_2(x_0) = b_2, \ \ldots, \ y_n(x_n) = b_n.$$

Here the b_i are certain constants and x_0 is in the interval $[a, b]$.

Let $A = [a_{ij}]$, the *coefficient matrix* of the system and write

$$Y = \begin{pmatrix} y_1 \\ y_2 \\ \vdots \\ y_n \end{pmatrix}.$$

Then we define the *derivative* of Y to be

$$Y' = \begin{pmatrix} y_1' \\ y_2' \\ \vdots \\ y_n' \end{pmatrix}.$$

With this notation the given system of differential equations can be written in matrix form

$$Y' = AY.$$

By a *solution* of this equation we shall mean any column vector Y of n functions in $D[a, b]$ which satisfies the equation. The set of all solutions is a subspace of the vector space of all n-column vectors of differentiable functions; this is called the *solution space*. It can be shown that *the dimension of the solution space equals* n, so that there are n linearly independent solutions, and every solution is a linear combination of them.

If a set of n initial conditions is given, there is in fact a unique solution of the system satisfying these conditions. For an account of the theory of systems of differential equations the reader may consult a book on differential equations such as [15] or [16]. Here we are concerned with methods of finding solutions, not with questions of existence and uniqueness of solutions.

Suppose that the coefficient matrix A is diagonalizable, so there is an invertible matrix S such that $D = S^{-1}AS$ is diagonal, with diagonal entries $d_1, \ldots, d_n$ say. Here of course the d_i are the eigenvalues of A. Define

$$U = S^{-1}Y.$$

Then $Y = SU$ and $Y' = SU'$ since S has constant entries. Substituting for Y and Y' in the equation $Y' = AY$, we obtain $SU' = ASU$, or

$$U' = (S^{-1}AS)U = DU.$$

This is a system of linear differential equations for $u_1, \ldots, u_n$, the entries of U. It has the very simple form

$$\begin{cases} u_1' & = d_1 u_1 \\ u_2' & = d_2 u_2 \\ \quad\cdot & \qquad \cdot \\ u_n' & = d_n u_n \end{cases}$$

The equation $u_i' = d_i u_i$ is easy to solve since its differential form is

$$d(\ln u_i) = d_i.$$

Thus its general solution is $u_i = c_i e^{d_i x}$ where c_i is a constant. The general solution of the system of linear differential equations for $u_1, \ldots, u_n$ is therefore

$$u_1 = c_1 e^{d_1 x}, \quad \ldots, u_n = c_n e^{d_n x}.$$

To find the original functions y_i, simply use the equation $Y = SU$ to get

$$y_i = \sum_{j=1}^{n} s_{ij} u_j = \sum_{j=1}^{n} s_{ij} c_j e^{d_j x}.$$

Since we know how to find S, this procedure provides an effective method of solving systems of first order linear differential equations in the case where the coefficient matrix is diagonalizable.

Example 8.3.1

Consider a long tube divided into four regions along which heat can flow. The regions on the extreme left and right are kept at $0°C$, while the walls of the tube are insulated. It is assumed that the temperature is uniform within each region. Let $y(t)$ and $z(t)$ be the temperatures of the regions A and B at time t. It is known that the rate at which each region cools equals the sum of the temperature differences with the surrounding media. Find a system of linear differential equations for $y(t)$ and $z(t)$ and solve it.

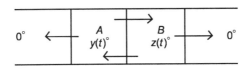

According to the law of cooling

$$\begin{cases} y' &= (z - y) + (0 - y) \\ z' &= (y - z) + (0 - z) \end{cases}$$

Thus we are faced with the linear system of differential equations

$$\begin{cases} y' = -2y + z \\ z' = y - 2z \end{cases}$$

Here

$$A = \begin{pmatrix} -2 & 1 \\ 1 & -2 \end{pmatrix} \text{ and } Y = \begin{pmatrix} y \\ z \end{pmatrix}.$$

Now the matrix A is diagonalizable; indeed

$$D = S^{-1}AS = \begin{pmatrix} -1 & 0 \\ 0 & -3 \end{pmatrix} \text{ where } S = \begin{pmatrix} 1 & 1 \\ 1 & -1 \end{pmatrix}.$$

Setting $U = S^{-1}Y$, we obtain from $Y' = AY$ the equation $U' = DU$. This yields two very simple differential equations

$$\begin{cases} u_1' = -u_1 \\ u_2' = -3u_2 \end{cases}$$

where u_1 and u_2 are the entries of U. Hence $u_1 = ce^{-t}$ and $u_2 = de^{-3t}$, with arbitrary constants c and d. Finally

$$Y = SU = \begin{pmatrix} ce^{-t} + de^{-3t} \\ ce^{-t} - de^{-3t} \end{pmatrix}.$$

The general solution of the original system of differential equations is therefore

$$\begin{cases} y = ce^{-t} + de^{-3t} \\ z = ce^{-t} - de^{-3t} \end{cases}$$

Thus the temperatures of both regions A and B tend to zero as $t \to \infty$.

In the next example complex eigenvalues arise, which causes a change in the procedure.

Example 8.3.2

Solve the linear system of differential equations

$$\begin{cases} y_1' = y_1 - y_2 \\ y_2' = y_1 + y_2 \end{cases}$$

The coefficient matrix here is

$$A = \begin{pmatrix} 1 & -1 \\ 1 & 1 \end{pmatrix},$$

which has complex eigenvalues $1 + i$ and $1 - i$; we are using the familiar notation $i = \sqrt{-1}$ here. The corresponding eigenvectors are

$$\begin{pmatrix} i \\ 1 \end{pmatrix} \text{ and } \begin{pmatrix} -i \\ 1 \end{pmatrix},$$

respectively. Let S be the 2×2 matrix which has these vectors as its columns; then $S^{-1}AS = D$, the diagonal matrix with diagonal entries $1 + i$ and $1 - i$. If we write $U = S^{-1}Y$, the system of equations becomes $U' = DU$, that is,

$$\begin{cases} u_1' = (1+i)u_1 \\ u_2' = (1-i)u_2 \end{cases}$$

where u_1 and u_2 are the entries of U.

The first equation has the solution $u_1 = e^{(1+i)x}$, while the second has the obvious solution $u_2 = 0$. Using these values for u_1 and u_2, we obtain a complex solution of the system of differential equations

$$Y = SU = \begin{pmatrix} ie^{(1+i)x} \\ e^{(1+i)x} \end{pmatrix}.$$

Of course we are looking for real solutions, but these are in fact at hand. For the real and imaginary parts of Y will also be solutions of the system $Y' = AY$. Thus we obtain two real solutions from the single complex solution Y, by taking the real and imaginary parts of Y; these are respectively

$$Y_1 = \begin{pmatrix} -e^x \sin x \\ e^x \cos x \end{pmatrix} \quad \text{and} \quad Y_2 = \begin{pmatrix} e^x \cos x \\ e^x \sin x \end{pmatrix}.$$

Now Y_1 and Y_2 are easily seen to be linearly independent solutions; therefore the general solution of the system is obtained by taking an arbitrary linear combination of these:

$$Y = c_1 Y_1 + c_2 Y_2 = e^x \begin{pmatrix} -c_1 \sin x + c_2 \cos x \\ c_1 \cos x + c_2 \sin x \end{pmatrix}$$

where c_1 and c_2 are arbitrary real constants. Hence

$$\begin{cases} y_1 = e^x(-c_1 \sin x + c_2 \cos x) \\ y_2 = e^x(c_1 \cos x + c_2 \sin x) \end{cases}$$

Of course the success of the method employed in the last two examples depended entirely upon the fact that A is diagonalizable. However, should this not be the case, one can still treat the system of differential equations by triangularizing the coefficient matrix and solving the resulting triangular system using back substitution, rather as was done for systems of linear recurrences in 8.2.

Example 8.3.3
Solve the linear system of differential equations

$$\begin{cases} y_1' = y_1 + y_2 \\ y_2' = -y_1 + 3y_2 \end{cases}$$

In this case the coefficient matrix

$$A = \begin{pmatrix} 1 & 1 \\ -1 & 3 \end{pmatrix}$$

is not diagonalizable, but it can be triangularized. In fact it was shown in Example 8.1.6 that

$$T = S^{-1}AS = \begin{pmatrix} 2 & 1 \\ 0 & 2 \end{pmatrix}$$

where $S = \begin{pmatrix} 1 & 0 \\ 1 & 1 \end{pmatrix}$. Put $U = S^{-1}Y$ and write u_1, u_2 for the entries of U. Then $Y = SU$ and $Y' = SU'$. The equation $Y' = AY$ now becomes $U' = TU$. This yields the triangular system

$$\begin{cases} u_1' = 2u_1 + u_2 \\ u_2' = 2u_2 \end{cases}$$

Solving the second equation, we find that $u_2 = c_2 e^{2x}$ with c_2 an arbitrary constant. Now substitute for u_2 in the first equation to get

$$u_1' - 2u_1 = c_2 e^{2x}.$$

This is a first order linear equation which can be solved by a standard method: multiply both sides of the equation by the "integrating factor"

$$e^{\int -2dx} = e^{-2x}.$$

The equation then becomes $(u_1 e^{-2x})' = c_2$, whence $u_1 e^{-2x} = c_2 x + c_1$, with c_1 another arbitrary constant. Thus $u_1 = c_2 x e^{2x} + c_1 e^{2x}$. To find the original functions y_1 and y_2, we form the product

$$Y = SU = e^{2x} \begin{pmatrix} c_1 + c_2 x \\ c_1 + c_2(x+1) \end{pmatrix}.$$

Thus the general solution of the system is

$$\begin{cases} y_1 = (c_1 + c_2 x)e^{2x}, \\ y_2 = (c_1 + c_2(x+1))e^{2x} \end{cases}$$

Finally, suppose that initial conditions $y_1(0) = 1$ and $y_2(0) = 0$ are given. We can find the correct values of c_1 and c_2 by substituting $t = 0$ in the expressions for y_1 and y_2, to get $c_1 = 1$ and $c_2 = -1$. The required solution is $y_1 = (1 - x)e^{2x}$ and $y_2 = -x\,e^{2x}$.

The next application is one of a military nature.

Example 8.3.4
Two armored divisions A and B engage in combat. At time t their respective numbers of tanks are $a(t)$ and $b(t)$. The rate at which tanks in a division are destroyed is proportional to the number of intact enemy tanks at that instant. Initially A and B have a_0 and b_0 tanks where $a_0 > b_0$. Predict the outcome of the battle.

According to the information given, the functions a and b satisfy the linear system

$$\begin{cases} a' &= -kb \\ b' &= -ka \end{cases}$$

where k is some positive constant. Here the coefficient matrix is

$$A = \begin{pmatrix} 0 & -k \\ -k & 0 \end{pmatrix}.$$

The characteristic equation is $x^2 - k^2 = 0$, so the eigenvalues are k and $-k$ and A is diagonalizable. It turns out that

$$S^{-1}AS = D = \begin{pmatrix} k & 0 \\ 0 & -k \end{pmatrix},$$

where $S = \begin{pmatrix} 1 & 1 \\ -1 & 1 \end{pmatrix}$. If we set $Y = \begin{pmatrix} a \\ b \end{pmatrix}$, the system of differential equations becomes $Y' = AY$. On writing $U = S^{-1}Y$, we get $U' = DU$. This is the system

$$\begin{cases} u' &= ku \\ v' &= -kv \end{cases}$$

where $U = \begin{pmatrix} u \\ v \end{pmatrix}$. Hence $u = ce^{kx}$ and $v = de^{-kx}$, with c and d arbitrary constants. The general solution is $Y = SU$, which yields

$$\begin{cases} a = ce^{kt} + de^{-kt} \\ b = -ce^{kt} + de^{-kt} \end{cases}$$

Now the initial conditions are $a(0) = a_0$ and $b(0) = b_0$, so

$$\begin{cases} c + d = a_0 \\ -c + d = b_0 \end{cases}$$

Solving we obtain $c = (a_0 - b_0)/2$, $d = (a_0 + b_0)/2$. Therefore the numbers of tanks surviving at time t in Divisions A and B are respectively

$$\begin{cases} a = \left(\dfrac{a_0 - b_0}{2}\right)e^{kt} + \left(\dfrac{a_0 + b_0}{2}\right)e^{-kt} \\ b = -\left(\dfrac{a_0 - b_0}{2}\right)e^{kt} + \left(\dfrac{a_0 + b_0}{2}\right)e^{-kt} \end{cases}$$

It is more convenient to write $a(t)$ and $b(t)$ in terms of the hyperbolic functions $\cosh(x) = \frac{1}{2}(e^x + e^{-x})$ and $\sinh(x) = \frac{1}{2}(e^x - e^{-x})$. Then the solution becomes

$$\begin{cases} a = a_0\cosh(kt) - b_0\sinh(kt) \\ b = b_0\cosh(kt) - a_0\sinh(kt) \end{cases}$$

Now Division B will have lost all its tanks when $b = 0$, i.e., after time

$$t = \frac{1}{k}\tanh^{-1}\left(\frac{b_0}{a_0}\right).$$

Observe also that

$$a^2 - b^2 = a_0^2 - b_0^2$$

because of the identity $\cosh^2(kt) - \sinh^2(kt) = 1$. Therefore at the time when Division B has lost all of its tanks, Division A still has a tanks where $a^2 - 0 = a_0^2 - b_0^2$. Hence the number of tanks that Division A has left at the end of the battle is

$$\sqrt{a_0^2 - b_0^2}.$$

Not surprisingly, since it had more tanks to start with, Division A wins the battle.

However, there is a way in which Division B could conceivably win. Suppose that

$$\frac{1}{\sqrt{2}}a_0 \;<\; b_0 \;<\; a_0.$$

Suppose further that Division A consists of two columns with equal numbers of tanks, and that Division B manages to attack one column of Division A before the other column can come to its aid. Since $b_0 > \frac{1}{2}a_0$, Division B defeats the first column of Division A, and it still has $\sqrt{b_0^2 - \frac{1}{4}a_0^2}$ tanks left. Then Division B attacks the second column and wins with

$$\sqrt{b_0^2 - \frac{1}{4}a_0^2 - \frac{1}{4}a_0^2} = \sqrt{b_0^2 - \frac{1}{2}a_0^2}$$

tanks left.

Thus Division B wins the battle despite having fewer tanks than Division A: but it must have more than $a_0/\sqrt{2}$ or 71% of the strength of the larger division for the plan to work. This explains the frequent success of the "divide and conquer strategy".

Higher order equations

Systems of linear differential equations of order 2 or more can be converted to first order systems by introducing additional functions. Once again the procedure is similar to that adopted for systems of linear recurrences.

Example 8.3.5
Solve the second order system

$$\begin{cases} y_1'' = -2y_2 + y_1' + 2y_2' \\ y_2'' = 2y_1 + 2y_1' - y_2' \end{cases}$$

The system may be converted to a first order system by introducing two new functions

$$y_3 = y_1' \quad \text{and} \quad y_4 = y_2'.$$

Thus $y_1'' = y_3'$ and $y_2'' = y_4'$. The given system is therefore equivalent to the first order system

$$\begin{cases} y_1' = y_3 \\ y_2' = y_4 \\ y_3' = -2y_2 + y_3 + 2y_4 \\ y_4' = 2y_1 + 2y_3 - y_4 \end{cases}$$

The coefficient matrix here is

$$A = \begin{pmatrix} 0 & 0 & 1 & 0 \\ 0 & 0 & 0 & 1 \\ 0 & -2 & 1 & 2 \\ 2 & 0 & 2 & -1 \end{pmatrix}.$$

Its eigenvalues turn out to be $1, -1, 2, -2$, with corresponding eigenvectors

$$\begin{pmatrix} 1 \\ 2 \\ 1 \\ 2 \end{pmatrix}, \begin{pmatrix} 2 \\ -1 \\ -2 \\ 1 \end{pmatrix}, \begin{pmatrix} 1 \\ 1 \\ 2 \\ 2 \end{pmatrix}, \begin{pmatrix} 1 \\ -1 \\ -2 \\ 2 \end{pmatrix}.$$

Therefore, if S denotes the matrix with these vectors as its columns, we have $S^{-1}AS = D$, the diagonal matrix with diagonal entries $1, -1, 2, -2$. Now write $U = S^{-1}Y$. Then the equation $Y' = AY$ becomes $U' = (S^{-1}AS)U = DU$, which is equivalent to

$$u_1' = u_1, \quad u_2' = -u_2, \quad u_3' = 2u_3, \quad u_4' = -2u_4.$$

Solving these simple equations, we obtain

$$u_1 = c_1e^x, \quad u_2 = c_2e^{-x}, \quad u_3 = c_3e^{2x}, \quad u_4 = c_4e^{-2x}.$$

The functions y_1 and y_2 may now be read off from the equation $Y = SU$ to give the general solution

$$\begin{cases} y_1 &= \quad c_1e^x \quad + 2c_2e^{-x} \quad + c_3e^{2x} \quad + c_4e^{-2x} \\ y_2 &= \quad 2c_1e^x \quad - c_2e^{-x} \quad + c_3e^{2x} \quad - c_4e^{-2x} \end{cases}$$

Exercises 8.3

1. Find the general solutions of the following systems of linear differential equations:

(a) $\begin{cases} y_1' = -y_1 + y_2 \\ y_2' = 2y_1 - 3y_2 \end{cases}$ (b) $\begin{cases} y_1' = 3y_1 - 2y_2 \\ y_2' = -2y_1 + 3y_2 \end{cases}$

(c) $\begin{cases} y_1' = y_1 + y_2 + y_3 \\ y_2' = \qquad y_2 \\ y_3' = \qquad y_2 + y_3 \end{cases}$

2. Find the general solution (in real terms) of the system of differential equations

$$\begin{cases} y_1' = \quad y_1 + y_2 \\ y_2' = -2y_1 + 3y_2 \end{cases}$$

Then find a solution satisfying the initial conditions $y_1(0) = 1$, $y_2(0) = 2$.

3. By triangularizing the coefficient matrix solve the system of differential equations

$$\begin{cases} y_1' = 5y_1 + 3y_2 \\ y_2' = -3y_1 - y_2 \end{cases}$$

Then find a solution satisfying the initial conditions $y_1(0) = 0$, $y_2(0) = 2$.

4. Solve the second order linear system

$$\begin{cases} y_1'' = 2y_1 + y_2 + y_1' + y_2' \\ y_2'' = -5y_1 + 2y_2 + 5y_1' - y_2' \end{cases}$$

5. Given a system of n (homogeneous) linear differential equations of order k, how would you convert this to a system of first order equations? How many equations will there be in the first order system?

6. Describe a general method for solving a system of second order linear differential equations of the form $Y'' = AY$, where A is diagonalizable.

7. Solve the systems of differential equations

$$\text{(a)} \begin{cases} y_1'' = y_1 - y_2 \\ y_2'' = 3y_1 + 5y_2 \end{cases} \quad \text{(b)} \begin{cases} y_1'' = -4y_2 \\ y_2'' = y_1 + 5y_2 \end{cases}$$

[Note that the general solution of the differential equation $u'' = a^2 u$ is $u = c_1\cosh(ax) + c_2\sinh(ax)$].

8. (*The double pendulum*) A string of length $2l$ is hung from a rigid support. Two weights each of mass m are attached to the midpoint and lower end of the string, which is then allowed to execute small vibrations subject to gravity only. Let y_1 and y_2 denote the horizontal displacements of the two weights from the equilibrium position at time t.

(a) (optional) By using Newton's Second Law of Motion, show that y_1 and y_2 satisfy the differential equations

$y_1'' = a^2(-3y_1 + y_2)$, $y_2'' = a^2(y_1 - y_2)$ where $a = \sqrt{g/l}$ and g is the acceleration due to gravity.

(b) Solve the linear system in (a) for y_1 and y_2. [Note: the general solution of the differential equation $y'' + a^2 y = 0$ is $y = c_1 \cos ax + c_2 \sin ax$].

9. In Example 8.3.4 assume that Division A consists of m equal columns. Suppose that Division B is able to attack each column of A in turn. Show that Division B will win the battle provided that $b_0 > \frac{a_0}{\sqrt{m}}$.

Chapter Nine

MORE ADVANCED TOPICS

This chapter is intended to serve as an introduction to some of the more advanced parts of linear algebra. The most important result of the chapter is the Spectral Theorem, which asserts that every real symmetric matrix can be diagonalized by means of a suitable real orthogonal matrix. This result has applications to quadratic forms, bilinear forms, conics and quadrics, which are described in 9.2 and 9.3. The final section gives an elementary account of the important topic of Jordan normal form, a subject not always treated in a book such as this.

9.1 Eigenvalues and Eigenvectors of Symmetric and Hermitian Matrices

In this section we continue the discussion of diagonalizability of matrices, which was begun in 8.1, with special regard to real symmetric matrices. More generally, a square complex matrix A is called *hermitian* if

$$A = A^*,$$

that is, $A = (\bar{A})^T$. Thus hermitian matrices are the complex analogs of real symmetric matrices. It will turn out that the eigenvalues and eigenvectors of such matrices have remarkable properties not possessed by complex matrices in general. The first indication of special behavior is the fact that their eigenvalues are always real, while the eigenvectors tend to be orthogonal.

Theorem 9.1.1

Let A be a hermitian matrix. Then:

(a) *the eigenvalues of A are all real;*

(b) *eigenvectors of A associated with distinct eigenvalues are orthogonal.*

Proof

Let c be an eigenvalue of A with associated eigenvector X, so that $AX = cX$. Taking the complex transpose of both sides of this equation and using 7.1.7, we obtain $X^*A = \bar{c}X^*$ since $A = A^*$. Now multiply both sides of this equation on the right by X to get $X^*AX = \bar{c}X^*X = \bar{c}\|X\|^2$: remember here that X^*X equals the square of the length of X. But $(X^*AX)^* = X^*A^*X^{**} = X^*AX$; thus the scalar X^*AX equals its complex conjugate and so it is real. It follows that $\bar{c}\|X\|^2$ is real. Since lengths of vectors are always real, we deduce that $\bar{c}$, and hence c, is real, which completes the proof of (a).

To prove (b) take two eigenvectors X and Y associated with distinct eigenvalues c and d. Thus $AX = cX$ and $AY = dY$. Then $Y^*AX = Y^*(cX) = cY^*X$, and in the same way $X^*AY = dX^*Y$. However, by 7.1.7 again, $(X^*AY)^* = Y^*A^*X = Y^*AX$. Therefore $(dX^*Y)^* = cY^*X$, or $dY^*X = cY^*X$ because d is real by the first part of the proof. This means that $(c-d)Y^*X = 0$, from which it follows that $Y^*X = 0$ since $c \neq d$. Thus X and Y are orthogonal.

Suppose now that $\{X_1, \ldots, X_r\}$ is a set of linearly independent eigenvectors of the $n \times n$ hermitian matrix A, and that r is chosen as large as possible. We can multiply X_i by $1/\|X_i\|$ to produce a unit vector; thus we may assume that each X_i is a unit vector. By 9.1.1 $\{X_1, \ldots, X_r\}$ is an orthonormal set. Now write $U = (X_1 \ldots X_r)$, an $n \times r$ matrix. Then U has the property

$$AU = (AX_1 \ \ldots \ AX_r) = (c_1X_1 \ \ldots \ c_rX_r),$$

where $c_1, \ldots, c_r$ are the eigenvectors corresponding to $X_1, \ldots, X_r$ respectively. Hence

$$AU = (X_1 X_2 \ldots X_r) \begin{pmatrix} c_1 & 0 & 0 & \cdots & 0 \\ 0 & c_2 & 0 & \cdots & 0 \\ . & . & . & \cdots & . \\ 0 & 0 & 0 & \cdots & c_r \end{pmatrix} = UD,$$

where D is the diagonal matrix with diagonal entries $c_1, \ldots, c_r$. Since the columns of U form an orthonormal set, $U^* U = I_r$.

In general $r \leq n$, but should it be the case that $r = n$, then U is $n \times n$ and we have $U^{-1} = U^*$, so that U is unitary (see 7.3). Therefore $U^* AU = D$ and A is diagonalized by the matrix U. In other words, if there exist n mutually orthogonal eigenvectors of A, then A can be diagonalized by a unitary matrix. The outstanding question is, of course, whether there are always that many linearly independent eigenvectors. We shall shortly see that this is the case.

A key result must first be established.

Theorem 9.1.2 (*Schur's Theorem*)
Let A be an arbitrary square complex matrix. Then there is a unitary matrix U such that $U^* AU$ is upper triangular. Moreover, if A is a real symmetric matrix, then U can be chosen real and orthogonal.

Proof
Let A be an $n \times n$ matrix. The proof is by induction on n. Of course, if $n = 1$, then A is already upper triangular, so let $n > 1$. There is an eigenvector X_1 of A, with associated eigenvalue c_1 say. Here we can choose X_1 to be a unit vector in $\mathbf{C}^n$. Using 5.1.4 we adjoin vectors to X_1 to form a basis of $\mathbf{C}^n$. Then the Gram-Schmidt procedure (in the complex case) may be applied to produce an orthonormal basis $X_1, \ldots, X_n$ of $\mathbf{C}^n$; note that X_1 is a member of this basis.

Let U_0 denote the matrix $(X_1 \ldots X_n)$; then U_0 is unitary since its columns form an orthonormal set. Now

$$U_0^* A X_1 = U_0^* (c_1 X_1) = c_1 (U_0^* X_1).$$

Also $X_i^* X_1 = 0$ if $i > 1$, while $X_1^* X_1 = 1$. Hence

$$U_0^* A X_1 = c_1 \begin{pmatrix} X_1^* X_1 \\ \vdots \\ X_n^* X_1 \end{pmatrix} = \begin{pmatrix} c_1 \\ 0 \\ \vdots \\ 0 \end{pmatrix}.$$

Since

$$U_0^* A U_0 = U_0^* A(X_1 \ldots X_n) = (U_0^* A X_1 \ U_0^* A X_2 \ \ldots \ U_0^* A X_n),$$

we deduce that

$$U_0^* A U_0 = \begin{pmatrix} c_1 & B \\ 0 & A_1 \end{pmatrix},$$

where A_1 is a matrix with $n - 1$ rows and columns and B is an $(n - 1)$-row vector.

We now have the opportunity to apply the induction hypothesis on n; there is a unitary matrix U_1 such that $U_1^* A_1 U_1 = T_1$ is upper triangular. Put

$$U_2 = \begin{pmatrix} 1 & 0 \\ 0 & U_1 \end{pmatrix},$$

which is surely a unitary matrix. Then let $U = U_0 U_2$; this also unitary since $U^* U = U_2^* (U_0^* U_0) U_2 = U_2^* U_2 = I$. Finally

$$U^* A U = U_2^* (U_0^* A U_0) U_2 = U_2^* \begin{pmatrix} c_1 & B \\ 0 & A_1 \end{pmatrix} U_2,$$

which equals

$$\begin{pmatrix} 1 & 0 \\ 0 & U_1^* \end{pmatrix} \begin{pmatrix} c_1 & B \\ 0 & A_1 \end{pmatrix} \begin{pmatrix} 1 & 0 \\ 0 & U_1 \end{pmatrix} = \begin{pmatrix} c_1 & BU_1 \\ 0 & U_1^* A_1 U_1 \end{pmatrix}.$$

This shows that

$$U^* AU = \begin{pmatrix} c_1 & BU_1 \\ 0 & T_1 \end{pmatrix},$$

an upper triangular matrix, as required.

If the matrix A is real symmetric, the argument shows that there is a real orthogonal matrix S such that $S^T AS$ is diagonal. The point to keep in mind here is that the eigenvalues of A are real by 9.1.1, so that A has a real eigenvector.

The crucial theorem on the diagonalization of hermitian matrices can now be established.

Theorem 9.1.3 (*The Spectral Theorem*)
Let A be a hermitian matrix. Then there is a unitary matrix U such that $U^* AU$ is diagonal. If A is a real symmetric matrix, then U may be chosen to be real and orthogonal.

Proof
By 9.1.2 there is a unitary matrix U such that $U^* AU = T$ is upper triangular. Then $T^* = U^* A^* U = U^* AU = T$, so T is hermitian. But T is upper triangular and T^* is lower triangular, so the only way that T and T^* can be equal is if all the off-diagonal entries of T are zero, that is, T is diagonal.

The case where A is real symmetric is handled by the same argument.

Corollary 9.1.4
If A is an $n \times n$ hermitian matrix, there is an orthonormal basis of $\mathbf{C}^n$ which consists entirely of eigenvectors of A. If in addition A is real, there is an orthonormal basis of $\mathbf{R}^n$ consisting of eigenvectors of A.

Proof

By 9.1.3 there is a unitary matrix U such that $U^*AU = D$ is diagonal, with diagonal entries $d_1, \ldots, d_n$ say. If $X_1, \ldots, X_n$ are the columns of U, then the equation $AU = UD$ implies that $AX_i = d_i X_i$ for $i = 1, \ldots, n$. Therefore the X_i are eigenvectors of A, and since U is unitary, they form an orthonormal basis of $\mathbf{C}^n$. The argument in the real case is similar.

This justifies our hope that an $n \times n$ hermitian matrix always has enough eigenvectors to form an orthonormal basis of $\mathbf{C}^n$. Notice that this will be the case even if the eigenvalues of A are not all distinct.

The following constitutes a practical method of diagonalizing an $n \times n$ hermitian matrix A by means of a unitary matrix. For each eigenvalue find a basis for the corresponding eigenspace. Then apply the Gram-Schmidt procedure to get an orthonormal basis of each eigenspace. These bases are then combined to form an orthonormal set, say $\{X_1, \ldots, X_n\}$. By 9.1.4 this will be a basis of $\mathbf{C}^n$. If U is the matrix with columns $X_1, \ldots, X_n$, then U is hermitian and U^*AU is diagonal, as was shown in the discussion preceding 9.1.2. The same procedure is effective for real symmetric matrices.

Example 9.1.1

Find a real orthogonal matrix which diagonalizes the matrix

$$A = \begin{pmatrix} 1 & 2 \\ 2 & 1 \end{pmatrix}.$$

The eigenvalues of A are 3 and -1, (real of course), and corresponding eigenvectors are

$$\begin{pmatrix} 1 \\ 1 \end{pmatrix} \text{ and } \begin{pmatrix} -1 \\ 1 \end{pmatrix}.$$

These are orthogonal; to get an orthonormal basis of $\mathbf{R}^2$, replace them by the unit eigenvectors

$$\frac{1}{\sqrt{2}} \begin{pmatrix} 1 \\ 1 \end{pmatrix} \text{ and } \frac{1}{\sqrt{2}} \begin{pmatrix} -1 \\ 1 \end{pmatrix}.$$

Finally let

$$S = \frac{1}{\sqrt{2}} \begin{pmatrix} 1 & -1 \\ 1 & 1 \end{pmatrix},$$

which is an orthogonal matrix. The theory predicts that

$$S^T A S = \begin{pmatrix} 3 & 0 \\ 0 & -1 \end{pmatrix},$$

as is easily verified by matrix multiplication.

Example 9.1.2
Find a unitary matrix which diagonalizes the hermitian matrix

$$A = \begin{pmatrix} 3/2 & i/2 & 0 \\ -i/2 & 3/2 & 0 \\ 0 & 0 & 1 \end{pmatrix},$$

where $i = \sqrt{-1}$.

The eigenvalues are found to be 1, 2, 1, with associated unit eigenvectors

$$\begin{pmatrix} -i/\sqrt{2} \\ 1/\sqrt{2} \\ 0 \end{pmatrix}, \begin{pmatrix} 1/\sqrt{2} \\ -i/\sqrt{2} \\ 0 \end{pmatrix}, \begin{pmatrix} 0 \\ 0 \\ 1 \end{pmatrix}.$$

Therefore

$$U^* A U = \begin{pmatrix} 1 & 0 & 0 \\ 0 & 2 & 0 \\ 0 & 0 & 1 \end{pmatrix},$$

where U is the unitary matrix

$$\frac{1}{\sqrt{2}} \begin{pmatrix} -i & 1 & 0 \\ 1 & -i & 0 \\ 0 & 0 & \sqrt{2} \end{pmatrix}.$$

Normal matrices

We have seen that every $n \times n$ hermitian matrix A has the property that there is an orthonormal basis of $\mathbf{C}^n$ consisting of eigenvectors of A. It was also observed that this property immediately leads to A being diagonalizable by a unitary matrix, namely the matrix whose columns are the vectors of the orthonormal basis. We shall consider what other matrices have this useful property.

A complex matrix A is called *normal* if it commutes with its complex transpose,

$$A^* A = A A^*.$$

Of course for a real matrix this says that A commutes with its transpose A^T. Clearly hermitian matrices are normal; for if $A = A^*$, then certainly A commutes with A^*. What is the connection between normal matrices and the existence of an orthonormal basis of eigenvectors? The somewhat surprising answer is given by the next theorem.

Theorem 9.1.5
Let A be a complex $n \times n$ matrix. Then A is normal if and only if there is an orthonormal basis of $\mathbf{C}^n$ consisting of eigenvectors of A.

Proof
First of all suppose that $\mathbf{C}^n$ has an orthonormal basis of eigenvectors of A. Then, as has been noted, there is a unitary matrix U such that $U^* A U = D$ is diagonal. This leads

to $A = UDU^*$ because $U^* = U^{-1}$. Next we perform a direct computation to show that A commutes with its complex transpose:

$$AA^* = UDU^*UD^*U^* = UDD^*U^*,$$

and in the same way

$$A^*A = UD^*U^*UDU^* = UD^*DU^*.$$

But diagonal matrices always commute, so $DD^* = D^*D$. It follows that $AA^* = A^*A$, so that A is normal.

It remains to show that if A is normal, then there is an orthonormal basis of $\mathbf{C}^n$ consisting entirely of eigenvectors of A. From 9.1.2 we know that there is a unitary matrix U such that $U^*AU = T$ is upper triangular. The next observation is that T is also normal. This too is established by a direct computation:

$$T^*T = U^*A^*UU^*AU = U^*(A^*A)U.$$

In the same way $TT^* = U^*(AA^*)U$. Since $A^*A = AA^*$, it follows that $T^*T = TT^*$.

Now equate the $(1, 1)$ entries of T^*T and TT^*; this yields the equation

$$|t_{11}|^2 = |t_{11}|^2 + |t_{12}|^2 + \cdots + |t_{1n}|^2,$$

which implies that $t_{12}, \ldots, t_{1n}$ are all zero. By looking at the $(2, 2), (3, 3), \ldots, (n, n)$ entries of T^*T and TT^*, we see that all the other off-diagonal entries of T vanish too. Thus T is actually a diagonal matrix.

Finally, since $AU = UT$, the columns of U are eigenvectors of A, and they form an orthonormal basis of $\mathbf{C}^n$ because U is unitary. This completes the proof of the theorem.

The last theorem provides us with many examples of diagonalizable matrices: for example, complex matrices which are unitary or hermitian are automatically normal, as are real symmetric and real orthogonal matrices. Any matrix of these types can therefore be diagonalized by a unitary matrix.

Exercises 9.1

1. Find unitary or orthogonal matrices which diagonalize the following matrices:

$$\text{(a } \begin{pmatrix} 3 & 1 \\ 1 & 3 \end{pmatrix}; \quad \text{(b)} \begin{pmatrix} 2 & 1 & 1 \\ 1 & 3 & -2 \\ 1 & -2 & 3 \end{pmatrix};$$

$$\text{(c)} \begin{pmatrix} 3 & \sqrt{-1} \\ -\sqrt{-1} & 3 \end{pmatrix}.$$

2. Suppose that A is a complex matrix with real eigenvalues which can be diagonalized by a unitary matrix. Prove that A must be hermitian.

3. Show that an upper triangular matrix is normal if and only if it is diagonal.

4. Let A be a normal matrix. Show that A is hermitian if and only if all its eigenvalues are real.

5. A complex matrix A is called *skew-hermitian* if $A^* = -A$. Prove the following statements:
   (a) a skew-hermitian matrix is normal;
   (b) the eigenvalues of a skew-hermitian matrix are purely imaginary, that is, of the form $a\sqrt{-1}$ where a is real;
   (c) a normal matrix is skew-hermitian if all its eigenvalues are purely imaginary.

6. Let A be a normal matrix. Prove that A is unitary if and only if all its eigenvalues c satisfy $|c| = 1$.

7. Let X be any unit vector in $\mathbf{C}^n$ and put $A = I_n - 2XX^*$. Prove that A is both hermitian and unitary. Deduce that $A = A^{-1}$.

8. Give an example of a normal matrix which is not hermitian, skew-hermitian or unitary. [Hint: use Exercises 4, 5, and 6].

9. Let A be a real orthogonal $n \times n$ matrix. Prove that A is similar to a matrix with blocks down the diagonal each of which is I_l, $-I_m$, or else a matrix of the form

$$\begin{pmatrix} \cos\theta & -\sin\theta \\ \sin\theta & \cos\theta \end{pmatrix}$$

where $0 < \theta < 2\pi$, and $\theta \neq \pi$. [Hint: by Exercise 6 the eigenvalues of A have modulus 1; also A is similar to a diagonal matrix whose diagonal entries are the eigenvalues].

9.2 Quadratic Forms

A *quadratic form* in the real variables $x_1, \ldots, x_n$ is a polynomial in $x_1, \ldots, x_n$ with real coefficients in which every term has degree 2. For example, the expression $ax^2 + 2bxy + cy^2$ is a quadratic form in x and y. Quadratic forms occur in many contexts; for example, the equations of a conic in the plane and a quadric surface in three-dimensional space involve quadratic forms.

We begin by observing that the quadratic form

$$q = ax^2 + 2bxy + cy^2$$

in x and y can be written as a product of two vectors and a symmetric matrix,

$$q = (x\ \ y)\begin{pmatrix} a & b \\ b & c \end{pmatrix}\begin{pmatrix} x \\ y \end{pmatrix} = \begin{pmatrix} x \\ y \end{pmatrix}^T \begin{pmatrix} a & b \\ b & c \end{pmatrix}\begin{pmatrix} x \\ y \end{pmatrix}.$$

In general any quadratic form q in $x_1, \ldots, x_n$ can be written in this form. For let q be given by the equation

$$q = \sum_{i=1}^{n}\sum_{j=1}^{n} a_{ij}x_i x_j$$

where the a_{ij} are real numbers. Setting $A = [a_{ij}]_{n,n}$ and writing X for the column vector with entries $x_1, \ldots, x_n$, we see from the definition of matrix products that q may be written in the form

$$q = X^T A X.$$

Thus the quadratic form q is determined by the real matrix A.

At this point we make the crucial observation that nothing is lost if we assume that A is *symmetric*. For, since $X^T A X$ is scalar, q may also be written as $(X^T A X)^T = X^T A^T X$; therefore

$$q = \frac{1}{2}(X^T A X + X^T A^T X) = X^T \left(\frac{1}{2}(A + A^T) \right) X.$$

It follows that A can be replaced by the symmetric matrix $\frac{1}{2}(A + A^T)$. For this reason it will in future be tacitly assumed that *the matrix associated with a quadratic form is symmetric*.

The observation of the previous paragraph allows us to apply the Spectral Theorem to an arbitrary quadratic form. The conclusion is that a quadratic form can be written in terms of squares only.

Theorem 9.2.1
Let $q = X^T A X$ be an arbitrary quadratic form. Then there is a real orthogonal matrix S such that $q = c_1 x_1'^2 + \cdots + c_n x_n'^2$ where $x_1', \ldots, x_n'$ are the entries of $X' = S^T X$ and $c_1, \ldots, c_n$ are the eigenvalues of the matrix A.

Proof
By 9.1.3 there is a real orthogonal matrix S such that $S^T A S = D$ is diagonal, with diagonal entries $c_1, \ldots, c_n$ say. Define X' to be $S^T X$; then $X = S X'$. Substituting for X, we find that

$$q = X^T A X = (S X')^T A (S X') = (X')^T (S^T A S) X'$$
$$= (X')^T D X'.$$

Multiplying out the final matrix product, we find that $q = c_1 x_1'^2 + \cdots + c_n x_n'^2$.

Application to conics and quadrics

We recall from the analytical geometry of two dimensions that a *conic* is a curve in the plane with equation of the second degree, the general form being

$$ax^2 + 2bxy + cy^2 + dx + ey + f = 0$$

where the coefficients are real numbers. This can be written in the matrix form

$$X^T A X + (d\ e)X + f = 0$$

where

$$X = \begin{pmatrix} x \\ y \end{pmatrix} \text{ and } A = \begin{pmatrix} a & b \\ b & c \end{pmatrix}.$$

So there is a quadratic form in x and y involved in this conic. Let us examine the effect on the equation of the conic of applying the Spectral Theorem.

Let S be a real orthogonal matrix such that $S^T A S = \begin{pmatrix} a' & 0 \\ 0 & c' \end{pmatrix}$ where a' and c' are the eigenvalues of A. Put $X' = S^T X$ and denote the entries of X' by x', y'; then $X = SX'$ and the equation of the conic takes the form

$$(X')^T \begin{pmatrix} a' & 0 \\ 0 & c' \end{pmatrix} X' + (d\ e)SX' + f = 0,$$

or equivalently,

$$a'x'^2 + c'y'^2 + d'x' + e'y' + f = 0$$

for certain real numbers d' and e'. Thus the advantage of changing to the new variables x' and y' is that no "cross term" in $x'y'$ appears in the quadratic form.

There is a good geometrical interpretation of this change of variables: it corresponds to a rotation of axes to a new set of coordinates x' and y'. Indeed, by Examples 6.2.9 and 7.3.7, any real 2×2 orthogonal matrix represents either a rotation or a reflection in $\mathbf{R}^2$; however a reflection will not arise in the present instance: for if it did, the equation of the conic would have had no cross term to begin with. By Example 7.3.7 the orthogonal matrix S has the form

$$\begin{pmatrix} \cos\theta & -\sin\theta \\ \sin\theta & \cos\theta \end{pmatrix}$$

where θ is the angle of rotation. Since $X' = S^T X$, we obtain the equations

$$\begin{cases} x' = x\cos\theta + y\sin\theta \\ y' = -x\sin\theta + y\cos\theta \end{cases}$$

The effect of changing the variables from x, y to x', y' is to rotate the coordinate axes to axes that are parallel to the axes of the conic, the so-called *principal axes*.

Finally, by completing the square in x' and y' as necessary, we can obtain the standard form of the conic, and identify it as an *ellipse*, *parabola*, *hyperbola* (or degenerate form). This final move amounts to a translation of axes. So our conclusion is that the equation of any conic can be put in standard form by a rotation of axes followed by a translation of axes.

Example 9.2.1

Identify the conic $x^2 + 4xy + y^2 + 3x + y - 1 = 0$.

The matrix of the quadratic form $x^2 + 4xy + y^2$ is

$$A = \begin{pmatrix} 1 & 2 \\ 2 & 1 \end{pmatrix}.$$

It was shown in Example 9.1.1 that the eigenvalues of A are 3 and -1 and that A is diagonalized by the orthogonal matrix

$$S = \frac{1}{\sqrt{2}} \begin{pmatrix} 1 & -1 \\ 1 & 1 \end{pmatrix}.$$

Put $X' = S^T X$ where X' has entries x' and y'; then $X = SX'$ and we read off that

$$\begin{cases} x & = \dfrac{1}{\sqrt{2}}(x' - y') \\ y & = \dfrac{1}{\sqrt{2}}(x' + y') \end{cases}$$

So here $\theta = \pi/4$ and the correct rotation of axes for this conic is through angle $\pi/4$ in an anticlockwise direction. Substituting for x and y in the equation of the conic, we get

$$3x'^2 - y'^2 + 2\sqrt{2}x' - \sqrt{2}y' - 1 = 0.$$

From this we can already see that the conic is a hyperbola. To obtain the standard form, complete the square in x' and y':

$$3(x' + \frac{\sqrt{2}}{3})^2 - (y' + \frac{1}{\sqrt{2}})^2 = \frac{7}{6}.$$

Hence the equation of the hyperbola in standard form is

$$3x''^2 - y''^2 = 7/6,$$

where $x'' = x' + \sqrt{2}/3$ and $y'' = y' + 1/\sqrt{2}$. This is a hyperbola whose center is at the point where $x' = -\sqrt{2}/3$ and $y' = -1/\sqrt{2}$; thus the xy - coordinates of the center of the hyperbola are $(1/6, -5/6)$. The axes of the hyperbola are the lines $x'' = 0$ and $y'' = 0$, that is, $x + y = -2/3$ and $x - y = 1$.

Quadrics

A *quadric* is a surface in three-dimensional space whose equation has degree 2 and therefore has the form

$$ax^2 + by^2 + cz^2 + 2dxy + 2eyz + 2fzx + gx + hy + iz + j = 0.$$

Let A be the symmetric matrix

$$\begin{pmatrix} a & d & f \\ d & b & e \\ f & e & c \end{pmatrix}.$$

Then the equation of the quadric may be written in the form

$$X^T AX + (g\ h\ i)X + j = 0.$$

where X is the column with entries x, y, z.

Recall from analytical geometry that a quadric is one the following surfaces: an *ellipsoid*, a *hyperboloid*, a *paraboloid*, a *cone*, a *cylinder* (or a degenerate form). The type of a quadric can be determined by a rotation to principal axes, just as for conics. Thus the procedure is to find a real orthogonal matrix S such that $S^T AX = D$ is diagonal, with entries a', b', c' say. Put $X' = S^T X$. Then $X = SX'$ and $X^T AX = (X')^T DX'$: the equation of the quadric becomes

$$(X')^T DX' + (g\ h\ i)SX' + j = 0,$$

which is equivalent to

$$a'x'^2 + b'y'^2 + c'z'^2 + g'x' + h'y' + i'z' + j = 0.$$

Here a', b', c' are the eigenvalues of A, while g', h', i' are certain real numbers. By completing the square in x', y', z' as necessary, we shall obtain the equation of the quadric in standard form; it will then be possible to recognise its type and position. The last step represents a translation of axes.

Example 9.2.2
Identify the quadric surface

$$x^2 + y^2 + z^2 + 2xy + 2yz + 2zx - x + 2y - z = 0.$$

The matrix of the relevant quadratic form is

$$A = \begin{pmatrix} 1 & 1 & 1 \\ 1 & 1 & 1 \\ 1 & 1 & 1 \end{pmatrix}$$

and the equation of the quadric in matrix form is

$$X^T AX + (-1 \quad 2 - 1)X = 0.$$

We diagonalize A by means of an orthogonal matrix. The eigenvalues of A are found to be 0, 0, 3, with corresponding unit eigenvectors

$$\begin{pmatrix} 1/\sqrt{2} \\ -1/\sqrt{2} \\ 0 \end{pmatrix}, \quad \begin{pmatrix} 0 \\ 1/\sqrt{2} \\ -1/\sqrt{2} \end{pmatrix}, \quad \begin{pmatrix} 1/\sqrt{3} \\ 1/\sqrt{3} \\ 1/\sqrt{3} \end{pmatrix}.$$

The first two vectors generate the eigenspace corresponding to the eigenvalue 0. We need to find an orthonormal basis of this subspace; this can be done either by using the Gram-Schmidt procedure or by guessing. Such a basis turns out to be

$$\frac{1}{\sqrt{2}} \begin{pmatrix} 1 \\ -1 \\ 0 \end{pmatrix}, \quad \frac{1}{\sqrt{6}} \begin{pmatrix} 1 \\ 1 \\ -2 \end{pmatrix}.$$

Therefore A is diagonalized by the orthogonal matrix

$$S = \begin{pmatrix} 1/\sqrt{2} & 1/\sqrt{6} & 1/\sqrt{3} \\ -1/\sqrt{2} & 1/\sqrt{6} & 1/\sqrt{3} \\ 0 & -2/\sqrt{6} & 1/\sqrt{3} \end{pmatrix}.$$

The matrix S represents a rotation of axes. Put $X' = S^T X$; then $X = SX'$ and

$$X^T A X = (X')^T (S^T A S) X' = (X')^T D X',$$

where D is the diagonal matrix with diagonal entries 0, 0, 3. The equation of the quadric becomes

$$X'^T D X' + (-1 \quad 2 \quad -1) S X' = 0$$

or

$$z'^2 = \frac{1}{\sqrt{2}} x' - \frac{1}{\sqrt{6}} y'.$$

This is a parabolic cylinder whose axis is the line with equations $y' = \sqrt{3} x'$, $z' = 0$.

Definite quadratic forms

Consider once again a quadratic form $q = X^T A X$ in real variables $x_1, \ldots, x_n$, where A is a real symmetric matrix. In some applications it is the sign of q that is significant.

The quadratic form q is said to be *positive definite* if $q > 0$ whenever $X \neq 0$. Similarly, q is called *negative definite* if $q < 0$ whenever $X \neq 0$. If, however, q can take both positive and negative values, then q is said to be *indefinite*. The terms positive definite, negative definite and indefinite can also be applied to a real symmetric matrix A, according to the behavior of the corresponding quadratic form $q = X^T A X$.

For example, the expression $2x^2 + 3y^2$ is positive unless $x = 0 = y$, so this is a positive definite quadratic form, while

$-2x^2 - 3y^2$ is clearly negative definite. On the other hand, the form $2x^2 - 3y^2$ can take both positive and negative values, so it is indefinite.

In these examples it was easy to decide the nature of the quadratic form since it contained only squared terms. However, in the case of a general quadratic form, it is not possible to decide the nature of the form by simple inspection. The diagonalization process for symmetric matrices allows us to reduce the problem to a quadratic form whose matrix is diagonal, and which therefore involves only squared terms. From this it is apparent that it is the *signs* of the eigenvalues of the matrix A that are important. The definitive result is

Theorem 9.2.2
Let A be a real symmetric matrix and let $q = X^T A X$: then
    (a) *q is positive definite if and only if all the eigenvalues of A are positive;*
    (b) *q is negative definite if and only if all the eigenvalues of A are negative;*
    (c) *q is indefinite if and only if A has both positive and negative eigenvalues.*

Proof
There is a real orthogonal matrix S such that $S^T A S = D$ is diagonal, with diagonal entries $c_1, \ldots, c_n$, say. Put $X' = S^T X$; then $X = SX'$ and

$$q = X^T A X = (X')^T (S^T A S) X' = (X')^T D X',$$

so that q takes the form

$$q = c_1 x_1'^2 + c_2 x_2'^2 + \cdots + c_n x_n'^2$$

where $x_1', \ldots, x_n'$ are the entries of X'. Thus q, considered as a quadratic form in $x_1', \ldots, x_n'$, involves only squares. Now observe that as X varies over the set of all non-zero vectors in $\mathbf{R}^n$, so does $X' = S^T X$. This is because $S^T = S^{-1}$ is invertible. Therefore $q > 0$ for all non-zero X if and only if $q > 0$ for all non-zero X'. In this way we see that it is sufficient to discuss the behavior of q as a quadratic form in $x_1', \ldots, x_n'$. Clearly q will be positive definite as such a form precisely when $c_1, \ldots, c_n$ are all positive, with a corresponding statement for negative definite: but q is indefinite if there are positive and negative c_i's. Finally $c_1, \ldots, c_n$ are just the eigenvalues of A, so the assertion of the theorem is proved.

Let us consider in greater detail the important case of a quadratic form q in two variables x and y, say $q = ax^2 + 2bxy + cy^2$; the associated symmetric matrix is

$$A = \begin{pmatrix} a & b \\ b & c \end{pmatrix}.$$

Let the eigenvalues of A be d_1 and d_2. Then by 8.1.3 we have the relations $\det(A) = d_1 d_2$ and $\operatorname{tr}(A) = d_1 + d_2$; hence

$$d_1 d_2 = ac - b^2 \text{ and } d_1 + d_2 = a + c.$$

Now according to 9.2.2 the form q is positive definite if and only if d_1 and d_2 are both positive. This happens precisely when $ac > b^2$ and $a > 0$. For these conditions are certainly necessary if d_1 and d_2 are to be positive, while if the conditions hold, a and c must both be positive since the inequality $ac > b^2$ shows that a and c have the same sign.

In a similar way we argue that the conditions for A to be negative definite are $ac > b^2$ and $a < 0$. Finally, q is indefinite if and only if $ac < b^2$: for by 9.2.2 the condition for q to be indefinite is that d_1 and d_2 have opposite signs, and this is equivalent to the inequality $d_1 d_2 < 0$. Therefore we have the following result.

Corollary 9.2.3

Let $q = ax^2 + 2bxy + cy^2$ be a quadratic form in x and y. Then:

    (a) *q is positive definite if and only if $ac > b^2$ and $a > 0$;*
    (b) *q is negative definite if and only if $ac > b^2$ and $a < 0$;*
    (c) *q is indefinite if and only if $ac < b^2$.*

Example 9.2.3

Let $q = -2x^2 + xy - 3y^2$. Here we have $a = -2$, $b = 1/2$, $c = -3$. Since $ac - b^2 > 0$ and $a < 0$, the quadratic form is negative definite, by 9.2.3.

The status of a quadratic form in three or more variables can be determined by using 9.2.2.

Example 9.2.4

Let $q = -2x^2 - y^2 - 2z^2 + 6xz$ be a quadratic form in x, y, z. The matrix of the form is

$$A = \begin{pmatrix} -2 & 0 & 3 \\ 0 & -1 & 0 \\ 3 & 0 & -2 \end{pmatrix},$$

which has eigenvalues $-5, -1, 1$. Hence q is indefinite.

Next we record a very different criterion for a matrix to be positive definite. While it is not a practical test, it has a very striking form.

Theorem 9.2.4

Let A be a real symmetric matrix. Then A is positive definite if and only if $A = B^T B$ for some invertible real matrix B.

Proof

Suppose first that $A = B^T B$ with B an invertible matrix. Then the quadratic form $q = X^T A X$ can be rewritten as

$$q = X^T B^T B X = (BX)^T BX = \|BX\|^2.$$

If $X \neq 0$, then $BX \neq 0$ since B is invertible. Hence $\|BX\|$ is positive if $X \neq 0$. It follows that q, and hence A, is positive definite.

Conversely, suppose that A is positive definite, so that all its eigenvalues are positive. Now there is a real orthogonal matrix S such that $S^T A S = D$ is diagonal, with diagonal entries $d_1, \ldots, d_n$ say. Here the d_i are the eigenvalues of A, so all of them are positive. Define $\sqrt{D}$ to be the real diagonal matrix with diagonal entries $\sqrt{d_1}, \ldots, \sqrt{d_n}$. Then we have $A = (S^T)^{-1}DS^T = SDS^T$ since $S^T = S^{-1}$, and hence

$$A = S(\sqrt{D}\sqrt{D})S^T = (\sqrt{D}S^T)^T(\sqrt{D}S^T).$$

Finally, put $B = \sqrt{D}S^T$ and observe that B is invertible since both S and $\sqrt{D}$ are.

Application to local maxima and minima

A well-known use of quadratic forms is to determine if a critical point of a function of several variables is a local maximum or a local minimum. We recall briefly the nature of the problem; for a detailed account the reader is referred to a textbook on calculus such as [18].

Let f be a function of independent real variables $x_1, \ldots, x_n$ whose first order partial derivatives exist in some region R. A point $P(a_1, \ldots, a_n)$ of R is called a *local maximum* (*minimum*) of f if within some neighborhood of P the function f assumes its largest (smallest) value at P. A basic result states that *if P is a local maximum or minimum of f lying inside R, then all the first order partial derivatives of f vanish at P:*

$$f_{x_i}(a_1, \ldots, a_n) = 0 \text{ for } i = 1, \ldots, n.$$

A point at which all these partial derivatives are zero is called a *critical point* of f. Thus every local maximum or minimum is a critical point of f. However there may be critical points which are not local maxima or minima, but are *saddle points* of f.

For example, the function $f(x, y) = x^2 - y^2$ has a saddle point at the origin, as shown in the diagram.

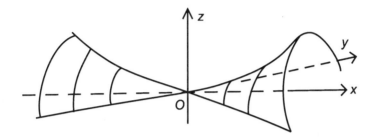

The problem is to devise a test which can distinguish local maxima and minima from saddle points. Such a test is furnished by the criterion for a quadratic form to be positive definite, negative definite or indefinite.

For simplicity we assume that f is a function of two variables x and y. Assume further that f and its partial derivatives of degree at most three are continuous inside a region R of the plane, and that (x_0, y_0) is a critical point of f in R.

Apply Taylor's Theorem to the function f at the point (x_0, y_0), keeping in mind that $f_x(x_0, y_0) = 0 = f_y(x_0, y_0)$. If h and k are sufficiently small, then $f(x_0 + h, y_0 + k) - f(x_0, y_0)$ equals

$$\frac{1}{2}\left(h^2 f_{xx}(x_0, y_0) + 2hk f_{xy}(x_0, y_0) + k^2 f_{yy}(x_0, y_0)\right) + S :$$

here S is a remainder term which is a polynomial of degree 3 or higher in h and k. Write $a = f_{xx}(x_0, y_0)$, $b = f_{xy}(x_0, y_0)$ and $c = f_{yy}(x_0, y_0)$; then

$$f(x_0 + h, y_0 + k) - f(x_0, y_0) = \frac{1}{2}(ah^2 + 2bhk + ck^2) + S.$$

Here S is small compared to the other terms of the sum if h and k are small.

Let $q = ax^2 + 2bxy + cy^2$. If q is negative definite, then $f(x_0 + h, y_0 + k) \leq f(x_0, y_0)$ when h and k are small and P is a local maximum. On the other hand, if q is positive definite, then P is a local minimum since $f(x_0 + h, y_0 + k) \geq f(x_0, y_0)$ for sufficiently small h and k. Finally, should q be indefinite, the expression $f(x_0+h, y_0+k) - f(x_0, y_0)$ can be both positive and negative, so P is neither a local maximum nor a local minimum, but a saddle point.

Thus the crucial quadratic form which provides us with a test for P to be a local maximum or minimum arises from the matrix

$$H = \begin{pmatrix} f_{xx} & f_{xy} \\ f_{xy} & f_{yy} \end{pmatrix}.$$

If the matrix $H(x_0, y_0)$ is positive definite or negative definite, then f will have a local minimum or local maximum respectively at P. If, however, $H(x_0, y_0)$ is indefinite, then P will be a saddle point of f. Combining this result with 9.2.3, we obtain

Theorem 9.2.5

Let f be a function of x and y and assume that f and its partial derivatives of order ≤ 3 are continuous in some region containing the critical point $P(x_0, y_0)$. Let $D = f_{xx} f_{yy} - f_{xy}^2$:

(a) *If $D(x_0, y_0) > 0$ and $f_{xx}(x_0, y_0) < 0$, then P is a local maximum of f;*

(b) *If $D(x_0, y_0) > 0$ and $f_{xx}(x_0, y_0) > 0$, then P is a local minimum of f;*

(c) *If $D < 0$, then P is a saddle point of f.*

The argument just given for a function of two variables can be applied to a function f of n variables $x_1, \ldots, x_n$. The relevant quadratic form in this case is obtained from the matrix

$$H = \begin{pmatrix} f_{x_1x_1} & f_{x_1x_2} & \cdots & f_{x_1x_n} \\ f_{x_2x_1} & f_{x_2x_2} & \cdots & f_{x_2x_n} \\ \cdot & \cdot & \cdots & \cdot \\ f_{x_nx_1} & f_{x_nx_2} & \cdots & f_{x_nx_n} \end{pmatrix},$$

which is called the *hessian* of the function f. Notice that the hessian matrix is symmetric since $f_{x_ix_j} = f_{x_jx_i}$, provided that f and all its derivatives of order ≤ 3 are continuous.

The fundamental theorem may now be stated.

Theorem 9.2.6
Let f be a function of independent variables $x_1, \ldots, x_n$. Assume that f and its partial derivatives of order ≤ 3 are continuous in a region containing a critical point $P(a_1, a_2, \ldots, a_n)$. Let H be the the hessian of f.

(a) *If $H(a_1, \ldots, a_n)$ is positive definite, then P is a local minimum of f;*
(b) *if $H(a_1, \ldots, a_n)$ is negative definite, then P is a local maximum of f;*
(c) *if $H(a_1, \ldots, a_n)$ is indefinite, then P is a saddle point of f.*

Example 9.2.5
Consider the function $f(x, y) = (x^2 - 2x) \cos y$. It has a single critical point $(1, \pi)$ since this is the only point where both first derivatives vanish. To decide the nature of this point we compute the hessian of f as

$$H = \begin{pmatrix} 2\cos y & -(2x - 2)\sin y \\ -(2x - 2)\sin y & -(x^2 - 2x)\cos y \end{pmatrix}.$$

Hence $H(1, \pi) = \begin{pmatrix} -2 & 0 \\ 0 & -1 \end{pmatrix}$, which is clearly negative definite. Thus the point in question is a local maximum of f.

Notice that the test given in 9.2.6 will fail to decide the nature of the critical point P if at P the matrix H is not positive definite, negative definite or indefinite: for example, H might equal 0 at P.

Extremal values of a quadratic form

Consider a quadratic form in variables $x_1, \ldots, x_n$

$$q = X^T A X,$$

where as usual A is a real symmetric $n \times n$ matrix and X is the column consisting of $x_1, \ldots, x_n$. Suppose that we want to find the maximum and minimum values of q when X is subject to a restriction. One possible restriction is that

$$||X|| = a$$

for some $a > 0$, that is, $x_1^2 + \cdots + x_n^2 = a^2$. Thus we are looking for the maximum and minimum values of q on the n-sphere with radius a and center the origin in $\mathbf{R}^n$. One could use calculus to attack this problem, but it is simpler to employ diagonalization.

There is a real orthogonal $n \times n$ matrix S such that $S^T A S = D$, where D is the diagonal matrix with the eigenvalues of A, say $d_1, \ldots, d_n$, on its diagonal. Put $Y = S^{-1}X$: thus we have $X = SY$ and

$$q = X^T A X = Y^T S^T A S Y = Y^T D Y = d_1 y_1^2 + \cdots + d_n y_n^2,$$

where $y_1, \ldots, y_n$ are the entries of Y.

In addition we find that

$$X^T X = Y^T (S^T S) Y = Y^T Y,$$

since $S^T = S^{-1}$. Therefore our problem may be reformulated as follows: find the maximum and minimum values of the expression $d_1 y_1^2 + \cdots + d_n y_n^2$ subject to $y_1^2 + \cdots + y_n^2 = a^2$. But this is easily answered. For assume that m and M are respectively the smallest and the largest eigenvalues of A. Then

$$q = d_1 y_1^2 + \cdots + d_n y_n^2 \leq M(y_1^2 + \cdots + y_n^2) = Ma^2,$$

and

$$q = d_1 y_1^2 + \cdots + d_n y_n^2 \geq m(y_1^2 + \cdots + y_n^2) = ma^2.$$

Suppose that the largest eigenvalue M occurs for k different y_i's; then we can take each of the corresponding y_i's to be equal to $a/\sqrt{k}$ and all other y_i's to be 0. Then $y_1^2 + \cdots + y_n^2 = a^2$ and the value of q at this point is exactly

$$Mk(a/\sqrt{k})^2 = Ma^2.$$

It follows that the largest value of q on the n-sphere really is Ma^2. By a similar argument the smallest value of q on the n-sphere is ma^2. We state this conclusion as:

Theorem 9.2.7
The minimum and maximum values of the quadratic form $q = X^T A X$ for $\|X\| = a > 0$ are respectively ma^2 and Ma^2 where m and M are the smallest and largest eigenvalues of the real symmetric matrix A.

We conclude with a geometrical example.

Example 9.2.6
The equation of an ellipsoid with center the origin is given as $X^T A X = c$, where A is a real symmetric 3×3 matrix and c is a positive constant. Find the radius of the largest sphere with center the origin which lies entirely within the ellipsoid.

By a rotation to principal axes we can write the equation of the ellipsoid in the form $dx'^2 + ey'^2 + fz'^2 = c$, where the eigenvalues d, e, f of A are positive. Hence the equation of the ellipsoid takes the standard form

$$\frac{x'^2}{c/d} + \frac{y'^2}{c/e} + \frac{z'^2}{c/f} = 1.$$

Clearly the sphere will lie entirely inside the ellipsoid provided that its radius a does not exceed the length of any of the semi-axes: thus a cannot be larger than any of

$$\sqrt{\frac{c}{d}}, \ \sqrt{\frac{c}{e}}, \ \sqrt{\frac{c}{f}}.$$

Therefore the condition on a is that $a \leq \sqrt{\frac{c}{M}}$, where M is the biggest of the eigenvalues d, e, f. Thus the largest sphere which is contained entirely within the ellipsoid has radius

$$\sqrt{\frac{c}{M}}.$$

Exercises 9.2

1. Determine if the following quadratic forms are positive definite, negative definite or indefinite:
    (a) $2x^2 - 2xy + 3y^2$;
    (b) $x^2 - 3xz - 2y^2 + z^2$;
    (c) $x^2 + y^2 + \frac{1}{2}xz + yz$.

2. Determine if the following matrix is positive definite, negative definite or indefinite:

$$\begin{pmatrix} 1 & 1 & 2 \\ 1 & 2 & 1 \\ 2 & 1 & 1 \end{pmatrix}.$$

3. A quadratic form $q = X^T A X$ is called *positive semidefinite* if $q \geq 0$ for all X. The definition of *negative semidefinite* is similar. Prove that q is positive semidefinite if and only if all the eigenvalues of A are ≥ 0, and negative semidefinite if and only if all the eigenvalues are ≤ 0.

4. Let A be a positive definite $n \times n$ matrix and let S be a real invertible $n \times n$ matrix. Prove that $S^T A S$ is also positive definite.

5. Let A be a real symmetric matrix. Prove that A is negative definite if and only if it has the form $-(B^T B)$ for some invertible matrix B.

6. Identify the following conics:
  (a) $14x^2 - 16xy + 5y^2 = 6$; (b) $2x^2 + 4xy + 2y^2 + x - 3y = 1$.

7. Identify the following quadrics:
  (a) $2x^2 + 2y^2 + 3z^2 + 4yz = 3$; (b) $2x^2 + 2y^2 + z^2 + 4xz = 4$.

8. Classify the critical points of the following functions as local maxima, local minima or saddle points:
  (a) $x^2 + 2xy + 2y^2 + 4x$;
  (b) $(x + y)^3 + (x - y)^3 - 12(3x + y)$;
  (c) $x^2 + y^2 + 3z^2 - xy + 2xz - z$.

9. Find the smallest and largest values of the quadratic form $q = 2x^2 + 2y^2 + 3z^2 + 4yz$ when the point (x, y, z) is required to lie on the sphere with radius 1 and center the origin.

10. Let $X^T A X = c$ be the equation of an ellipsoid with center the origin, where A is a real symmetric 3×3 matrix and c is a positive constant. Show that the radius of the smallest sphere with center the origin which contains the ellipsoid is $\sqrt{\frac{c}{m}}$, where m is the smallest eigenvalue of A.

11. Show that $5x^2 + 2xy + 2y^2 + 5z^2 = 1$ is the equation of an ellipsoid with center the origin. Then find the radius of the smallest and largest sphere with center the origin which contains, respectively is contained in, the ellipsoid.

9.3 Bilinear Forms

Roughly speaking, a bilinear form is a scalar-valued linear function of two vector variables. One type of a bilinear form which we have already met is an inner product on a real vector space. It will be seen that there is a close connection between bilinear forms and quadratic forms.

Let V be a vector space over a field of scalars F and write

$$V \times V$$

for the set of all pairs $(\mathbf{u}, \mathbf{v})$ of vectors from V. Then a *bilinear form* on V is a function

$$f : V \times V \to F,$$

that is, a rule assigning to each pair of vectors $(\mathbf{u}, \mathbf{v})$ a scalar $f(\mathbf{u}, \mathbf{v})$, which satisfies the following requirements:

(i) $f(\mathbf{u}_1 + \mathbf{u}_2, \mathbf{v}) = f(\mathbf{u}_1, \mathbf{v}) + f(\mathbf{u}_2, \mathbf{v})$;
(ii) $f(\mathbf{u}, \mathbf{v}_1 + \mathbf{v}_2) = f(\mathbf{u}, \mathbf{v}_1) + f(\mathbf{u}, \mathbf{v}_2)$;
(iii) $f(c\mathbf{u}, \mathbf{v}) = cf(\mathbf{u}, \mathbf{v})$;
(iv) $f(\mathbf{u}, c\mathbf{v}) = cf(\mathbf{u}, \mathbf{v})$.

These rules must hold for all vectors $\mathbf{u}, \mathbf{u}_1, \mathbf{u}_2, \mathbf{v}, \mathbf{v}_1, \mathbf{v}_2$ in V and all scalars c in F. The effect of the four defining properties is to make $f(\mathbf{u}, \mathbf{v})$ "linear" in both the variables $\mathbf{u}$ and $\mathbf{v}$.

As has been mentioned, an inner product $< \; >$ on a real vector space is a bilinear form f in which

$$f(\mathbf{u}, \mathbf{v}) = \; < \mathbf{u}, \mathbf{v} > .$$

Indeed the defining properties of the inner product guarantee this.

A very important example of a bilinear form arises whenever a square matrix is given.

Example 9.3.1

Let A be an $n \times n$ matrix over a field F. A function $f : F^n \times F^n \to F$ is defined by the rule

$$f(X, Y) = X^T A Y.$$

That f is a bilinear form on F^n follows from the usual rules of matrix algebra. The importance of this example stems from the fact that it is typical of bilinear forms on finite-dimensional vector spaces in a sense that will now be made precise.

Matrix representation of bilinear forms

Suppose that $f : V \times V \to F$ is a bilinear form on a vector space V of dimension n over a field F. Choose an ordered basis $\mathcal{B} = \{\mathbf{v}_1, \ldots, \mathbf{v}_n\}$ of V and define a_{ij} to be the scalar $f(\mathbf{v}_i, \mathbf{v}_j)$. Thus we can associate with f the $n \times n$ matrix

$$A = [a_{ij}].$$

Now let $\mathbf{u}$ and $\mathbf{v}$ be arbitrary vectors of V and write them in terms of the basis as $\mathbf{u} = \sum_{i=1}^{n} b_i \mathbf{v}_i$ and $v = \sum_{j=1}^{n} c_j \mathbf{v}_j$; then the coordinate vectors of $\mathbf{u}$ and $\mathbf{v}$ with respect to the given basis are

$$[\mathbf{u}]_{\mathcal{B}} = \begin{pmatrix} b_1 \\ \vdots \\ b_n \end{pmatrix} \quad \text{and} \quad [\mathbf{v}]_{\mathcal{B}} = \begin{pmatrix} c_1 \\ \vdots \\ c_n \end{pmatrix}.$$

The linearity properties of f can be used to compute $f(\mathbf{u}, \mathbf{v})$ in terms of the matrix A.

$$f(\mathbf{u}, \mathbf{v}) = f\left(\sum_{i=1}^{n} b_i \mathbf{v}_i, \sum_{j=1}^{n} c_j \mathbf{v}_j\right) = \sum_{i=1}^{n} b_i f\left(\mathbf{v}_i, \sum_{j=1}^{n} c_j \mathbf{v}_j\right)$$

$$= \sum_{i=1}^{n} \sum_{j=1}^{n} b_i f(\mathbf{v}_i, \mathbf{v}_j) c_j.$$

Since $f(\mathbf{v}_i, \mathbf{v}_j) = a_{ij}$, this becomes

$$f(\mathbf{u}, \mathbf{v}) = \sum_{i=1}^{n} \sum_{j=1}^{n} b_i a_{ij} c_j,$$

from which we obtain the fundamental equation

$$f(\mathbf{u}, \mathbf{v}) = ([\mathbf{u}]_{\mathcal{B}})^T A [\mathbf{v}]_{\mathcal{B}}.$$

Thus the bilinear form f is represented with respect to the basis $\mathcal{B}$ by the $n \times n$ matrix A whose (i, j) entry is $f(\mathbf{v}_i, \mathbf{v}_j)$. The values of f can be computed using the above rule. In particular, if f is a bilinear form on F^n and the standard basis of F^n is used, then $f(X, Y) = X^T A Y$.

Conversely, if we start with a matrix A and define f by means of the equation $f(\mathbf{u}, \mathbf{v}) = ([\mathbf{u}]_{\mathcal{B}})^T A [\mathbf{v}]_{\mathcal{B}}$, then it is easy to verify that f is a bilinear form on V and that the matrix representing f with respect to the basis $\mathcal{B}$ is A.

Now suppose we decide to use another ordered basis $\mathcal{B}'$: what will be the effect on the matrix A? Let S be the invertible matrix which describes the change of basis $\mathcal{B}' \rightarrow \mathcal{B}$. Thus $[\mathbf{u}]_{\mathcal{B}} = S[\mathbf{u}]_{\mathcal{B}'}$, according to 6.2.4. Therefore

$$f(\mathbf{u}, \mathbf{v}) = (S[\mathbf{u}]_{\mathcal{B}'})^T A (S[\mathbf{v}]_{\mathcal{B}'}) = ([\mathbf{u}]_{\mathcal{B}'})^T (S^T A S)[\mathbf{v}]_{\mathcal{B}'},$$

which shows that the matrix $S^T A S$ represents f with respect to the basis $\mathcal{B}'$.

At this point we recognize that a new relation between matrices has arisen: a matrix B is said to be *congruent* to a matrix A if there is an invertible matrix S such that

$$B = S^T A S.$$

While there is an analogy between congruence and similarity of matrices, in general similar matrices need not be congruent, nor congruent matrices similar.

The point that has emerged from the preceding discussion is that matrices which represent the same bilinear form with respect to different bases of the vector space are congruent. This result is to be compared with the fact that the matrices representing the same linear transformation are similar.

The conclusions of the the last few paragraphs are summarized in the following basic theorem.

Theorem 9.3.1

(i) *Let f be a bilinear form on an n-dimensional vector space V over a field F and let $\mathcal{B} = \{\mathbf{v}_1, \ldots, \mathbf{v}_n\}$ be an ordered basis of V. Define A to be the $n \times n$ matrix whose (i,j) entry is $f(\mathbf{v}_i, \mathbf{v}_j)$; then*

$$f(\mathbf{u}, \mathbf{v}) = ([\mathbf{u}]_\mathcal{B})^T A [\mathbf{v}]_\mathcal{B},$$

and A is the $n \times n$ matrix representing f with respect to $\mathcal{B}$.

(ii) *If $\mathcal{B}'$ is another ordered basis of V, then f is represented with respect to $\mathcal{B}'$ by the matrix $S^T A S$ where S is the invertible matrix describing the basis change $\mathcal{B}' \to \mathcal{B}$.*

(iii) *Conversely, if A is any $n \times n$ matrix over F, a bilinear form on V is defined by the rule $f(\mathbf{u}, \mathbf{v}) = ([\mathbf{u}]_\mathcal{B})^T A [\mathbf{v}]_\mathcal{B}$. It is represented by the matrix A with respect to the basis $\mathcal{B}$.*

Symmetric and skew-symmetric bilinear forms

A bilinear form f on a vector space V is called *symmetric* if its values are unchanged by reversing the arguments, that is, if

$$f(\mathbf{u}, \mathbf{v}) = f(\mathbf{v}, \mathbf{u})$$

for all vectors $\mathbf{u}$ and $\mathbf{v}$. Similarly, f is said to be *skew-symmetric* if

$$f(\mathbf{u}, \mathbf{v}) = -f(\mathbf{v}, \mathbf{u})$$

is always valid. Notice the consequence, $f(\mathbf{u}, \mathbf{u}) = \mathbf{0}$ for all vectors $\mathbf{u}$. For example, any real inner product is a symmetric bilinear form; on the other hand, the form defined by the rule

$$f(\begin{pmatrix} x_1 \\ x_2 \end{pmatrix}, \begin{pmatrix} y_1 \\ y_2 \end{pmatrix}) = x_1 y_2 - x_2 y_1$$

is an example of a skew-symmetric bilinear form on $\mathbf{R}^2$. As the reader may suspect, there are connections with symmetric and skew-symmetric matrices.

Theorem 9.3.2
Let f be a bilinear form on a finite-dimensional vector space V and let A be a matrix representing f with respect to some basis of V. Then f is symmetric if and only if A is symmetric and f is skew-symmetric if and only if A is skew-symmetric.

Proof
Let A be symmetric. Then, remembering that $[\mathbf{u}]^T A[\mathbf{v}]$ is scalar, we have

$$f(\mathbf{u}, \mathbf{v}) = [\mathbf{u}]^T A[\mathbf{v}] = ([\mathbf{u}]^T A[\mathbf{v}])^T = [\mathbf{v}]^T A^T[\mathbf{u}] = [\mathbf{v}]^T A[\mathbf{u}]$$
$$= f(\mathbf{v}, \mathbf{u}).$$

Therefore f is symmetric. Conversely, suppose that f is symmetric, and let the ordered basis in question be $\{\mathbf{v}_1, \ldots, \mathbf{v}_n\}$. Then $a_{ij} = f(\mathbf{v}_i, \mathbf{v}_j) = f(\mathbf{v}_j, \mathbf{v}_i) = a_{ji}$, so that A is symmetric.

The proof of the skew-symmetric case is similar and is left as an exercise.

Symmetric bilinear forms and quadratic forms

Let f be a bilinear form on $\mathbf{R}^n$ given by $f(X, Y) = X^T A Y$. Then f determines a quadratic form q where

$$q = f(X, X) = X^T A X.$$

Conversely, if q is a quadratic form in $x_1, \ldots, x_n$, we can define a corresponding symmetric bilinear form f on $\mathbf{R}^n$ by means of the rule

$$f(X, Y) = \frac{1}{2}\{q(X + Y) - q(X) - q(Y)\}$$

where X and Y are the column vectors consisting of $x_1, \ldots, x_n$ and $y_1, \ldots, y_n$. To see that f is bilinear, first write $q(X) = X^T A X$ with A symmetric; then we have

$$
\begin{aligned}
f(X, Y) &= \tfrac{1}{2}\{(X + Y)^T A(X + Y) - X^T A X - Y^T A Y\} \\
&= \tfrac{1}{2}(X^T A Y + Y^T A X) \\
&= X^T A Y,
\end{aligned}
$$

since $X^T A Y = (X^T A Y)^T = Y^T A X$. This shows that f is bilinear.

It is readily seen that the correspondence $q \to f$ just described is a bijection from quadratic forms to symmetric bilinear forms on $\mathbf{R}^n$.

Theorem 9.3.3
There is a bijection from the set of quadratic forms in n variables to the set of symmetric bilinear forms on $\mathbf{R}^n$.

From past experience we would expect to get significant information about symmetric bilinear forms by using the Spectral Theorem. In fact what is obtained is a canonical or standard form for such bilinear forms.

Theorem 9.3.4

Let f be a symmetric bilinear form on an n-dimensional real vector space V. Then there is a basis $\mathcal{B}$ of V such that

$$f(\mathbf{u}, \mathbf{v}) = u_1 v_1 + \cdots + u_k v_k - u_{k+1} v_{k+1} - \cdots - u_l v_l$$

where $u_1, \ldots, u_n$ and $v_1, \ldots, v_n$ are the entries of the coordinate vectors $[\mathbf{u}]_\mathcal{B}$ and $[\mathbf{v}]_\mathcal{B}$ respectively and k and l are integers satisfying $0 \le k \le l \le n$.

Proof

Let f be represented by a matrix A with respect some basis $\mathcal{B}'$ of V. Then A is symmetric. Hence there is an orthogonal matrix S such that $S^T A S = D$ is diagonal, say with diagonal entries $d_1, \ldots, d_n$; of course these are the eigenvalues of A. Here we can assume that $d_1, \ldots, d_k > 0$, while $d_{k+1}, \ldots, d_l < 0$ and $d_{l+1} = \cdots = d_n = 0$, by reordering the basis if necessary. Let E be the $n \times n$ diagonal matrix whose diagonal entries are the real numbers

$$1/\sqrt{d_1}, \ \ldots, \ 1/\sqrt{d_k}, \ 1/\sqrt{-d_{k+1}}, \ \ldots, 1/\sqrt{-d_l}, \ 1, \ldots, 1.$$

Then

$$(SE)^T A(SE) = E^T(S^T AS)E = EDE,$$

and the final product is the matrix

$$B = \begin{pmatrix} I_k & | & 0 & | & 0 \\ -- & | & -- & | & -- \\ 0 & | & -I_{l-k} & | & 0 \\ -- & | & -- & | & -- \\ 0 & | & 0 & | & 0 \end{pmatrix}.$$

Now the matrix SE is invertible, so its inverse determines a change of basis from $\mathcal{B}'$ to say $\mathcal{B}$. Then f will be represented by the matrix B with respect to the basis $\mathcal{B}$. Finally, $f(\mathbf{u}, \mathbf{v}) =$

$([\mathbf{u}]_\mathcal{B})^T B[\mathbf{v}]_\mathcal{B}$, so the result follows on multiplying the matrices together.

Example 9.3.2
Find the canonical form of the symmetric bilinear form on $\mathbf{R}^2$ defined by $f(X, Y) = x_1 y_1 + 2x_1 y_2 + 2x_2 y_1 + x_2 y_2$.

The matrix of the bilinear form with respect to the standard basis is

$$A = \begin{pmatrix} 1 & 2 \\ 2 & 1 \end{pmatrix},$$

which, by Example 9.1.1, has eigenvalues 3 and -1, and is diagonalized by the matrix

$$S = \frac{1}{\sqrt{2}} \begin{pmatrix} 1 & -1 \\ 1 & 1 \end{pmatrix};$$

then

$$f(X, Y) = X^T A Y = (X')^T S^T A S \, Y' = (X')^T \begin{pmatrix} 3 & 0 \\ 0 & -1 \end{pmatrix} Y',$$

so that

$$f(X, Y) = 3x_1' \, y_1' - x_2' y_2'.$$

Here $x_1' = \frac{1}{\sqrt{2}}(x_1 + x_2)$ and $x_2' = \frac{1}{\sqrt{2}}(-x_1 + x_2)$, with corresponding formulas in y.

To obtain the canonical form of f, put $x_1'' = \sqrt{3}x_1'$, $y_1'' = \sqrt{3}y_1'$, and $x_2'' = x_2'$, $y_2'' = y_2'$. Then

$$f(X, Y) = x_1'' y_1'' - x_2'' y_2'',$$

which is the canonical form specified in 9.3.4.

Eigenvalues of congruent matrices

Since congruent matrices represent the same symmetric bilinear form, it is natural to expect that such matrices should have some common properties, as similar matrices do. However, whereas similar matrices have the same eigenvalues, this is not true of congruent matrices. For example, the matrix

$$\begin{pmatrix} 2 & 0 \\ 0 & -3 \end{pmatrix}$$

has eigenvalues 2 and -3, but the congruent matrix

$$\begin{pmatrix} 1 & 0 \\ 1 & 1 \end{pmatrix} \begin{pmatrix} 2 & 0 \\ 0 & -3 \end{pmatrix} \begin{pmatrix} 1 & 1 \\ 0 & 1 \end{pmatrix} = \begin{pmatrix} 2 & 2 \\ 2 & -1 \end{pmatrix}$$

has eigenvalues -2 and 3.

Notice that, although the eigenvalues of these congruent matrices are different, the numbers of positive and negative eigenvalues are the same for each matrix. This is an instance of a general result.

Theorem 9.3.5 (*Sylvester's Law of Inertia*)
Let A be a real symmetric $n \times n$ matrix and S an invertible $n \times n$ matrix. Then A and $S^T A S$ have the same numbers of positive, negative and zero eigenvalues.

Proof
Assume first of all that A is invertible; this is the essential case. Recall that by 7.3.6 it is possible to write S in the form QR where Q is real orthogonal and R is real upper triangular with positive diagonal entries; this was a consequence of the Gram-Schmidt process.

The idea of the proof is to obtain a continuous chain of matrices leading from S to the orthogonal matrix Q; the point of this is that $Q^T A Q = Q^{-1} A Q$ certainly has the same eigenvalues as A. Define

$$S(t) = tQ + (1-t)S,$$

where $0 \le t \le 1$. Thus $S(0) = S$ while $S(1) = Q$. Now write $U = tI + (1-t)R$, so that $S(t) = QU$. Next U is an upper triangular matrix and its diagonal entries are $t + (1 - t)r_{ii}$; these cannot be zero since $r_{ii} > 0$ and $0 \leq t \leq 1$. Hence U is invertible, while Q is certainly invertible since it is orthogonal. It follows that $S(t) = QU$ is invertible; thus $\det(S(t)) \neq 0$.

Now consider $A(t) = S(t)^T A S(t)$; since

$$\det(A(t)) = \det(A)\det(S(t))^2 \neq 0,$$

it follows that $A(t)$ cannot have zero eigenvalues. Now as t goes from 0 to 1, the eigenvalues of $A(0) = S^T A S$ gradually change to those of $A(1) = Q^T AQ$, that is, to those of A. But in the process no eigenvalue can change sign because the eigenvalues that appear are continuous functions of t and they are never zero. Consequently the numbers of positive and negative eigenvalues of $S^T A S$ are equal to those of A.

Finally, what if A is singular? In this situation the trick is to consider the matrix $A + \epsilon I$, which may be thought of as a "perturbation" of A. Now $A + \epsilon I$ will be invertible provided that ϵ is sufficiently small and positive: for $\det(A + xI)$ is a polynomial of degree n in x, so it vanishes for at most n values of x. The previous argument shows that the result is true for $A + \epsilon I$ if ϵ is small and positive; then by taking the limit as $\epsilon \to 0$, we can deduce the result for A.

It follows from this theorem that the numbers of positive and negative signs that appear in the canonical form of 9.3.4 are uniquely determined by the bilinear form and do not depend on the particular basis chosen.

Example 9.3.3

Show that the matrices $\begin{pmatrix} 2 & 1 \\ 1 & 2 \end{pmatrix}$ and $\begin{pmatrix} 1 & 2 \\ 2 & 1 \end{pmatrix}$ are not congruent.

All one need do here is note that the first matrix has eigenvalues $1, 3$, while the second has eigenvalues $3, -1$. Hence by 9.3.5 they cannot be congruent.

Skew-symmetric bilinear forms

Having seen that there is a canonical form for symmetric bilinear forms on real vector spaces, we are led to enquire if something similar can be done for skew-symmetric bilinear forms. By 9.3.2 this is equivalent to trying to describe all skew-symmetric matrices up to congruence. The theorem that follows provides a solution to this problem.

Theorem 9.3.6

Let f be a skew-symmetric bilinear form on an n-dimensional vector space V over either $\mathbf{R}$ or $\mathbf{C}$. Then there is an ordered basis of V with the form $\{\mathbf{u}_1, \mathbf{v}_1, \ldots, \mathbf{u}_k, \mathbf{v}_k, \mathbf{w}_1, \ldots, \mathbf{w}_{n-2k}\}$, where $0 \leq 2k \leq n$, such that

$$f(\mathbf{u}_i, \mathbf{v}_i) = 1 = -f(\mathbf{v}_i, \mathbf{u}_i), \ \ i = 1, \ldots, k$$

and f vanishes on all other pairs of basis elements.

Let us examine the consequence of this theorem before setting out to prove it. If we use the basis provided by the theorem, the bilinear form f is represented by the matrix

$$\left(\begin{array}{ccccc|ccc} 0 & 1 & \cdots & 0 & 0 & 0 & \cdots & 0 \\ -1 & 0 & \cdots & 0 & 0 & 0 & \cdots & 0 \\ \cdot & \cdot & \cdots & \cdot & \cdot & \cdot & \cdots & \cdot \\ 0 & 0 & \cdots & 0 & 1 & 0 & \cdots & 0 \\ 0 & 0 & \cdots & -1 & 0 & 0 & \cdots & 0 \\ \hline 0 & 0 & \cdots & 0 & 0 & 0 & \cdots & 0 \\ \cdot & \cdot & \cdots & \cdot & \cdot & \cdot & \cdots & \cdot \\ 0 & 0 & \cdots & 0 & 0 & 0 & \cdots & 0 \end{array} \right)$$

where the number of blocks of the type $\begin{pmatrix} 0 & 1 \\ -1 & 0 \end{pmatrix}$ is k. This allows us to draw an important conclusion about skew-symmetric matrices.

Corollary 9.3.7
A skew-symmetric $n \times n$ matrix A over $\mathbf{R}$ or $\mathbf{C}$ is congruent to a matrix M of the above form.

This is because the bilinear form f given by $f(X, Y) = X^T A Y$ is skew-symmetric and hence is represented with respect to a suitable basis by a matrix of type M; thus A must be congruent to M.

Proof of 9.3.6
Let $\mathbf{z}_1, \ldots, \mathbf{z}_n$ be any basis of V. If $f(\mathbf{z}_i, \mathbf{z}_j) = 0$ for all i and j, then $f(\mathbf{u}, \mathbf{v}) = 0$ for all vectors $\mathbf{u}$ and $\mathbf{v}$, so that f is the zero bilinear form and it is represented by the zero matrix. This is the case $k = 0$. So assume that $f(\mathbf{z}_i, \mathbf{z}_j)$ is not zero for some i and j. Since the basis can be reordered, we may suppose that $f(\mathbf{z}_1, \mathbf{z}_2) = a \neq 0$. Then $f(a^{-1}\mathbf{z}_1, \mathbf{z}_2) = a^{-1} f(\mathbf{z}_1, \mathbf{z}_2) = 1$. Now replace $\mathbf{z}_1$ by $a^{-1}\mathbf{z}_1$; the effect is to make $f(\mathbf{z}_1, \mathbf{z}_2) = 1$, and of course $f(\mathbf{z}_2, \mathbf{z}_1) = -1$ since f is skew-symmetric.
Next put $b_i = f(\mathbf{z}_1, \mathbf{z}_i)$ where $i > 2$. Then

$$f(\mathbf{z}_1, \mathbf{z}_i - b\mathbf{z}_2) = f(\mathbf{z}_1, \mathbf{z}_i) - bf(\mathbf{z}_1, \mathbf{z}_2) = b - b = 0.$$

This suggests that we modify the basis further by replacing $\mathbf{z}_i$ by $\mathbf{z}_i - b\mathbf{z}_2$ for $i > 2$; notice that this does not disturb linear independence, so we still have a basis of V. The effect of this substitution is make

$$f(\mathbf{z}_1, \mathbf{z}_i) = 0 \text{ for } i = 3, \ldots, n.$$

Next we have to address the possibility that $f(\mathbf{z}_2, \mathbf{z}_i)$ may be non-zero when $i > 2$; let $c = f(\mathbf{z}_2, \mathbf{z}_i)$. Then

$$f(\mathbf{z}_2, \mathbf{z}_i + c\mathbf{z}_1) = f(\mathbf{z}_2, \mathbf{z}_i) + cf(\mathbf{z}_2, \mathbf{z}_1) = c + c(-1) = 0.$$

This suggests that the next step should be to replace $\mathbf{z}_i$ by $\mathbf{z}_i + c\mathbf{z}_1$ where $i > 2$; again we need to observe that $\mathbf{z}_1, \ldots, \mathbf{z}_n$

will still be a basis of V. Also important is the remark that this substitution will not nullify what has already been achieved; the reason is that when $i > 2$

$$f(\mathbf{z}_1, \mathbf{z}_i + c\mathbf{z}_1) = f(\mathbf{z}_1, \mathbf{z}_i) + cf(\mathbf{z}_1, \mathbf{z}_1) = 0.$$

We have now reached the point where

$$f(\mathbf{z}_1, \mathbf{z}_2) = 1 = -f(\mathbf{z}_2, \mathbf{z}_1) \text{ and } f(\mathbf{z}_1, \mathbf{z}_i) = 0 = f(\mathbf{z}_2, \mathbf{z}_i),$$

for all $i > 2$. Now we rename our first two basis elements, writing $\mathbf{u}_1 = \mathbf{z}_1$ and $\mathbf{v}_1 = \mathbf{z}_2$.

So far the matrix representing f has the form

$$\begin{pmatrix} 0 & 1 & | & 0 & 0 \\ -1 & 0 & | & 0 & 0 \\ -- & -- & | & -- & -- \\ 0 & 0 & | & B & \end{pmatrix}$$

where B is a skew-symmetric matrix with $n - 2$ rows and columns. We can now repeat the argument just given for the subspace with basis $\{\mathbf{z}_3, \ldots, \mathbf{z}_n\}$; it follows by induction on n that there is a basis for this subspace with respect to which f is represented by a matrix of the required form. Indeed let $\mathbf{u}_2, \ldots, \mathbf{u}_k, \mathbf{v}_2, \ldots, \mathbf{v}_k, \mathbf{w}_1, \ldots, \mathbf{w}_{n-2k}$ be this basis. By adjoining $\mathbf{u}_1$ and $\mathbf{v}_1$, we obtain a basis of V with respect to which f is represented by a matrix of the required form.

Example 9.3.4

Find the canonical form of the skew-symmetric matrix

$$A = \begin{pmatrix} 0 & 0 & 2 \\ 0 & 0 & -1 \\ -2 & 1 & 0 \end{pmatrix}.$$

We need to carry out the procedure indicated in the proof of the theorem. Let $\{E_1, E_2, E_3\}$ be the standard basis of $\mathbf{R}^3$.

The matrix A determines a skew-symmetric bilinear form f with the properties $f(E_1, E_3) = 2 = -f(E_3, E_1)$, $f(E_3, E_2) = 1 = -f(E_2, E_3)$, $f(E_1, E_2) = 0 = f(E_2, E_1)$.

The first step is to reorder the basis as $\{E_1, E_3, E_2\}$; this is necessary since $f(E_1, E_2) = 0$ whereas $f(E_1, E_3) \neq 0$. Now replace $\{E_1, E_3, E_2\}$ by $\{\frac{1}{2}E_1, E_3, E_2\}$, noting that $f(\frac{1}{2}E_1, E_3) = 1 = -f(E_3, \frac{1}{2}E_1)$. Next $f(E_3, E_2) = 1$, so we replace E_2 by

$$E_2 + f(E_3, E_2)\frac{1}{2}E_1 = \frac{1}{2}E_1 + E_2.$$

Note that $f(\frac{1}{2}E_1, \frac{1}{2}E_1 + E_2) = 0 = f(E_3, \frac{1}{2}E_1 + E_2)$.

The procedure is now complete. The bilinear form is represented with respect to the new ordered basis

$$\{\frac{1}{2}E_1,\ E_3,\ \frac{1}{2}E_1 + E_2\}$$

by the matrix

$$M = \begin{pmatrix} 0 & 1 & 0 \\ -1 & 0 & 0 \\ 0 & 0 & 0 \end{pmatrix},$$

which is in canonical form. The change of basis from $\{\frac{1}{2}E_1, E_3, \frac{1}{2}E_1 + E_2\}$ to the standard ordered basis is represented by the matrix

$$S = \begin{pmatrix} 1/2 & 0 & 1/2 \\ 0 & 0 & 1 \\ 0 & 1 & 0 \end{pmatrix}.$$

The reader should now verify that $S^T A S$ equals M, the canonical form of A, as predicted by the proof of 9.3.6.

Exercises 9.3

1. Which of the following functions f are bilinear forms?
   (a) $f(X, Y) = X - Y$ on $\mathbf{R}^n$;
   (b) $f(X, Y) = X^T Y$ on $\mathbf{R}^n$;
   (c) $f(g, h) = \int_a^b g(x)h(x)\, dx$ on $C[a, b]$.

2. Let f be the bilinear form on $\mathbf{R}^2$ which is defined by the equation $f(X, Y) = 2x_1 y_2 - 3x_2 y_1$. Write down the matrices which represent f with respect to (a) the standard basis, and (b) the basis $\{\begin{pmatrix} 0 \\ 1 \end{pmatrix} \begin{pmatrix} 1 \\ 1 \end{pmatrix}\}$.

3. If f and g are two bilinear forms on a vector space V, define their sum $f + g$ by the rule $f + g(u, v) = f(u, v) + g(u, v)$; also define the scalar multiple cf by the equation $cf(u, v) = c(f(u, v))$. Prove that with these operations the set of all bilinear forms on V becomes a vector space V'. If V has dimension n, what is the dimension of V'?

4. Prove that every bilinear form on a real or complex vector space is the sum of a symmetric and a skew-symmetric bilinear form.

5. Find the canonical form of the symmetric bilinear form on $\mathbf{R}^2$ given by $f(X, Y) = 3x_1 y_1 + x_1 y_2 + x_2 y_1 + 3x_2 y_2$.

6. Let f be a bilinear form on $\mathbf{R}^n$. Prove that f is an inner product on $\mathbf{R}^n$ if and only if f is symmetric and the corresponding quadratic form is positive definite.

7. Test each of the following bilinear forms to see if it is an inner product:
   (a) $f(X, Y) = 3x_1 y_1 + x_1 y_2 + x_2 y_1 + 5x_2 y_2$;
   (b) $f(X, Y) = 2x_1 y_1 + x_1 y_2 + x_1 y_3 + x_2 y_1 + 3x_2 y_2 - 2x_2 y_3 + x_3 y_1 - 2x_3 y_2 + 3x_3 y_3$.

8. Find the canonical form of the skew-symmetric matrix

$$A = \begin{pmatrix} 0 & 2 & -1 \\ -2 & 0 & 1 \\ 1 & -1 & 0 \end{pmatrix}$$

and also find an invertible matrix S such that $S^T A S$ equals the canonical form.

9. (a) If A is a square matrix and S is an invertible matrix, prove that A and $S^T A S$ have the same rank.

(b) Deduce that the rank of a skew-symmetric matrix equals twice the number of 2×2 blocks in the canonical form of the matrix. Conclude that the canonical form is unique.

10. Call a skew-symmetric bilinear form f on a vector space V *non-isotropic* if for every non-zero vector $\mathbf{v}$ there is another vector $\mathbf{w}$ in V such that $f(\mathbf{v}, \mathbf{w}) \neq 0$. Prove that a finite-dimensional real or complex vector space which has a non-isotropic skew-symmetric bilinear form must have even dimension.

9.4 Minimum Polynomials and Jordan Normal Form

The aim of this section is to introduce the reader to one of the most famous results in linear algebra, the existence of what is known as Jordan normal form of a matrix. This is a canonical form which applies to any square complex matrix. The existence of Jordan normal form is often presented as the climax of a series of difficult theorems; however the simplified approach adopted here depends on only elementary facts about vector spaces. We begin by introducing the important concept of the minimum polynomial of a linear operator or matrix.

The minimum polynomial

Let T be a linear operator on an n-dimensional vector space V over some field of scalars F. We show that T must satisfy some polynomial equation with coefficients in F. At this point the reader needs to keep in mind the definitions of sum, scalar multiple and product for linear operators introduced in 6.3. For any vector $\mathbf{v}$ of V, the set $\{\mathbf{v}, T(\mathbf{v}), \ldots, T^n(\mathbf{v})\}$ contains $n + 1$ vectors and so it must be linearly dependent by 5.1.1. Consequently there are scalars $a_0, a_1, \ldots, a_n$, not all of them zero, such that

$$a_0\mathbf{v} + a_1 T(\mathbf{v}) + \cdots + a_n T^n(\mathbf{v}) = \mathbf{0}.$$

Let us write $f_\mathbf{v}$ for the polynomial $a_0 + a_1 x + \cdots + a_n x^n$. Then

$$f_\mathbf{v}(T) = a_0 1 + a_1 T + \cdots + a_n T^n,$$

where 1 denotes the identity linear operator. Therefore

$$f_\mathbf{v}(T)(\mathbf{v}) = a_0\mathbf{v} + a_1 T(\mathbf{v}) + \cdots + a_n T^n(\mathbf{v}) = \mathbf{0}.$$

Now let $\{\mathbf{v}_1, \ldots, \mathbf{v}_n\}$ be a basis of the vector space V and define f to be the product of the polynomials $f_{\mathbf{v}_1}, f_{\mathbf{v}_2}, \ldots, f_{\mathbf{v}_n}$. Then

$$f(T)(\mathbf{v}_i) = f_{\mathbf{v}_1}(T) \cdots f_{\mathbf{v}_n}(T)(\mathbf{v}_i) = \mathbf{0}$$

for each $i = 1, \ldots, n$. This is because $f_{\mathbf{v}_i}(T)(\mathbf{v}_i) = \mathbf{0}$ and the $f_{\mathbf{v}_j}(T)$ commute, since powers of T commute by Exercise 6.3.13. Therefore $f(T)$ is the zero linear transformation on V, that is,

$$f(T) = 0.$$

Here of course f is a polynomial with coefficients in F.

Having seen that T satisfies a polynomial equation, we can select a polynomial f in x over F of *smallest degree* such that $f(T) = 0$. In addition, we may suppose that f is *monic*, that is, the highest power of x in f has its coefficient equal to 1. This polynomial f is called a *minimum polynomial* of T.

Suppose next that g is an arbitrary polynomial with coefficients in F. Using long division, just as in elementary algebra, we can divide g by f to obtain a quotient q and a remainder r; both of these will be polynomials in x over F. Thus $g = fq + r$, and either $r = 0$ or the degree of r is less than that of f. Then we have

$$g(T) = f(T)q(T) + r(T) = r(T)$$

since $f(T) = 0$. Therefore $g(T) = 0$ if and only if $r(T) = 0$. But, remembering that f was chosen to be of smallest degree subject to $f(T) = 0$, we can conclude that $r(T) = 0$ if and only if $r = 0$, that is, g is divisible by f. Thus the polynomials that vanish at T are precisely those that are divisible by the polynomial f.

If g is another monic polynomial of the same degree as f such that $g(T) = 0$, then in fact g must equal f. For g is divisible by f and has the same degree as f, which can only mean that g is a constant multiple of f. However g is monic, so it actually equals f. Therefore the minimum polynomial of T is the unique monic polynomial f of smallest degree such that $f(T) = 0$.

These conclusions are summed up in the following result.

Theorem 9.4.1

Let T be a linear operator on a finite-dimensional vector space over a field F with a minimum polynomial f. Then the only polynomials g with coefficients in F such that $g(T) = 0$ are the multiples of f. Hence f is the unique monic polynomial of smallest degree such that $f(T) = 0$ and T has a unique minimum polynomial.

So far we have introduced the minimum polynomial of a linear operator, but it is to be expected that there will be a corresponding concept for matrices. The *minimum poly-nomial of a square matrix A* over a field F is defined to be the monic polynomial f with coefficients in F of least degree such that $f(A) = 0$. The existence of f is assured by 9.4.1 and the relationship between linear operators and matrices. Clearly the minimum polynomial of a linear operator equals the minimum polynomial of any representing matrix. There is of course an exact analog of 9.4.1 for matrices.

Example 9.4.1

What is the minimum polynomial of the following matrix?

$$A = \begin{pmatrix} 2 & 1 & 1 \\ 0 & 2 & 0 \\ 0 & 0 & 2 \end{pmatrix}.$$

In the first place we can see directly that $(A - 2I_3)^2 = 0$. Therefore the minimum polynomial f must divide the poly-nomial $(x - 2)^2$, and there are two possibilities, $f = x - 2$ and $f = (x - 2)^2$. However f cannot equal $x - 2$ since $A - 2I \neq 0$. Hence the minimum polynomial of A is $f = (x - 2)^2$.

Example 9.4.2

What is the minimum polynomial of a diagonal matrix D?

Let $d_1, \ldots, d_r$ be the *distinct* diagonal entries of D. Again there is a fairly obvious polynomial equation that is satisfied by the matrix, namely

$$(A - d_1 I) \cdots (A - d_r I) = 0.$$

So the minimum polynomial divides $(x - d_1) \cdots (x - d_r)$ and hence is the product of certain of the factors $x - d_i$. However, we cannot miss out even one of these factors; for the product of all the $A - d_j I$ for $j \neq i$ is not zero since $d_j \neq d_i$. It follows that the minimum polynomial of D is the product of all the factors, that is, $(x - d_1) \cdots (x - d_r)$.

In the computation of minimum polynomials the next result is very useful.

Lemma 9.4.2
Similar matrices have the same minimum polynomial.

The quickest way to see this is to recall that similar matrices represent the same linear operator, and hence their minimum polynomials equal the minimum polynomial of the linear operator. Thus, by combining Lemma 9.4.2 and Example 9.4.2, we can find the minimum polynomial of any diagonalizable complex matrix.

Example 9.4.3
Find the minimum polynomial of the the matrix $\begin{pmatrix} 1 & 2 \\ 2 & 1 \end{pmatrix}$.

By Example 9.1.1 the matrix is similar to $\begin{pmatrix} 3 & 0 \\ 0 & -1 \end{pmatrix}$. Hence the minimum polynomial of the given matrix is $(x-3)(x+1)$.

In Chapter Eight we encountered another polynomial associated with a matrix or linear operator, namely the characteristic polynomial. It is natural to ask if there is a connection between these two polynomials. The answer is provided by a famous theorem.

Theorem 9.4.3 (*The Cayley-Hamilton Theorem*)
Let A be an $n \times n$ matrix over $\mathbf{C}$. If p is the characteristic polynomial of A, then $p(A) = 0$. Hence the minimum polynomial of A divides the characteristic polynomial of A.

Proof
According to 8.1.8, the matrix A is similar to an upper triangular matrix T; thus we have $S^{-1}AS = T$ with S invertible. By 9.4.2 the matrices A and T have the same minimum polynomial, and we know from 8.1.4 that they have the same

characteristic polynomial. Therefore it is sufficient to prove
the statement for the triangular matrix T. From Example 8.1.2
we know that the characteristic polynomial of T is

$$(t_{11} - x) \cdots (t_{nn} - x).$$

On the other hand, direct matrix multiplication shows that
$(t_{11}I - T) \cdots (t_{nn}I - T) = 0$: the reader may find it helpful to
check this statement for $n = 2$ and 3. The result now follows
from 9.4.1.

At this juncture the reader may wonder if the minimum
polynomial is really of much interest, given that it is a divisor
of the more easily calculated characteristic polynomial. But in
fact there are features of a matrix that are easily recognized
from its minimum polynomial, but which are unobtainable
from the characteristic polynomial. One such feature is diagonalizability.

Example 9.4.4

Consider for example the matrices I_2 and $\begin{pmatrix} 1 & 1 \\ 0 & 1 \end{pmatrix}$: both of
these have characteristic polynomial $(x - 1)^2$, but the first
matrix is diagonalizable while the second is not. Thus the
characteristic polynomial alone cannot tell us if a matrix is
diagonalizable. On the other hand, the two matrices just considered have different minimum polynomials, $x - 1$ and $(x - 1)^2$
respectively.

This example raises the possibility that it is the minimum polynomial which determines if a matrix is diagonalizable. The next theorem confirms this.

Theorem 9.4.4

Let A be an $n \times n$ matrix over **C**. *Then A is diagonalizable if
and only if its minimum polynomial splits into a product of n
distinct linear factors.*

Proof

Assume first that A is diagonalizable, so that $S^{-1}AS = D$, a diagonal matrix, for some invertible S. Then A and D have the same minimum polynomials by 9.4.2. Let $d_1, \ldots, d_r$ be the *distinct* diagonal entries of D; then Example 9.4.2 shows that the minimum polynomial of D is $(x - d_1)\cdots(x - d_r)$, which is a product of distinct linear factors.

Conversely, suppose that A has minimum polynomial

$$f = (x - d_1)\cdots(x - d_r)$$

where $d_1, \ldots, d_r$ are distinct complex numbers. Define g_i to be the polynomial obtained from f by deleting the factor $x - d_i$. Thus

$$g_i = \frac{f}{x - d_i}.$$

Next we recall the method of *partial fractions*, which is useful in calculus for integrating rational functions. This tells us that there are constants $b_1, \ldots, b_r$ such that

$$\frac{1}{f} = \sum_{i=1}^{r} \frac{b_i}{x - d_i}.$$

Multiplying both sides of this equation by f, we obtain

$$1 = b_1 g_1 + \cdots + b_r g_r$$

by definition of g_i.

At this point we prefer to work with linear operators, so we introduce the linear operator T on $\mathbf{C}^n$ defined by $T(X) = AX$. It follows from the above equation that $b_1 g_1(T) + \cdots + b_r g_r(T)$ is the identity function. Hence

$$X = b_1 g_1(T)(X) + \cdots + b_r g_r(T)(X)$$

for any vector X. Let V_i denote the set of all elements of the form $g_i(T)X$ with X a vector in $\mathbf{C}^n$. Then V_i is a subspace and the above equation for X tells us that

$$\mathbf{C}^n = V_1 + \cdots + V_r.$$

Now in fact $\mathbf{C}^n$ is the *direct* sum of the subspaces V_i, which amounts to saying that the intersection of a V_i and the sum of the remaining V_j, with $j \neq i$, is zero. To see why this is true, take a vector X in the intersection. Observe that $g_i(T)g_j(T) = 0$ if $i \neq j$ since every factor $x - d_k$ is present in the polynomial $g_i g_j$. Therefore $g_k(T)(X) = 0$ for all k. Since $X = \sum_{k=1}^{n} b_k g_k(T)(X)$, it follows that $X = 0$. Hence $\mathbf{C}^n$ is the direct sum

$$\mathbf{C}^n = V_1 \oplus \cdots \oplus V_r.$$

Now the effect of T on vectors in V_i is merely to multiply them by d_i since $(T - d_i)g_i(T) = f(T) = 0$. Therefore, if we choose bases for each subspace $V_1, \ldots, V_r$ and combine them to form a basis of $\mathbf{C}^n$, then T will be represented by a diagonal matrix. Consequently A is similar to a diagonal matrix.

Example 9.4.5
The matrix

$$\begin{pmatrix} 2 & 1 & 1 \\ 0 & 2 & 0 \\ 0 & 0 & 2 \end{pmatrix}$$

has minimum polynomial $(x-2)^2$, as we saw in Example 9.4.1. Since this is not a product of distinct linear factors, the matrix cannot be diagonalized.

Example 9.4.6

The $n \times n$ upper triangular matrix

$$
A = \begin{pmatrix}
c & 1 & 0 & \cdots & 0 & 0 \\
0 & c & 1 & \cdots & 0 & 0 \\
0 & 0 & c & \cdots & 0 & 0 \\
\cdot & \cdot & \cdot & \cdots & \cdot & \cdot \\
0 & 0 & 0 & \cdots & c & 1 \\
0 & 0 & 0 & \cdots & 0 & c
\end{pmatrix}
$$

has minimum polynomial $(x-c)^n$; this is because $(A-cI)^n = 0$, but $(A-cI)^{n-1} \neq 0$. Hence A is diagonalizable if and only if $n = 1$. Notice that the characteristic polynomial of A equals $(c-x)^n$.

Jordan normal form

We come now to the definition of the Jordan normal form of a square complex matrix. The basic components of this are certain complex matrices called Jordan blocks, of the type considered in Example 9.4.6. In general an $n \times n$ *Jordan block* is a matrix of the form

$$
J = \begin{pmatrix}
c & 1 & 0 & \cdots & 0 & 0 \\
0 & c & 1 & \cdots & 0 & 0 \\
0 & 0 & c & \cdots & 0 & 0 \\
\cdot & \cdot & \cdot & \cdots & \cdot & \cdot \\
0 & 0 & 0 & \cdots & c & 1 \\
0 & 0 & 0 & \cdots & 0 & c
\end{pmatrix}
$$

for some scalar c. Thus J is an upper triangular $n \times n$ matrix with constant diagonal entries, a superdiagonal of 1's, and zeros elsewhere. By Example 9.4.6 the minimum and characteristic polynomials of J are $(x-c)^n$ and $(c-x)^n$ respectively.

We must now take note of the essential property of the matrix J. Let $E_1, \ldots, E_n$ be the vectors of the standard basis of $\mathbf{C}^n$. Then matrix multiplication shows that $JE_1 = cE_1$, and $JE_i = cE_i + E_{i-1}$ where $1 < i \leq n$.

In general, if A is any complex $n \times n$ matrix, we call a sequence of vectors $X_1, \ldots, X_r$ in $\mathbf{C}^n$ a *Jordan string* for A if it satisfies the equations

$$AX_1 = cX_1 \text{ and } AX_i = cX_i + X_{i-1}$$

where c is a scalar and $1 < i \leq r$. Thus every $n \times n$ Jordan block determines a Jordan string of length n.

Now suppose there is a basis of $\mathbf{C}^n$ which consists of Jordan strings for the matrix A. Group together basis elements in the same string. Then the linear operator on $\mathbf{C}^n$ given by $T(X) = AX$ is represented with respect to this basis of Jordan strings by a matrix which has Jordan blocks down the diagonal:

$$N = \begin{pmatrix} J_1 & 0 & \cdots & 0 \\ 0 & J_2 & \cdots & 0 \\ \cdot & \cdot & \cdots & \cdot \\ 0 & 0 & \cdots & J_k \end{pmatrix}.$$

Here J_i is a Jordan block, say with c_i on the diagonal. This is because of the effect produced on the basis elements when they are multiplied on the left by A.

Our conclusion is that A is similar to the matrix N, which is called the *Jordan normal form* of A. Notice that the diagonal elements c_i of N are just the eigenvalues of A. Of course we still have to establish that a basis consisting of Jordan strings always exists; only then can we conclude that every matrix has a Jordan normal form.

Theorem 9.4.5 (*Jordan Normal Form*)
Every square complex matrix is similar to a matrix in Jordan normal form.

Proof
Let A be an $n \times n$ complex matrix. We have to establish the existence of a basis of $\mathbf{C}^n$ consisting of Jordan strings for A. This is done by induction on n; if $n = 1$, any non-zero vector

qualifies as a Jordan string of length 1, so we can assume that $n > 1$.

Since A is complex, it has an eigenvalue c. Thus the matrix $A' = A - cI$ is singular, and so its column space C has dimension $r < n$. Recall from Example 6.3.2 that C is the image of the linear operator on $\mathbf{C}^n$ which sends X to $A'X$. Restriction of this linear operator to C produces a linear operator which is represented by an $r \times r$ matrix. Since $r < n$, we may assume by induction hypothesis on n that C has a basis which is a union of Jordan strings for A. Let the ith such string be written X_{ij}, $j = 1, \ldots, l_i$; thus $AX_{i1} = c_i X_{i1}$ and in addition $AX_{ij} = c_i X_{ij} + X_{ij-1}$ for $1 < j \le l_i$. Then $A'X_{i1} = 0$ and $A'X_{ij} = X_{ij-1}$ if $j > 1$.

Next let D denote the intersection of C with N, the null space of A', and set $p = \dim(D)$. We need to identify the elements of D. Now any element of C has the form

$$Y = \sum_i \sum_j a_{ij} X_{ij}$$

where a_{ij} is a complex number. Assume that Y is in D, and thus in N, the null space of A'. Suppose that $a_{ij} \ne 0$ and let j be as large as possible with this property for the given i. If $j > 1$, then the equations $A'X_{i1} = 0$ and $A'X_{ik} = X_{ik-1}$ will prevent $A'Y$ from being zero. Hence $j = 1$. It follows that the X_{i1} form a basis of D, so there are exactly p of these X_{i1}.

Every vector in C is of the form $A'Y$ for some Y, since C is the image space of the linear operator sending X to $A'X$. For each i write the vector X_{il_i} in the form $X_{il_i} = A'Y_i$, for some Y_i, $i = 1, \ldots, p$. There are p of these Y_i. Finally, N has dimension $n - r$, so we can adjoin a further set of $n - r - p$ vectors to the X_{i1} to get a basis for N, say $Z_1, \ldots, Z_{n-r-p}$.

Altogether we have a total of $r + p + (n - r - p) = n$ vectors

$$X_{ij}, \; Y_k, \; Z_m.$$

We now assert that these vectors form a basis of $\mathbf{C}^n$ which consists of Jordan strings of A. Certainly

$$AY_k = (A' + cI)Y_k = A'Y_k + cY_k = cY_k + X_{kl_k}.$$

Thus the Jordan string $X_{k1}, \ldots, X_{kl_k}$ has been extended by adjoining Y_k. Also $AZ_m = cZ_m$ since Z_m belongs to the null space of A'; thus Z_m is a Jordan string of A with length 1. Hence the vectors in question constitute a set of Jordan strings of A.

What remains to be done is to prove that the vectors X_{ij}, Y_k, Z_m form a basis of $\mathbf{C}^n$, and by 5.1.9 it is enough to show that they are linearly independent. To accomplish this, we assume that e_{ij}, f_k, g_m are scalars such that

$$\sum \sum e_{ij} X_{ij} + \sum f_k Y_k + \sum g_m Z_m = 0.$$

Multiplying both sides of this equation on the left by A', we get

$$\sum \sum e_{ij}(0 \text{ or } X_{ij-1}) + \sum f_k X_{kl_k} = 0.$$

Now X_{kl_k} does not appear among the terms of the first sum in the above equation since $j - 1 < l_k$. Hence $f_k = 0$ for all k. Thus

$$\sum g_m Z_m = -\sum \sum e_{ij} X_{ij},$$

which therefore belongs to D. Hence $e_{ij} = 0$ if $j > 1$, and $\sum g_m Z_m = -\sum e_{i1} X_{i1}$. This can only mean that $g_m = 0$ and $e_{i1} = 0$ since the X_{i1} and Z_m are linearly independent. Hence the theorem is established.

Corollary 9.4.6

Every complex $n \times n$ matrix is similar to an upper triangular matrix with zeros above the superdiagonal.

This follows at once from the theorem since every Jordan block is an upper triangular matrix of the specified type.

Example 9.4.7

Put the matrix

$$A = \begin{pmatrix} 3 & 1 & 0 \\ -1 & 1 & 0 \\ 0 & 0 & 2 \end{pmatrix}$$

in Jordan normal form.

We follow the method of the proof of 9.4.5. The eigenvalues of A are 2, 2, 2, so define

$$A' = A - 2I = \begin{pmatrix} 1 & 1 & 0 \\ -1 & -1 & 0 \\ 0 & 0 & 0 \end{pmatrix}.$$

The column space C of A' is generated by the single vector $X = \begin{pmatrix} 1 \\ -1 \\ 0 \end{pmatrix}$. Note that $AX = 2X$, so X is a Jordan string of length 1 for A. Also the null space N of A' is generated by X and the vector

$$\begin{pmatrix} 0 \\ 0 \\ 1 \end{pmatrix}.$$

Thus $D = C \cap N = C$ is generated by X. The next step is to write X in the form $A'Y$; in fact we can take

$$Y = \begin{pmatrix} 1 \\ 0 \\ 0 \end{pmatrix}.$$

Thus the second basis element is Y. Finally, put $Z = \begin{pmatrix} 0 \\ 0 \\ 1 \end{pmatrix}$,

so that $\{X, Z\}$ is a basis for N. Then

$$A'X = 0, \ A'Y = X, \ A'Z = 0$$

and hence

$$AX = 2X, \ AY = 2Y + X \text{ and } AZ = 2Z.$$

It is now evident that $\{X, Y, Z\}$ is a basis of $\mathbf{C}^3$ consisting of the two Jordan strings X, Y, and Z. Therefore the Jordan form of A has two blocks and is

$$N = \left(\begin{array}{cc|c} 2 & 1 & 0 \\ 0 & 2 & 0 \\ \hline 0 & 0 & 2 \end{array}\right).$$

As an application of Jordan form we establish an interesting connection between a matrix and its transpose.

Theorem 9.4.7
Every square complex matrix is similar to its transpose.

Proof
Let A be a square matrix with complex entries, and write N for the Jordan normal form of A. Thus $S^{-1}AS = N$ for some invertible matrix S by 9.4.5. Now

$$N^T = S^T A^T (S^{-1})^T = S^T A^T (S^T)^{-1},$$

so N^T is similar to A^T. It will be sufficient if we can prove that N and N^T are similar. The reason for this is the transitive property of similarity: if P is similar to Q and Q is similar to R, then P is similar to R.

Because of the block decomposition of N, it is enough to prove that any Jordan block J is similar to its transpose. But this can be seen directly. Indeed, if P is the permutation matrix with a line of 1's from top right to bottom left, then matrix multiplication shows that $P^{-1}JP = J^T$.

Another use of Jordan form is to determine which matrices satisfy a given polynomial equation.

Example 9.4.8

Find up to similarity all complex $n \times n$ matrices A satisfying the equation $A^2 = I$.

Let N be the Jordan normal form of A, and write $N = S^{-1}AS$. Then $N^2 = S^{-1}A^2S$. Hence $A^2 = I$ if and only if $N^2 = I$. Since N consists of a string of Jordan blocks down the diagonal, we have only to decide which Jordan blocks J can satisfy $J^2 = I$. This is easily done. Certainly the diagonal entries of J will have to be 1 or -1. Furthermore, matrix multiplication reveals that $J^2 \neq I$ if J has two or more rows. Hence the block J must be 1×1. Thus N is a diagonal matrix with all its diagonal entries equal to $+1$ or -1. After reordering the rows and columns, we get a matrix of the form

$$N = \begin{pmatrix} I_r & 0 \\ 0 & -I_s \end{pmatrix}$$

where $r + s = n$. Therefore $A^2 = 1$ if and only if A is similar to a matrix with the form of N.

Next we consider the relationship between Jordan normal form and the minimum and characteristic polynomials. It will emerge that knowledge of Jordan form permits us to write down the minimum polynomial immediately. Since in principle we know how to find the Jordan form – by using the method of Example 9.4.7 – this leads to a systematic way of computing minimum polynomials, something that was lacking previously.

Let A be a complex $n \times n$ matrix whose distinct eigenvalues are $c_1, \ldots, c_r$. For each c_i there are corresponding Jordan blocks in the Jordan normal form N of A which have c_i on their principal diagonals, say $J_{i1}, \ldots, J_{il_i}$; let n_{ij} be the number of rows of J_{ij}. Of course A and N have the same minimum and characteristic polynomials since they are similar matrices. Now J_{ij} is an $n_{ij} \times n_{ij}$ upper triangular matrix with c_i on the principal diagonal, so its characteristic polynomial is just $(c_i - x)^{n_{ij}}$. The characteristic polynomial p of N is clearly the product of all of these polynomials: thus

$$p = \prod_{i=1}^{r} (c_i - x)^{m_i} \quad \text{where} \quad m_i = \sum_{j=1}^{l_i} n_{ij}.$$

The minimum polynomial is a little harder to find. If f is any polynomial, it is readily seen that $f(N)$ is the matrix with the blocks $f(J_{ij})$ down the principal diagonal and zeros elsewhere. Thus $f(N) = 0$ if and only if all the $f(J_{ij}) = 0$. Hence the minimum polynomial of N is the least common multiple of the minimum polynomials of the blocks J_{ij}. But we saw in Example 9.4.6 that the minimum polynomial of the Jordan block J_{ij} is $(x - c_i)^{n_{ij}}$. It follows that the minimum polynomial of N is

$$f = \prod_{i=1}^{n} (x - c_i)^{k_i}$$

where k_i is the largest of the n_{ij} for $j = 1, \ldots, l_i$.

These conclusions, which amount to a method of computing minimum polynomials from Jordan normal form, are summarized in the next result.

Theorem 9.4.8
Let A be an $n \times n$ complex matrix and let $c_1, \ldots, c_r$ be the distinct eigenvalues of A. Then the characteristic and minimum polynomials of A are

$$\prod_{i=1}^{n} (c_i - x)^{m_i} \quad \text{and} \quad \prod_{i=1}^{n} (x - c_i)^{k_i}$$

respectively, where m_i is the sum of the numbers of columns in Jordan blocks with eigenvalue c_i and k_i is the number of columns in the largest such Jordan block.

Example 9.4.9
Find the minimum polynomial of the matrix

$$A = \begin{pmatrix} 3 & 1 & 0 \\ -1 & 1 & 0 \\ 0 & 0 & 2 \end{pmatrix}.$$

The Jordan form of A is

$$N = \left(\begin{array}{cc|c} 2 & 1 & 0 \\ 0 & 2 & 0 \\ \hline 0 & 0 & 2 \end{array} \right)$$

by Example 9.4.7. Here 2 is the only eigenvalue and there are two Jordan blocks, with 2 and 1 columns. The minimum polynomial of A is therefore $(x - 2)^2$. Of course the characteristic polynomial is $(2 - x)^3$.

Application of Jordan form to differential equations

In 8.3 we studied systems of first order linear differential equations for functions $y_1, y_2, \ldots, y_n$ of a variable x. Such a system takes the matrix form

$$Y' = AY.$$

Here Y is the column of functions $y_1, \ldots, y_n$ and A is an $n \times n$ matrix with constant coefficients. Since any such matrix A is similar to a triangular matrix (by 8.1.8), it is possible to change to a system of linear differential equations for a new set of functions which has a triangular coefficient matrix. This new system can then be solved by back substitution,

as in Example 8.3.3. However this method can be laborious
for large n and Jordan form provides a simpler alternative
method.

Returning to the system $Y' = AY$, we know that there is
a non-singular matrix S such that $N = S^{-1}AS$ is in Jordan
normal form: say

$$N = \begin{pmatrix} J_1 & 0 & \cdots & 0 \\ 0 & J_2 & \cdots & 0 \\ \cdot & \cdot & \cdots & \cdot \\ 0 & 0 & \cdots & J_k \end{pmatrix}.$$

Here J_i is a Jordan block, say with d_i on the diagonal. Of
course the d_i are the eigenvalues of A. Now put $U = S^{-1}Y$,
so that the system $Y' = AY$ becomes $(SU)' = ASU$, or

$$U' = NU,$$

since $N = S^{-1}AS$. To solve this system of differential equations it is plainly sufficient to solve the subsystems $U_i' = J_iU_i$
for $i = 1, \ldots, k$ where U_i is the column of entries of U corresponding to the block J_i in N.

This observation effectively reduces the problem to one
in which the coefficient matrix is a Jordan block, let us say

$$A = \begin{pmatrix} d & 1 & 0 & \cdots & 0 & 0 \\ 0 & d & 1 & \cdots & 0 & 0 \\ \cdot & \cdot & \cdot & \cdots & \cdot & \cdot \\ 0 & 0 & 0 & \cdots & d & 1 \\ 0 & 0 & 0 & \cdots & 0 & d \end{pmatrix}.$$

Now the equations in the corresponding system have a
much simpler form than in the general triangular case:

$$\begin{cases} u_1' & = du_1 & + u_2 \\ u_2' & = & du_2 & + u_3 \\ \cdot & & \cdot & \cdot \\ u_{n-1}' & = & & du_{n-1} & + u_n \\ u_n' & = & & & du_n \end{cases}$$

The functions u_i can be found by solving a series of first order linear equations, starting from the bottom of the list. Thus $u'_n = du_n$ yields $u_n = c_{n-1}e^{dx}$ where c_{n-1} is a constant. The second last equation becomes

$$u'_{n-1} - du_{n-1} = c_{n-1}e^{dx},$$

which is first order linear with integrating factor e^{-dx}. Multiplying the equation by this factor, we get $(u_{n-1}e^{-dx})' = c_{n-1}$. Hence

$$u_{n-1} = (c_{n-2} + c_{n-1}x)e^{dx}$$

where c_{n-2} is another constant. The next equation yields

$$u'_{n-2} - du_{n-2} = (c_{n-2} + c_{n-1}x)e^{dx},$$

which is also first order linear with integrating factor e^{-dx}. It can be solved to give

$$u_{n-2} = (c_{n-3} + \frac{c_{n-2}}{1!}x + \frac{c_{n-1}}{2!}x^2)e^{dx},$$

where c_{n-2} is constant. Continuing in this manner, we find that the function u_{n-i} is given by

$$u_{n-i} = (c_{n-i-1} + \frac{c_{n-i}}{1!}x + \cdots + \frac{c_{n-1}}{i!}x^i)e^{dx},$$

where the c_i are constants. The original functions y_i can then be calculated by using the equation $Y = SU$.

Example 9.4.10
Solve the linear system of differential equations below using Jordan normal form:

$$\begin{cases} y'_1 &= 3y_1 + y_2 \\ y'_2 &= -y_1 + y_2 \\ y'_3 &= \qquad\qquad 2y_3 \end{cases}$$

Here the coefficient matrix is

$$A = \begin{pmatrix} 3 & 1 & 0 \\ -1 & 1 & 0 \\ 0 & 0 & 2 \end{pmatrix}.$$

The Jordan form of A was found in Example 9.4.7: we recall the results obtained there. There is a basis of $\mathbf{R}^3$ consisting of Jordan strings: this is $\{X, W, Z\}$, where

$$X = \begin{pmatrix} 1 \\ -1 \\ 0 \end{pmatrix}, \quad W = \begin{pmatrix} 1 \\ 0 \\ 0 \end{pmatrix} \text{ and } Z = \begin{pmatrix} 0 \\ 0 \\ 1 \end{pmatrix}.$$

Here $AX = 2X$, $AW = 2W + X$, $AZ = 2Z$.

The matrix which describes the change of basis from $\{X, W, Z\}$ to the standard basis is

$$S = \begin{pmatrix} 1 & 1 & 0 \\ -1 & 0 & 0 \\ 0 & 0 & 1 \end{pmatrix}.$$

By 6.2.6 the matrix which represents the linear operator arising from left multiplication by A, with respect to the basis $\{X, W, Z\}$, is

$$S^{-1}AS = J = \begin{pmatrix} 2 & 1 & 0 \\ 0 & 2 & 0 \\ 0 & 0 & 2 \end{pmatrix}.$$

Now put $U = S^{-1}Y$, so that $Y = SU$ and the system of equations becomes $U' = S^{-1}ASU = JU$, that is,

$$\begin{cases} u_1' = 2u_1 + u_2 \\ u_2' = 2u_2 \\ u_3' = 2u_3 \end{cases}$$

We solve this system, beginning with the last equation, and obtain $u_3 = c_2 e^{2x}$. Next $u_2' = 2u_2$, so that $u_2 = c_1 e^{2x}$. Finally we solve

$$u_1' - 2u_1 = c_1 e^{2x},$$

a first order linear equation, and find the solution to be

$$u_1 = (c_0 + c_1 x)e^{2x}.$$

Therefore

$$U = \begin{pmatrix} c_0 + c_1 x \\ c_1 \\ c_2 \end{pmatrix} e^{2x},$$

and since $Y = SU$, we obtain

$$Y = \begin{pmatrix} c_0 + c_1 + c_1 x \\ -c_0 - c_1 x \\ c_2 \end{pmatrix} e^{2x},$$

from which the values of the functions y_1, y_2, y_3 can be read off.

Exercises 9.4

1. Find the minimum polynomials of the following matrices *by inspection* :

(a) $\begin{pmatrix} 2 & 0 \\ 0 & 2 \end{pmatrix}$; (b) $\begin{pmatrix} 2 & 1 \\ 0 & 3 \end{pmatrix}$; (c) $\begin{pmatrix} 0 & 1 \\ 1 & 0 \end{pmatrix}$;

(d) $\begin{pmatrix} 3 & 0 & 0 \\ 0 & 2 & 1 \\ 0 & 0 & 2 \end{pmatrix}$.

2. Let A be an $n \times n$ matrix and S an invertible $n \times n$ matrix over a field F. If f is any polynomial over F, show that $f(S^{-1}AS) = S^{-1}f(A)S$. Use this result to give another proof of the fact that similar matrices have the same minimum polynomial (see 9.4.2).

3. Use 9.4.4 to prove that if an $n \times n$ complex matrix has n distinct eigenvalues, then the matrix is diagonalizable.

4. Show that the minimum polynomial of the companion matrix

$$A = \begin{pmatrix} 0 & 0 & -c \\ 1 & 0 & -b \\ 0 & 1 & -a \end{pmatrix}$$

is $x^3 + ax^2 + bx + c$. (See Exercise 8.1.6). [Hint: show that $uI + vA + wA^2 = 0$ implies that $u = v = w = 0$].

5. Find the Jordan normal forms of the following matrices:

(a) $\begin{pmatrix} 1 & 3 \\ 2 & 2 \end{pmatrix}$; (b) $\begin{pmatrix} 1 & 1 \\ -1 & 3 \end{pmatrix}$; (c) $\begin{pmatrix} 1 & -1 & 0 \\ 0 & 1 & 2 \\ 0 & 0 & 1 \end{pmatrix}$.

6. Read off the minimum polynomials from the Jordan forms in Exercise 5.

7. Find up to similarity all $n \times n$ complex matrices A satisfying $A = A^2$.

8. The same problem for matrices such that $A^2 = A^3$.

9. (*Uniqueness of Jordan normal form*) Let A be a complex $n \times n$ matrix with Jordan blocks J_{ij}, where J_{ij} is a block associated with the eigenvalue c_i. Prove that the number of $r \times r$ Jordan blocks $J_{i,j}$ for a given i equals $d_{r-1} - d_r$, where d_k is the dimension of the intersection of the column space of $(A - c_i I_n)^k$ and the null space of $A - c_i I_n$. Deduce that the blocks that appear in the Jordan normal form of A are unique up to order.

10. Using Exercise 9.4.4 as a model, suggest an $n \times n$ matrix whose minimal polynomial is $x^n + a_{n-1}x^{n-1} + \cdots + a_1 x + a_0$.

11. Use Jordan normal form to solve the following system of differential equations:

$$\begin{cases} y_1' = y_1 + y_2 + y_3 \\ y_2' = \qquad y_2 \\ y_3' = \qquad y_2 + y_3 \end{cases}$$

Chapter Ten

LINEAR PROGRAMMING

One of the great successes of linear algebra has been the construction of algorithms to solve certain optimization problems in which a linear function has to be maximized or minimized subject to a set of linear constraints. Typically the function is a profit or cost.Such problems are called *linear programming problems.*

The need to solve such problems was recognized during the Second World War, when supplies and labor were limited by wartime conditions. The pioneering work of George Danzig led to the creation of the Simplex Algorithm, which for over half a century has been the standard tool for solving linear programming problems. Our purpose here is to describe the linear algebra which underlies the simplex algorithm and then to show how it can be applied to solve specific problems.

10.1 Introduction to Linear Programming

We begin by giving some examples of linear programming problems.

Example 10.1.1 *(A productionproblem)*

A food company markets two products F_1 and F_2, which are made from two ingredients I_1 and I_2. To produce one unit of product F_j one requires a_{ij} units of ingredient I_i. The maximum amounts of I_1 and I_2 available are m_1 and m_2, respectively. The company makes a profit of p_i on each unit of product F_i sold. How many units of F_1 and F_2 should the company produce in order to maximize its profit without running out of ingredients?

370

Suppose the company decides to produce x_j units of product F_j. Then the profit on marketing the products will be $z = p_1x_1 + p_2x_2$. On the other hand, the production process will use $a_{11}x_1 + a_{12}x_2$ units of ingredient I_1 and $a_{21}x_2 + a_{22}x_2$ units of ingredient I_2. Therefore x_1 and x_2 must satisfy the constraints

$$a_{11}x_1 + a_{12}x_2 \leq m_1 \quad \text{and} \quad a_{21}x_1 + a_{22}x_2 \leq m_2.$$

Also x_1 and x_2 cannot be negative.

We therefore have to solve the following linear programming problem:

$$\text{maximize}: \quad z = p_1x_1 + p_2x_2$$

$$\text{subject to:} \quad \begin{cases} a_{11}x_1 + a_{12}x_2 & \leq m_1 \\ a_{21}x_1 + a_{22}x_2 & \leq m_2 \\ x_1, x_2 \geq 0 \end{cases}$$

Example 10.1.2 (*A transportationproblem*)

A company has m factories $F_1, \ldots, F_m$ and n warehouses $W_1, \ldots, W_n$. Factory F_i can produce at most r_i units of a certain product per week and warehouse W_j must be able to supply at least s_j units per week. The cost of shipping one unit from factory F_i to warehouse W_j is c_{ij}. How many units should be shipped from each factory to each warehouse per week in order to minimize the total transportation cost and yet still satisfy the requirements on the factories and warehouses?

Let x_{ij} be the number of units to be shipped from factory F_i to warehouse W_j per week. Then the total transportation cost for the week is

$$z = \sum_{i=1}^{m} \sum_{j=1}^{n} c_{ij}x_{ij}.$$

The condition on factory F_i is that $\sum_{j=1}^{n} x_{ij} \leq r_i$, while that on warehouse W_j is $\sum_{i=1}^{m} x_{ij} \geq s_j$. We are therefore faced with the following linear programming problem:

$$\text{minimize:} \quad z = \sum_{i=1}^{m} \sum_{j=1}^{n} c_{ij} x_{ij}$$

$$\text{subject} \quad \text{to} : \quad \begin{cases} \sum_{j=1}^{n} x_{ij} & \leq r_i, \ i = 1, \ldots, m \\ \sum_{i=1}^{m} x_{ij} & \geq s_j, \ j = 1, \ldots, n \\ x_{ij} \geq 0 \end{cases}$$

The general linear programming problem

After these examples we are ready to describe the general form of a linear programming problem.

Let $x_1, x_2, \ldots, x_n$ be variables. There is given a linear function of the variables

$$z = c_1 x_1 + c_2 x_2 + \cdots + c_n x_n,$$

called the *objective function*, which has to be maximized or minimized. The variables x_j are subject to a number of linear conditions, called the *constraints*, which take the form

$$a_{i1} x_1 + a_{i2} x_2 + \cdots + a_{in} x_n \ \leq \ \text{or} \ = \ \text{or} \ \geq \ b_i,$$

$i = 1, 2, \ldots, m$. In addition, certain of the variables may be *constrained*, i.e., they must take non-negative values. The general linear programming problem therefore takes the following form:

$$\text{maximize or minimize:} \quad z = c_1 x_1 + \cdots + c_n x_n$$

$$\text{subject to:} \quad \begin{cases} a_{i1} x_1 + \cdots + a_{in} x_n \ \leq \ \text{or} \ = \ \text{or} \ \geq \ b_i, \\ \qquad i = 1, 2, \ldots, m, \\ \qquad \text{certain} \ x_j \geq 0. \end{cases}$$

The understanding here is that a_{ij}, b_i, c_j are all known quantities. The object is to find $x_1, \ldots, x_n$ which optimize the objective function z, while satisfying the constraints. Evidently Examples 10.1.1 and 10.1.2 areproblems of this type.

Feasible and optimal solutions

It will be convenient to think of $X = (x_1, x_2, \ldots, x_n)^T$ in the above problem as a *point* in Euclidean space $\mathbf{R}^n$. If X satisfies all the constraints (including the conditions $x_j \geq 0$), then it is called a *feasible solution* of the problem. A feasible solution for which the objective function is maximum or minimum is said to be an *optimal solution*.

For a general linear programming problem there are three possible outcomes.

(i) There are no feasible solutions and thus the problem has nooptimal solutions.

(ii) Feasible solutions exist, but the objective function has arbitrarily large or small values at feasible solutions. Again thereare no optimal solutions.

(iii) The objective function has finite maximum or minimumvalues at feasible points. Then optimal solutions exist.

In a linear programming problem the object is to find an optimal solution or show that none exists.

Standard and canonical form

Since the general linear programming problem has a complex form, it is important to develop simpler types of problem which are equivalent to it. Here two linear programming problems are said to be *equivalent* if they have the same sets of feasible solutions and the same optimal solutions.

A linear programming problem is said to be in *standard form* if it is a maximization problem with all constraints inequalities and all variables constrained. It therefore has the general form

maximize: $z = c_1 x_1 + \cdots + c_n x_n$

subject to:

$$\begin{cases} a_{11}x_1 + a_{12}x_2 + \cdots + a_{1n}x_n \leq b_1 \\ a_{21}x_1 + a_{22}x_2 + \cdots + a_{2n}x_n \leq b_2 \\ \qquad \cdot \qquad\quad \cdot \quad \cdots \quad\ \cdot \qquad \cdot \\ a_{m1}x_1 + a_{m2}x_2 + \cdots + a_{mn}x_n \leq b_m \\ \qquad x_j \geq 0, \ j = 1, 2, \ldots, n \end{cases}$$

This problem can be written in matrix form: let $A = (a_{ij})_{mn}$, $B = (b_1\ b_2\ \cdots\ b_m)^T$, $C = (c_1\ c_2\ \cdots\ c_n)^T$ and $X = (x_1\ x_2\ \ldots x_n)^T$. Then the problem takes the form:

maximize : $z = C^T X$

subject to: $\begin{cases} AX \leq B \\ X \geq 0. \end{cases}$

Here a matrix inequality $U \leq V$ means that U and V are of the same size and $u_{ij} \leq v_{ij}$ for all i, j: there is a similar definition of $U \geq V$.

A second important type of linear programming problem is a maximization problem with all constraints equalities and allvariables constrained. The general form is:

$$\text{maximize}: \quad z = C^T X$$

$$\text{subject to:} \quad \begin{cases} AX &=& B \\ X &\geq& 0. \end{cases}$$

Such a linear programming problem is said to be in *canonical form*.

Changes to a linear programming problem

Our aim is to show that any linear programming problem is equivalent to one in standard form and to one in canonical form. To do this we need to consider what changes to a program will produce an equivalent program. There are four types of change that can be made.

Replace a minimization by a maximization

If the objective function in a linear program is $z = C^T X$, the minimum value of z occurs for the same X as the maximum value of $(-C)^T X$. Thus we can replace "minimize" by "maximize" and $C^T X$ by the new objective function $(-C)^T X$.

Reverse an inequality

The inequality $a_{i1}x_1 + \cdots + a_{in}x_n \geq b_i$ is clearly equivalent to $(-a_{i1})x_1 + \cdots + (-a_{in})x_n \leq -b_i$.

Replace an equality by two inequalities

The constraint $a_{i1}x_1 + \cdots + a_{in}x_n = b_n$ is equivalent tothe two inequalities

$$\begin{cases} a_{i1}x_1 + \cdots + a_{in}x_n &\leq& b_i \\ (-a_{i1})x_1 + \cdots + (-a_{in})x_n &\leq& -b_i \end{cases}$$

Elimination of an unconstrained variable

Suppose that the variable x_j is unconstrained, i.e., it can take negative values. The trick here is to replace x_j by two new variables x_j^+, x_j^- which are constrained. Write x_j in the form $x_j = x_j^+ - x_j^{-1}$ where $x_j^+, x_j^- \geq 0$. This is possible since any real number can be written as the difference between two positive numbers.

If we replace x_j by $x_j^+ - x_j^-$ in each constraint and in the objective function, and we add new constraints $x_j^+ \geq 0$, $x_j^- \geq 0$, then the resulting equivalent program will have fewer unconstrained variables.

By a sequence of operations of types I–IV a general linear programming problem may be transformed to an equivalence problem in standard form. Thus we have proved:

Theorem 10.1.1 *Every linear programming problem is equivalent to a program in standard form.*

Example 10.1.3

Put the following programming problem in standard form.

$$\text{minimize:} \quad z = 3x_1 + 2x_2 - x_3$$

$$\text{subject to:} \quad \begin{cases} x_1 + x_2 + 2x_3 & \geq 6 \\ x_1 + x_2 + 3x_3 & \leq 2 \\ x_1, x_3 \geq 0 \end{cases}$$

First of all change the minimization to a maximization and replace the constraints involving $=$ and $\geq$ by constraints involving $\leq$:

$$\text{maximize:} \quad z = -3x_1 - 2x_2 + x_3$$

$$\text{subject to:} \quad \begin{cases} -x_1 & - \; x_2 \; - \; 2x_3 & \le -6 \\ x_1 & + \; x_2 + \; x_3 & \le \;\;\; 4 \\ -x_1 & - \;\; x_2 - \;\; x_3 & \le -4 \\ x_1 & - \;\; x_2 + 3x_3 & \le \;\;\; 2 \\ & x_1, x_3 \ge 0 \end{cases}$$

Next write x_2, which is an unconstrained variable, in the form $x_2^+ - x_2^-$. This yields a problem in standard form:

$$\text{maximize:} \quad z = -3x_1 - 2x_2^+ + 2x_2^- + x_3$$

subject to:

$$\begin{cases} -x_1 & - \; x_2^+ \; + \; x_2^- \; - \; 2x_3 & \le -6 \\ x_1 & + \; x_2^+ \; - \; x_2^- \; + \; x_3 & \le \;\;\; 4 \\ -x_1 & + \; x_2^+ \; + \; x_2^- \; - \; x_3 & \le -4 \\ x_1 & - \; x_2^+ \; + \; x_2^- \; + \; 3x_3 & \le \;\;\; 2 \\ & x_1, x_2^+, x_2^-, x_3 \ge 0 \end{cases}$$

Slack variables

If we wish to transform a linear programming problem to canonical form, a method for converting inequalities into equalities is needed. This can be achieved by the introduction of what are called slack variables.

Consider a linear programming problem in standard form:

$$\text{maximize:} \quad z = C^T X$$

$$\text{subject to:} \quad \begin{cases} AX \le B \\ X \ge 0 \end{cases}$$

where A is $m \times n$ and the variables are $x_1, x_2, \ldots, x_n$. We introduce m new variables, $x_{n+1}, \ldots, x_{n+m}$, the so-called *slack variables*, and replace the ith constraint $a_{i1}x_1+\cdots+a_{in}x_n \leq b_i$ by the new constraint

$$a_{i1}x_1 + \cdots + a_{in}x_n + x_{n+i} = b_i,$$

for $i = 1, 2, \ldots, m$, together with $x_{i+n} \geq 0$, $i = 1, \ldots, m$. The effect is totransform the problem to an equivalent linear programming problem incanonical form:

$$\text{maximize:} \quad z = c_1 x_1 + \cdots + c_n x_n$$

subject to:

$$\begin{cases} a_{11}x_1 & + \cdots & + a_{1n}x_n & + x_{n+1} & = b_1 \\ a_{21}x_1 & + \cdots & + a_{2n}x_n & + x_{n+2} & = b_2 \\ \cdot & \cdots & \cdot & \cdot & \cdot \\ a_{m1}x_1 & + \cdots & + a_{mn}x_n & + x_{n+m} & = b_m \end{cases}$$

$$x_i \geq 0, \ i = 1, 2, \ldots, n + m.$$

Combining this observation with 10.1.1, we obtain:

Theorem 10.1.2 *Every linear programmingproblem is equivalent to one in canonical form.*

Exercises 10.1

1. A publishing house plans to issue three types of pamphlets P_1, P_2, P_3. Each pamphlet has to be printed and bound. The times in hours required to print and to bind one copy of pamphlet P_i are u_i and v_i respectively. The printing and binding machines can run for maximum times s and t hours per day respectively. The profit made on one pamphlet of type P_i is p_i. Let x_1, x_2, x_3 be thenumbers of pamphlets of the three types to be produced per day. Set up a linear program in x_1, x_2, x_3 which maximizes the profit$p$ per day and takes into account the times for which the machines are available.

2. A nutritionist is planning a lunch menu with two food types A and B. One ounce of A provides a_c units of carbohydrate, a_f units of fat and a_p units of protein: for B the figures are b_c, b_f, b_p, respectively. The costs of one unit of A and one unit of B are p and q respectively. The meal must provide at least m_c units of carbohydrate, m_f units of fat and m_p units of protein. Set up a linear program to determine how many ounces of A and B should be provided in the meal in order to minimize the cost e, while satisfying the dietary requirements.

3. Write the following linear programming problem in standard form:

$$\text{minimize:} \quad z = 2x_1 - x_2 - x_3 + x_4$$

$$\text{subject to:} \quad \begin{cases} x_1 + 2x_2 + x_3 - x_4 \geq 5 \\ 3x_1 + x_2 - x_3 + x_4 \leq 4 \\ x_1, x_2 \geq 0 \end{cases}$$

4. Write the linear programming problem in Exercise 10.4.3 in canonical form.

5. Consider the following linear programming problem in $x_1, x_2, \ldots, x_n$ with n constraints:

$$\text{maximize:} \quad z = C^T X$$

$$\text{subject to:} \quad \begin{cases} AX = B \\ X \geq 0 \end{cases}$$

where A is an $n \times n$ matrix with rank n.
    (a) Show that there is a feasible solution if and only if $A^{-1}B \geq 0$.
    (b) Show that if a feasible solution exists, it must be optimal.
    (c) If an optimal solution exists, what is the maximum value of z?

10.2　The Geometry of Linear Programming

Valuable insight into the nature of the linear programming problem is gained by adopting a geometrical point of view and regarding the problem as one about n-dimensional space.

We will identify an n-column vector X with a point

$$(x_1, x_2, \ldots, x_n)$$

in n-dimensional space and denote the latter by $\mathbf{R}^n$. The set of points X such that

$$a_1 x_1 + \cdots + a_n x_n = b,$$

where the real numbers a_i, b are not all zero, is called a *hyperplane* in $\mathbf{R}^n$. Thus a hyperplane in $\mathbf{R}^2$ is a line and a hyperplane in $\mathbf{R}^3$ is a plane.

Let $A = (a_1 \; a_2 \; \ldots \; a_n)$; thus the equation of the hyperplane is $AX = b$: let us call it H. Then H divides $\mathbf{R}^n$ into two *halfspaces*

$$H_1 = \{X \in \mathbf{R}^n \mid AX \leq b\}$$

and

$$H_2 = \{X \in \mathbf{R}^n \mid AX \geq b\}.$$

Clearly

$$\mathbf{R}^n = H_1 \cup H_2 \text{ and } H = H_1 \cap H_2.$$

In a linear programming program in $x_1, \ldots, x_n$, each constraint requires the point X to lie in a half space or a hyperplane. Thus the set of feasible solutions corresponds to the points lying in all of the half spaces or hyperplanes corresponding to the constraints. In this way we obtain a geometrical picture of the set of feasible solutions of the problem.

Example 10.2.1

Consider the simple linear programming problem in standard form:

$$\text{maximize: } z = x + y$$

$$\text{subject to: } \begin{cases} 2x + y \leq 3 \\ x + 2y \leq 3 \\ x, y \geq 0 \end{cases}$$

The set S of feasible solutions is the region of the plane which is bounded by the lines $2x + y = 3$, $x + 2y = 3$, $x = 0$, $y = 0$

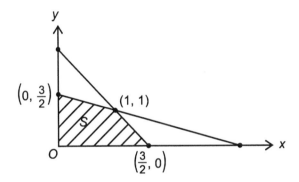

The objective function $z = x + y$ corresponds to a plane in 3-dimensional space. The problem is to find a point of S at which the height of the plane above the xy-plane is largest. Geometrically, it is clear that this point must be one of the "corner points" $(0,0), (\frac{3}{2},0), (1,1), (0,\frac{3}{2})$. The largest value of $z = x + y$ occurs at $(1,1)$. Therefore $x = 1 = y$ is an optimal solution of the problem.

The next step is to investigate the geometrical properties of the set of feasible solutions. This involves the concept of convexity.

Convex subsets

Let X_1 and X_2 be two distinct points in $\mathbf{R}^n$. The *line segment* $X_1 X_2$ joining X_1 and X_2 is defined to be the set of points

$$\{tX_1 + (1 - t)X_2 \mid 0 \le t \le 1\}.$$

For example, if $n \le 3$, the point $tX_1 + (1 - t)X_2$, where $0 \le t \le 1$, is a typical point lying between X_1 and X_2 on the line which joins them. To see this one has to notice that

$$X_2 - (tX_1 + (1 - t)X_2) = t(X_2 - X_1)$$

and

$$(tX_1 + (1 - t)X_2) - X_1 = (1 - t)(X_2 - X_1)$$

are parallel vectors.

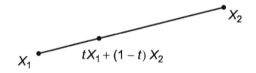

(Keep in mind that we are using X to denote both the point (x_1, x_2, x_3) and the column vector $(x_1 \ x_2 \ x_3)^T$.)

A non-empty subset S of $\mathbf{R}^n$ is called *convex* if, whenever X_1 and X_2 are points in S, every point on the line segment $X_1 X_2$ is also a point of S.

It is easy to visualize the situation in $\mathbf{R}^2$: for example, consider the shaded regions shown.

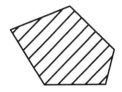 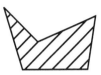

The interior of the left hand figure is clearly convex, but the interior of the right hand one is not.

The following property of convex sets is almost obvious.

Lemma 10.2.1
The intersection of a collection of convex subsets of $\mathbf{R}^n$ is either empty or convex.

Proof
Let $\{S_i \mid i \in I\}$ be a set of convex subsets of $\mathbf{R}^n$ and assume that $S = \bigcap_{i \in I} S_i$ is not empty. If S has only one element, then it is obviously convex. So assume X_1 and X_2 are distinct points of S and let $0 \leq t \leq 1$. Now X_1 and X_2 belong to S_i for all i, as must $tX_1 + (1 - t)X_2$ since S_i is convex. Hence $tX_1 + (1 - t)X_2 \in S$ and S is convex.

Our interest in convex sets is motivated by the following fundamental result.

Theorem 10.2.2
The set of all feasible solutions of a linear programming problem is either empty or convex.

Proof

By 10.1.1 we may assume that the linear programming problem is in standard form. Hence the set of feasible solutions is the intersection of a collection of half spaces. Because of 10.2.1 it is enough to prove that every half space H is convex.

For example, consider $H = \{X \in \mathbf{R}^n \mid AX \leq b\}$ where A is an n-row vector. Suppose that $X_1, X_2 \in H$ and $0 \leq t \leq 1$. Then

$$A(tX_1 + (1-t)X_2) = t(A\,X_1) + (1-t)A\,X_2 \leq tb + (1-t)b = b.$$

Hence $tX_1 + (1-t)X_2 \in H$ and H is convex.

The convex hull

Let $X_1, X_2, \ldots, X_m$ be vectors in $\mathbf{R}^n$. Then a vector of the form

$$\sum_{i=1}^{m} c_i X_i,$$

where

$$c_i \geq 0 \quad \text{and} \quad \sum_{i=1}^{m} c_i = 1,$$

is called a *convex combination* of $X_1, X_2, \ldots, X_m$. For example, when $m = 2$, every convex combination of X_1, X_2 has the form $t\,X_1 + (1-t)X_2$, where $0 \leq t \leq 1$. Thus the line segment $X_1 X_2$ consists of all the convex combinations of X_1 and X_2.

The set of all convex combinations of elements of a nonempty subset S of $\mathbf{R}^n$ is called the *convex hull* of S:

$$C(S).$$

For example, the convex hull of $\{X_1, X_2\}$, where $X_1 \neq X_2$, is just the line segment $X_1 X_2$.

The relation between the convex hull and convexity is made clear by the next result.

Theorem 10.2.3

Let S be a non-empty subset of $\mathbf{R}^n$. Then $C(S)$ is the smallest convex subset of $\mathbf{R}^n$ which contains S.

Proof

In the first place it is clear that $S \subseteq C(S)$. We show next that $C(S)$ is a convex set. Let $X, Y \in C(S)$; then we can write $X = \sum_{i=1}^{m} c_i X_i$ and $Y = \sum_{i=1}^{m} d_i X_i$, where $X_1, \ldots, X_m \in S$, $0 \le c_i, d_i \le 1$, and $\sum_{i=1}^{m} c_i = 1 = \sum_{i=1}^{m} d_i$. Then for any t satisfying $0 \le t \le 1$, we have

$$tX + (1-t)Y = \sum_{i=1}^{m} tc_i X_i + \sum_{i=1}^{m} (1-t)d_i X_i$$

$$= \sum_{i=1}^{m} (tc_i + (1-t)d_i) X_i.$$

Now

$$\sum_{i=1}^{m} (t\,c_i + (1-t)d_i) = t \left(\sum_{i=1}^{m} c_i \right) + (1-t) \left(\sum_{i=1}^{m} d_i \right)$$

$$= t + (1-t)$$

$$= 1.$$

Consequently $tX + (1-t)Y \in C(S)$ and $C(S)$ is convex.

Next suppose T is any convex subset of $\mathbf{R}^n$ containing S. We must show that $C(S) \subseteq T$; for then $C(S)$ will be the smallest convex subset containing S.

Let $X \in C(S)$ and write $X = \sum_{i=1}^{m} c_i X_i$, where $X_i \in S$, $c_i \ge 0$ and $\sum_{i=1}^{m} c_i = 1$. We will show that $X \in T$ by induction on $m > 1$, the claim being clearly true if $m = 1$. Now we have

$$X = (1 - c_m) \sum_{i=1}^{m-1} \left(\frac{c_i}{1 - c_m}\right) X_i + c_m X_m.$$

Next, since $\sum_{i-1}^{m-1} c_i = 1 - c_m$, we have

$$\sum_{i=1}^{m-1} \left(\frac{c_i}{1 - c_m}\right) = \frac{1 - c_m}{1 - c_m} = 1.$$

Also $0 \leq \frac{c_i}{1-c_m} \leq 1$ since $c_i \leq c_1 + \cdots + c_{m-1} = 1 - c_m$ for $1 \leq i \leq m - 1$. Hence

$$Y = \sum_{i=1}^{m-1} \left(\frac{c_i}{1 - c_m}\right) X_i \in T,$$

by the induction hypothesis on m. Finally,

$$X = (1 - c_m)Y + c_m X_m \in T,$$

since T is convex.

Extreme points

Let S be a convex subset of $\mathbf{R}^n$. A point of S is called an *extreme point* if it is not an interior point of any line segment joining two points of S. For example, the extreme points of the set of points in the polygon below are just the six vertices shown.

The extreme points of a convex set can be characterized in terms of convex combinations.

Theorem 10.2.4
Let S be a convex subset of $\mathbf{R}^n$ and let $X \in S$. Then X is an extreme point of S if and only if it is not a convex combination of other points of S.

Proof
Suppose X is not an extreme point of S; then

$$X = tY + (1-t)Z,$$

where $0 < t < 1$ and $Y, Z \in S$. Then X is certainly a convex combination of points of S, namely Y and Z.

Conversely, suppose that X is a convex combination of other points of S. We will show that X is not an extreme point of S. By assumption it is possible to write $X = \sum_{i=1}^{m} c_i X_i$

where $X_i \in S$, $X_i \neq X$, $0 < c_i \leq 1$ and $\sum_{i=1}^{m} c_i = 1$. Notice

that $c_j \neq 1$; for otherwise $\sum_{i=1,\ i \neq j}^{m} c_i = 0$ and $c_i = 0$ for all

$i \neq j$, so that $X = X_j$.

Just as in the proof of Theorem 10.2.3, we can write

$$X = (1 - c_m) \sum_{i=1}^{m-1} \left(\frac{c_i}{1 - c_m}\right) X_i + c_m X_m.$$

Also

$$\sum_{i=1}^{m-1} \left(\frac{c_i}{1 - c_m}\right) = \frac{1 - c_m}{1 - c_m} = 1$$

and $0 < \frac{c_i}{1-c_m} < 1$, since $c_i < c_1 + \cdots + c_{m-1} = 1 - c_m$ if $i < m$. It follows that

$$Y = \sum_{i=1}^{m-1} \left(\frac{c_i}{1 - c_m}\right) X_i \in C(S) = S$$

and $X = (1 - c_m)Y + c_m X_m$ is an interior point of the line
segment joining Y and X_m. Hence X is not an extreme point
of S, which completes the proof.

It is now time to explain the connection between optimal
solutions of a linear programming problem and the extreme
points of the set of feasible solutions.

Theorem 10.2.5 (*The Extreme Point Theorem*)
*Let S be the set of all feasible solutions of a linear program-
ming problem.*

(i) *If S is non-empty and bounded, then there is an
optimal solution.*

(ii) *If an optimal solution exists, then it occurs at an
extreme point of S.*

Here a subset S of $\mathbf{R}^n$ is said to be *bounded* if there exists
a positive number d such that $-d \le x_i \le d$, for $i = 1, 2, \ldots, n$
and all $(x_1, x_2, \ldots, x_n)$ in S.

Proof of Theorem 10.2.5
Suppose that we have a maximization problem. For simplicity
we will assume throughout that S is bounded and $n = 2$:
thus S can be visualized as a region of the plane bounded by
straight lines corresponding to the constraints.

Let $z = f(x, y) = cx + dy$ be the objective function: we
can assume c and d are not both 0. Since S is bounded and f
is continuous in S, a standard theorem from calculus can be
applied to show that f has an absolute maximum in S. This
establishes (i).

Next assume that there is an optimal solution. By another standard theorem, if $P(x, y)$ is a point of P which is an
absolute maximum of f, then either P is a critical point of
S or else it lies on the boundary of S. But f has no critical
points: for $f_x = c$, $f_y = d$, so f_x and f_y cannot both vanish.
Thus P lies on the boundary of S and so on a line. By the

same argument P cannot lie in the interior of the line. There-fore P is a point of intersection of two lines and hence it is an extreme point of S.

We can now summarize the possible situations for a linear programming problem with set of feasible solutions S.

(a) S is empty: the problem has no solutions;

(b) S is non-empty and bounded: in this case the problem has an optimal solution and it occurs at an extreme point of S.

(c) S is unbounded: here optimal solutions need not exist, but, if they do, they occur at extreme points of S.

We will see in 10.3 that if S is non-empty and bounded, then it has a finite number of extreme points. By computing the value of the objective function at each extreme point one can find an optimal solution of the problem. We conclude with two examples.

Example 10.2.2

$$\text{maximize:} \quad z = 2x + 3y$$

$$\text{subject to:} \quad \begin{cases} x + y & \geq \ 1 \\ x - y & \geq -1 \\ x, y \geq 0 \end{cases}$$

Here the set of feasible solutions S corresponds to the region of the xy-plane bounded by the lines $x + y = 1$, $x - y = -1$, $x = 0$, $y = 0$. Clearly it is unbounded and z can be arbitrary large at points in S. Thus no optimal solutions exist.

Example 10.2.3

$$\text{maximize:} \quad z = 1 - 12x - 3y$$

$$\text{subject to:} \quad \begin{cases} x + y & \geq \ 1 \\ x - y & \geq -1 \\ x, y \geq 0 \end{cases}$$

In this problem the set of feasible solutions is the same set S as in the previous example. However the maximum value of z in S occurs at $x = 0$, $y = 1$: this is an optimal solution of the problem.

Exercises 10.2

In Exercises 10.2.1–10.2.3 sketch the convex subset of all feasible solutions of a linear programming problem with the given constraints.

1. $\begin{cases} x - \ y & \leq -2 \\ 2x + \ y & \leq \ 3 \\ x, y \geq 0 \end{cases}$

2. $\begin{cases} x - \ 2y & \leq 3 \\ x + \ \ y & \leq 6 \\ x, y \geq 0 \end{cases}$

3. $\begin{cases} x + y + \ z & \leq 5 \\ x - y - \ z & \leq 0 \\ x, y, z \geq 0 \end{cases}$

4. Find all the extreme points in the programs of Exercises 10.2.1 and 10.2.2 .

5. Suppose the objective function in Exercise 10.2.2 is $z = 2x + 3y$. Find the optimal solution when z is to be maximized.

6. Let S be any *subspace* of $\mathbf{R}^n$. Prove that S is convex. Then give an example of a convex subset of $\mathbf{R}^2$ containing $(0, 0)$ which is *not* a subspace.

7. Let S be a convex subset of $\mathbf{R}^n$ and let T be a linear operator on $\mathbf{R}^n$. Define $T(S)$ to be $\{T(X) \mid X \in S\}$. Prove that $T(S)$ is convex.

8. Suppose that X_1 and X_2 are distinct feasible solutions of a linear programming problem in standard form. If the objective function has the same values at X_1 and X_2, prove that this is the value of the objective function at any point on the line segment joining X_1 and X_2.

10.3 Basic Solutions and Extreme Points

We have seen in 10.2 that the extreme points for a linear programming problem are the key to obtaining an optimal solution. In this section we describe a method for finding the extreme points which is the basis of the Simplex Algorithm.

Consider a linear programming problem in canonical form – remember that any problem can be put in this form:

$$\text{maximize:} \quad z = C^T X$$

$$\text{subject to:} \quad \begin{cases} AX = B \\ \quad X \geq 0 \end{cases}$$

Suppose that the problem has n variables $x_1, \ldots, x_n$ and m constraints, which means that A is an $m \times n$ matrix, while $X, C \in \mathbf{R}^n$ and $B \in \mathbf{R}^m$.

The linear system $AX = B$ must be consistent if there is to be any chance of a feasible solution, so we assume this to be the case; thus the matrix A and the augmented matrix $(A \mid B)$ have the same rank r. Hence the linear system $AX = B$ is equivalent to a system whose augmented matrix has rank r, with its final $m - r$ rows zero. These rows correspond to constraints of the form $0 = 0$, which are negligible. Therefore there is no loss in supposing that A is an $m \times n$ matrix with rank m; of course now $m \leq n$.

Since A has rank m, this matrix has m linearly independent columns, say $A_{j_1}, A_{j_2}, \ldots, A_{j_m}$, $(j_1 < j_2 < \cdots < j_m)$. Define

$$A' = (A_{j_1} \ A_{j_2} \ \ldots \ A_{j_m}),$$

which is an $m \times m$ matrix of rank m, so that $(A')^{-1}$ exists. The linear system

$$A'(x_{j_1} \ x_{j_2} \ \ldots \ x_{j_m})^T = B$$

therefore has a unique solution for $(x_{j_1} \ x_{j_2} \ \ldots \ x_{j_m})^T$, namely $(A')^{-1}B$.

This solution is in $\mathbf{R}^m$, not $\mathbf{R}^n$. To remedy this, define $X = (x_1 \ x_2 \ \ldots \ x_n)$ by putting $x_j = 0$ if $j \neq j_1, j_2, \ldots, j_m$. Then

$$AX = A(x_1 \ x_2 \ \ldots \ x_n)^T = x_{j_1} A_{j_1} + x_{j_2} A_{j_2} + \cdots + x_{j_m} A_{j_m}$$
$$= B.$$

Therefore X is a solution of $AX = B$ with the property that all entries of X, except perhaps those in positions $j_1, \ldots, j_m$, are zero. Such a solution is called a *basic solution* of the linear programming problem; if in addition all the x_{j_i} are non-negative, it is a *basic feasible solution*. The x_{j_i} are called the *basic variables*.

The next step is to relate the basic feasible solutions to the extreme points of a linear programming problem in canonical form.

Theorem 10.3.1

A basic feasible solution of a linear programming problem in canonical form is an extreme point of the set of feasible solutions.

Proof

Suppose that the linear programming problem is

$$\text{maximize: } z = C^T X$$

$$\text{subject to: } \begin{cases} AX = B \\ \quad X \geq 0 \end{cases}$$

and it has variables $x_1, \ldots, x_n$. Here A may be assumed to be an $m \times n$ matrix with rank m: as has been pointed out, this is no restriction. Then A has m linearly independent columns and, by relabeling the variables if necessary, we can assume these are the last m columns, say $A'_1, \ldots, A'_m$. Let

$$X = (0 \ \ldots \ 0 \ x'_1 \ \ldots \ x'_m)^T$$

be the corresponding basic solution. Assume that X is feasible, i.e., $x'_j \geq 0$ for $j = 1, 2, \ldots, m$. Our task is to prove that X is an extreme point of S, the set of all feasible solutions. Suppose X is not an extreme point of S; then

$$X = tU + (1-t)V,$$

where $0 < t < 1$, $U, V \in S$ and $X \neq U, V$. Write

$$U = (u_1 \ \ldots \ u_{n-m} \ u'_1 \ \ldots \ u'_m)^T$$

and

$$V = (v_1 \ \ldots \ v_{n-m} \ v'_1 \ \ldots \ v'_m)^T.$$

Equating the jth entries of X and $tU + (1-t)V$, we obtain

$$\begin{cases} tu_j + (1-t)v_j \ = 0, & 1 \leq j \leq n - m \\ tu'_j + (1-t)v'_j \ = x'_j, & 1 \leq j \leq m \end{cases}$$

Since $0 < t < 1$ and $u_j, v_j \geq 0$, the first equation shows that $u_j = 0 = v_j$ for $j = 1, \ldots, n - m$.

Since $U \in S$, we have $AU = B$, so that

$$u_1' A_1' + \cdots + u_m' A_m' = B$$

and

$$x_1' A_1' + \cdots + x_m' A_m' = B,$$

since $AX = B$. Therefore, on subtracting, we find that

$$(u_1' - x_1') A_1' + \cdots + (u_m' - x_m') A_m' = 0.$$

However $A_1', \ldots, A_m'$ are linearly independent, which means that $u_1' = x_1'$, $\ldots$, $u_m' = x_m'$, i.e., $U = X$, which is a contradiction.

The converse of this result is true.

Theorem 10.3.2
An extreme point of the set of feasible solutions of a linear programming problem in canonical form is a basic feasible solution.

Proof
Let the linear programming problem be

$$\text{maximize:} \quad z = C^T X$$

$$\text{subject to:} \quad \begin{cases} AX = B \\ X \geq 0 \end{cases}$$

where A is an $m \times n$ matrix of rank m. Let X be an extreme point of the set of feasible solutions.

Suppose that X has s non-zero entries and label the variables so that the last s entries of X are non-zero, say

$$X = (0 \ \ldots \ 0 \ x_1' \ \ldots \ x_s')^T.$$

Let A_j' be the column of A which corresponds to the entry x_j'. We will prove that $A_1', \ldots, A_s'$ are linearly independent.

Assume that $d_1 A'_1 + \cdots + d_s A'_s = 0$ where not all the d_j are 0. Let e be any positive number. Then

$$\sum_{j=1}^{s} (x'_j + ed_j) A'_j = \sum_{j=1}^{s} x'_j A'_j + e\left(\sum_{j=1}^{s} d_j A'_j\right)$$
$$= AX$$
$$= B.$$

In a similar fashion we have

$$\sum_{j=1}^{s} (x'_j - ed_j) A'_j = B.$$

Now define

$$\begin{cases} U = (0 \ \ldots 0 \ x'_1 + ed_1 \ \ldots \ x'_s + ed_s)^T \\ V = (0 \ \ldots 0 \ x'_1 - ed_1 \ \ldots \ x'_s - ed_s)^T \end{cases}$$

Then $AU = B = AV$.

Next choose e so that

$$0 < e < \frac{x'_j}{|d_j|}, \quad j = 1, 2, \ldots, s,$$

if $d_j \neq 0$. This choice of e ensures that $x'_j \pm ed_j > 0$ for $j = 1, 2, \ldots, s$. Hence $U \geq 0$ and $V \geq 0$, so that U and V are feasible solutions. However, $X = \frac{1}{2}U + \frac{1}{2}V$, which means that $X = U$ or V since X is an extreme point. But both of these are impossible because $e > 0$. It follows that $A'_1, \ldots, A'_s$ are linearly independent and X is a basic feasible solution.

We are now able to show that there are only finitely many extreme points in the set of feasible solutions.

Theorem 10.3.3

In a linear programming problem there are finitely many extreme points in the set of feasible solutions.

Proof

We assume that the linear programming problem is in canonical form:

$$\text{maximize: } z = C^T X$$

$$\text{subject to: } \begin{cases} AX = B \\ A \geq 0 \end{cases}$$

We can further assume here that A is an $m \times n$ matrix with rank m. Let X be an extreme point of S, the set of feasible solutions. Then X is a basic feasible solution by 10.3.2. In fact, if the non-zero entries of X are $x_{j_1}, \ldots, x_{j_s}$, the proof of the theorem shows that the corresponding columns of A, that is, $A_{j_1}, \ldots, A_{j_s}$, are linearly independent and thus $s \leq m$. In addition we have

$$x_{j_1} A_{j_1} + \cdots + x_{j_s} A_{j_s} = B.$$

By 2.2.1 this equation has a *unique* solution for $x_{j_1}, \ldots, x_{j_s}$. Therefore X is uniquely determined by $j_1, \ldots, j_s$. Now there are at most $\binom{n}{s}$ choices for $j_1, \ldots, j_s$, so the total number of extreme points is at most $\sum_{s=0}^{m} \binom{n}{s}$.

The last theorem shows that in order to find an optimal solution of a linear programming problem in canonical form, one can determine the finite set of basic feasible solutions and test the value of the objective function at each one. The simplex algorithm provides a practical method for doing this and is discussed in the next section.

In conclusion, we present an example of small order which illustrates how the basic feasible solutions can be determined.

Example 10.3.1

Consider the linear programming problem

$$\text{maximize: } z = 3x + 2y$$

$$\text{subject to: } \begin{cases} 2x - y \le 6 \\ 2x + y \le 10 \\ x, y \ge 0 \end{cases}$$

First transform the problem to canonical form by introducing slack variables u and v:

$$\text{maximize: } z = 3x + 2y$$

$$\text{subject to: } \begin{cases} 2x - y + u = 6 \\ 2x + y + v = 10 \\ x, y, u, v \ge 0 \end{cases}$$

The matrix form of the constraints is

$$\begin{pmatrix} 2 & -1 & 1 & 0 \\ 2 & 1 & 0 & 1 \end{pmatrix} \begin{pmatrix} x \\ y \\ u \\ v \end{pmatrix} = \begin{pmatrix} 6 \\ 10 \end{pmatrix}$$

The coefficient matrix has rank 2 and each pair of columns is linearly independent. Clearly there are $\binom{4}{2} = 6$ basic solutions, not all of them feasible. In each such solution two of the non-basic variables are zero. The basic solutions are listed in the table below:

x	y	u	v	type	z
0	0	6	10	feasible	0
0	10	16	0	feasible	20
3	0	0	4	feasible	9
5	0	−4	0	infeasible	15
4	2	0	0	feasible	16
0	−6	0	16	infeasible	−12

There are four basic feasible solutions, i.e., extreme points. The one that produces the largest value of z is $x = 0$, $y = 10$, giving $z = 20$. Thus $x = 0$, $y = 10$ is an optimal solution.

Exercises 10.3

In each of the following linear programming problems, transform the problem to canonical form and determine all the basic solutions. Classify these as infeasible or basic feasible, and then find the optimal solutions.

1.

maximize: $z = 3x - y$

subject to: $\begin{cases} x + 3y \leq 6 \\ x - y \leq 2 \\ x, y \geq 0 \end{cases}$

2.

maximize: $z = 2x + 3y$

subject to: $\begin{cases} 2x - y \leq 6 \\ 2x + y \leq 10 \\ x, y \geq 0 \end{cases}$

3.

maximize: $z = x_1 + x_2 + x_3$

subject to: $\begin{cases} 2x_1 - x_2 + 4x_3 \leq 12 \\ 4x_1 + 2x_2 + 5x_3 \leq 4 \\ x_1, x_2, x_3 \geq 0 \end{cases}$

4. A linear programming problem in standard form has m constraints and n variables. Prove that the number of extreme points is at most $\sum_{s=0}^{m} \binom{m+n}{s}$.

10.4 The Simplex Algorithm

We are now in a position to describe the simplex algorithm, which is a practical method for solving linear programming problems, based on the theory developed in the preceding sections. The method starts with a basic feasible solution and, by changing one basic variable at a time, seeks to find an optimal solution of the problem. It should be kept in mind that there are finitely many basic feasible solutions.

Consider a linear programming problem in standard form with variables $x_1, x_2, \ldots, x_n$ and m constraints:

$$\text{maximize:} \quad z = C^T X$$

$$\text{subject to:} \quad \begin{cases} AX \leq B \\ X \geq 0 \end{cases}$$

Thus A is an $m \times n$ matrix. For the present we will assume that $B \geq 0$, which is likely to be true in many applications: just what to do if this condition does not hold will be discussed later.

Convert the program to one in canonical form by introducing slack variables $x_{n+1}, \ldots, x_{n+m}$:

$$\text{maximize:} \quad z = C^T X$$

$$\text{subject to:} \quad \begin{cases} A'X = B \\ X \geq 0 \end{cases}$$

where

$$A' = (A \mid 1_m),$$

an $m \times (n + m)$ matrix. Also $X = (x_1 \ x_2 \ \ldots \ x_{n+m})^T$ and $C = (c_1 \ c_2 \ \ldots \ c_n \ 0 \ldots \ 0)^T$. Notice that A' has rank m since columns $n + 1, n + 2, \ldots, n + m$ are linearly independent.

Recall from 10.3 that the extreme points of the set of feasible solutions are exactly the basic feasible solutions. Also keep in mind that in a basic solution the non-basic variables all have the value 0.

The initial tableau

For the linear programming problem in canonical form above we have the solution

$$x_1 = x_2 = \cdots = x_n = 0, \ x_{n+1} = b_1, \ \ldots, \ x_{n+m} = b_m.$$

Since $B \geq 0$, this is a basic feasible solution in which the basic variables are the slack variables $x_{n+1}, \ldots, x_{n+m}$. The value of z at this point is 0 since $z = c_1 x_1 + \cdots + c_n x_n$.

The data are displayed in an array called the *initial tableau*.

	x_1	x_2	.	x_n	x_{n+1}	.	x_{n+m}	z	
x_{n+1}	a_{11}	a_{12}	.	a_{1n}	1	.	0	0	b_1
x_{n+2}	a_{21}	a_{22}	.	a_{2n}	0	.	0	0	b_2
.	.	.	.	.	.	.	.	.	.
x_{n+m}	a_{m1}	a_{m2}	.	a_{mn}	0	.	1	0	b_n
	$-c_1$	$-c_2$	.	$-c_n$	0	.	0	1	0

Here the rows in the array correspond to the basic variables, which appear on the left, while the columns correspond to all the variables, including z. The bottom row, which lies outside the main array and is called the *objective row*, displays the coefficients in the equation $-c_1 x_1 - \cdots - c_n x_n + z = 0$. The z-column is often omitted since it never changes during the algorithmic process. The right most column displays the current values of the basic variables, with the value of z in the lower right corner.

Entering and departing variables

Consider the initial tableau above. Suppose that all the entries in the objective row are non-negative. Then $c_j \leq 0$ and, since $z = \sum_{j=1}^{n} c_j x_j = 0$, if we change the value of one of the non-basic variables $x_1, \ldots, x_n$ by making it positive, the value of z will decrease or remain the same. Therefore the value of z cannot be increased from 0 and thus the solution is optimal.

On the other hand, suppose that the objective row contains a negative entry $-c_j$, so $c_j > 0$. Since $z = c_1 x_1 + \cdots + c_n x_n$ and $1 \leq j \leq n$, it may be possible to increase z by increasing x_j; however this must be done in a manner that does not violate any of the constraints.

Suppose that the most negative entry in the objective row is $-c_j$. The question of interest is: by how much can we increase the value of x_j? Since all other non-basic variables equal 0, the ith constraint requires that

$$a_{ij} x_j + x_{n+j} = b_i,$$

so that $x_{n+j} = b_i - a_{ij} x_j \geq 0$. Hence

$$a_{ij} x_j \leq b_i$$

for $i = 1, 2, \ldots, m$. Now if $a_{ij} \leq 0$, this imposes no restriction on x_j since $b_i \geq 0$. Thus if $a_{ij} \leq 0$ for all i, then x_j can be increased without limit, so *there are no optimal solutions*.

If $a_{ij} > 0$ for some i, on the other hand, we must ensure that

$$0 \leq x_j \leq \frac{b_i}{a_{ij}}.$$

The number

$$\frac{b_i}{a_{ij}}$$

is called a *θ-ratio* for x_j. Hence the value of x_j cannot be increased by more than the smallest non-negative θ-ratio of x_j; for otherwise one of the constraints will be violated.

Suppose that the smallest non-negative θ-ratio for x_j occurs in the ith row: this is called the *pivotal row*. One then applies row operations to the tableau, with the aim of making the ith entry of column j equal to 1 and all other entries of the column equal to 0. (This is called *pivoting* about (i, j) entry). The choice of i and j guarantees that no negative entries will appear in the right most column. Replace x_i (the *departing variable*) by x_j (the *entering variable*). Now x_j becomes a basic variable with value $\frac{b_i}{a_{ij}}$, replacing x_i. With this value of x_j, the value of z will increase by $b_i c_j / a_{ij}$, at least if $b_i > 0$.

After substituting x_j for x_i in the list of basic variables, we obtain the second tableau. This is treated in the same way as the first tableau, and if it is not optimal, one proceeds to a third tableau. If at some point in the procedure all the entries of the objective row become non-negative, an optimal solution has been reached and the algorithm stops.

Summary of the simplex algorithm

Assume that a linear programming problem is given in standard form

$$\text{maximize:} \quad z = C^T X$$

$$\text{subject to:} \quad \begin{cases} AX \leq B \\ X \geq 0 \end{cases}$$

where $B \geq 0$. Then the following procedure is to be applied.

1. Convert the program to canonical form by introducing slack variables. With the slack variables as basic variables, construct the initial tableau.

2. If no negative entries appear in the objective row, the solution is optimal. Stop.

3. Choose the column with the most negative entry in

the objective row. The variable for this column, say x_j, is the entering variable.

4. If all the entries in column j are negative, then there are no optimal solutions. Stop.

5. Find the row with the smallest non-negative θ-value of x_j. If this corresponds to x_i, then x_i is the departing variable.

6. Pivot about the (i, j) entry, i.e., apply row operations to the tableau to obtain 1 as the (i, j) entry, with all other entries in column j equal to 0.

7. Replace x_i by x_j in the tableau obtained in step 6. This is the new tableau. Return to Step 2.

Example 10.4.1

$$\text{maximize:} \quad z = 8x_1 + 9x_2 + 5x_3$$

$$\text{subject to:} \quad \begin{cases} x_1 + x_2 + 2x_3 \leq 2 \\ 2x_1 + 3x_2 + 4x_3 \leq 3 \\ 3x_1 + 3x_2 + x_3 \leq 4 \\ x_1, x_2, x_3 \geq 0 \end{cases}$$

Convert the problem to canonical form by introducing slack variables x_4, x_5, x_6:

$$\text{maximize:} \quad z = 8x_1 + 9x_2 + 5x_3$$

$$\text{subject to:} \quad \begin{cases} x_1 + x_2 + 2x_3 + x_4 = 2 \\ 2x_1 + 3x_2 + 4x_3 + x_5 = 3 \\ 3x_1 + 3x_2 + x_3 + x_6 = 4 \\ x_j \geq 0, \end{cases}$$

The initial basic feasible solution is $x_1 = x_2 = x_3 = 0$, $x_4 = 2$, $x_5 = 3$, $x_6 = 4$, with basic variables x_4, x_5, x_6. The initial tableau is:

	x_1	$*x_2$	x_3	x_4	x_5	x_6	
x_4	1	1	2	1	0	0	2
$**x_5$	2	3	4	0	1	0	3
x_6	3	3	1	0	0	1	4
	-8	-9	-5	0	0	0	0

Here the z-column has been suppressed. The initial basic feasible solution $x_4 = 2$, $x_5 = 3$, $x_6 = 4$ is not optimal since there are negative entries in the objective row; the most negative entry occurs in column 2, so x_2 is the entering variable, (indicated in the tableau by $*$).

The θ-values for x_2 are 2, 1, $\frac{4}{3}$, corresponding to x_4, x_5, x_6. The smallest (non-negative) θ-value is 1, so x_5 is the departing variable (indicated in the tableau by $**$). Now pivot about the $(2,2)$ entry to obtain the second tableau.

	$*x_1$	x_2	x_3	x_4	x_5	x_6	
x_4	$1/3$	0	$2/3$	1	$-1/3$	0	1
x_2	$2/3$	1	$4/3$	0	$1/3$	0	1
$**x_6$	1	0	-3	0	-1	1	1
	-2	0	7	0	3	0	9

The objective row still has a negative entry, so this is not optimal: the entering variable is x_1. The smallest θ-value for x_1 is 1, occurring for x_6, so this is the departing variable. Now pivot about the $(3,1)$ entry to get the third tableau.

	x_1	x_2	x_3	x_4	x_5	x_6	
x_4	0	0	$5/3$	1	0	$-1/3$	$2/3$
x_2	0	1	$10/3$	0	1	$-2/3$	$1/3$
x_1	1	0	-3	0	-1	1	1
	0	0	1	0	1	2	11

Since there are no negative entries in the objective row, this tableau is optimal. The optimal solution is therefore $x_1 = 1$, $x_2 = \frac{1}{3}$, $x_3 = 0$, giving $z = 11$.

The next example shows how the simplex method can detect a case where there are no optimal solutions.

Example 10.4.2

$$\text{maximize: } z = 5x_1 - 4x_2$$

$$\text{subject to: } \begin{cases} x_1 - x_2 \leq 2 \\ -2x_1 + x_2 \leq 2 \\ x_1, x_2 \geq 0 \end{cases}$$

Introduce slack variables x_3 and x_4 and pass to canonical form:

$$\text{maximize: } z = 5x_1 - 4x_2$$

$$\text{subject to: } \begin{cases} x_1 - x_2 + x_3 = 2 \\ -2x_1 + x_2 + x_4 = 2 \\ x_1, x_2, x_3, x_4 \geq 0 \end{cases}$$

The initial basic feasible solution is $x_1 = 0 = x_2$, $x_3 = 2$, $x_4 = 2$, with basic variables x_3, x_4. The initial tableau is therefore

	$*x_1$	x_2	x_3	x_4	
$**x_3$	1	-1	1	0	2
x_4	-2	1	0	1	2
	-5	4	0	0	0

The entering variable is x_1 and the departing variable x_3. The second tableau is:

	x_1	$*x_2$	x_3	x_4	
x_1	1	-1	1	0	2
x_4	0	-1	2	1	6
	0	-1	5	0	10

The next entering variable is x_2; however all the entries in the x_2-column are negative, which means that x_2 can be increased without limit. Therefore this problem has no optimal solution.

Geometrically, what happened here is that the set of feasible solutions is the *infinite* region of the plane lying between the lines $x_1 - x_2 = 2$, $-2x_1 + x_2 = 2$, $x_1 = 0$, $x_2 = 0$. In this region $z = 5x_1 - 4x_2$ can take arbitrarily large values.

Degeneracy

Up to this point we have not taken into account the possibility that the simplex algorithm may fail to terminate: in fact this could happen.

To see how it might occur, suppose that at some stage in the simplex algorithm the entering variable has two equal smallest non-negative θ-values. Then after pivoting one of the basic variables will have the value zero, a phenomenon called *degeneracy*. If in the next tableau the basic variable whose value was 0 is the departing variable, *the objective function will not increase in value*. This raises the possibility that at some point we might return to this tableau, in which event the simplex algorithm will run forever.

In practice the simplex algorithm very seldom fails to terminate. In any case there is a simple adjustment to the algorithm which avoids the possibility of non-termination. These adjustments involve different choices of entering and departing variables, as indicated below.

(i) To select the entering variable, choose the variable with a negative entry in the objective row which has the smallest subscript.

(ii) To select the departing variable choose the basic variable with smallest non-negative θ-value and smallest subscript.

This procedure is known as *Bland's Rule*. It can be shown

that the simplex method, when combined with Bland's Rule, will always terminate, even if degeneracy occurs.

The Two Phase Method

The reader may have noticed that our version of the simplex algorithm does not work if some constraints have negative numbers on the right side. We consider briefly how this situation can be remedied.

Consider a linear program in standard form:

$$\text{maximize: } z = C^T X$$

$$\text{subject to: } \begin{cases} AX \le B \\ X \ge 0 \end{cases}$$

where A is $m \times n$. As usual we introduce slack variables $x_{n+1}, \ldots, x_{n+m}$ to obtain a problem in canonical form:

$$\text{maximize: } z = C^T X$$

$$\text{subject to: } \begin{cases} A'X = B \\ X \ge 0 \end{cases}$$

where $A' = [A \mid 1_m]$. If some b_i is negative, we can multiply that constraint by -1 to get an entry $-b_i > 0$ on the right hand side. The problem now is that we do not have a basic feasible solution — for the obvious solution $x_{n+i} = b_i$ is not feasible. What is called for at this point is a general method for finding an initial basic feasible solution for any linear programming problem in canonical form.

Suppose we have a linear programming problem in canonical form:

$$\text{maximize: } z = C^T X$$

$$\text{subject to: } \begin{cases} AX = B \\ X \ge 0 \end{cases} \qquad \text{(I)}$$

where A is $m \times n$. The problem is to find an initial basic feasible solution. Once this is found, the simplex algorithm can be run. There is no loss in assuming that $B \geq 0$ since we can, if necessary, multiply a constraint by -1.

The method is to introduce new variables $y_1, y_2, \ldots, y_m$ called *artificial variables*. These are used to form the *auxiliary program*:

$$\text{maximize:} \quad z = -\sum_{i=1}^{m} y_i$$

$$\text{subject to:} \quad \begin{cases} AX + Y = B \\ X, Y \geq 0 \end{cases} \qquad \text{(II)}$$

If (II) has an optimal solution X, Y with $z = 0$, then all the y_i must equal 0 and thus $AX = B$. Hence X is a basic feasible solution of (I). On the other hand, if the optimal solution of (II) yields a negative value of z, there are no feasible solutions of (II) with $Y = 0$, i.e., there are no feasible solutions of (I). Thus if we can solve the problem (II), we will either find a basic feasible solution of (I) or else conclude that (I) has no feasible solutions.

But can we in fact solve the problem (II)? The answer is affirmative: for $X = 0$, $y_1 = b_1, \ldots, y_m = b_m$ is clearly a basic feasible solution of (II), so it can be used to form the initial tableau for problem (II). After solving (II), either we will have a basic feasible solution of (I) or we will know that no feasible solutions exist. In the former event the simplex algorithm can then be run for problem (I). This is known as the *Two Phase Method*.

We summarize the two phases for solving the linear programming problem (I).

Phase One
Apply the simplex method to the auxiliary program (II). If there is no optimal solution or if the optimal solution yields a negative value z, then there are no feasible solutions of (I). Stop. Otherwise a basic feasible solution to problem I is found.

Phase Two
Starting with the basic feasible solution obtained in Phase
One, use the simplex algorithm to find an optimal solution of
(I) or show that none exists.

In conclusion, the Two Phase Method can be applied to
any linear programming problem in canonical form.

Example 10.4.3

$$\text{maximize:} \quad z = 2x_1 - 2x_2 - 3x_3 + 2x_4$$

$$\text{subject to:} \quad \begin{cases} x_1 & + & 2x_2 & + & x_3 & + & x_4 & = 18 \\ 3x_1 & + & 6x_2 & + & 2x_3 & & & = 24 \\ & & & x_i \geq 0 & & & \end{cases}$$

This problem is given in canonical form. The Two Phase
Method will be applied, the first phase being to find a basic
feasible solution. To this end we set up the auxiliary problem:

$$\text{maximize:} \quad z = -y_1 - y_2 - y_3$$

$$\text{subject to:} \quad \begin{cases} x_1 & + & 2x_2 & + & 2x_3 & & & + & y_1 & & & = 12 \\ x_1 & + & 2x_2 & + & x_3 & + & x_4 & + & & y_2 & & = 18 \\ 3x_1 & + & 6x_2 & + & 2x_3 & & & + & & & y_3 & = 24 \\ & & & & x_i, y_j \geq 0 & & & & & & & \end{cases}$$

The initial tableau for this problem is:

	x_1	x_2	x_3	x_4	y_1	y_2	y_3	
y_1	1	2	2	0	1	0	0	12
y_2	1	2	1	1	0	1	0	18
y_3	3	6	2	0	0	0	1	24
	0	0	0	0	-1	-1	-1	-54

Here the initial basic feasible solution is $y_1 = 12$, $y_2 = 18$, $y_3 = 24$, with $z = -54$. But notice that the entries in the objective row corresponding to the basic variables are not 0; this is because z is expressed as $-y_1 - y_2 - y_3$. We need to replace y_1, y_2, y_3 by expressions in x_1, x_2, x_3, x_4 and thereby eliminate the offending entries. Note that $-y_1 = x_1 + 2x_2 + 2x_3 - 12$, $-y_2 = x_1 + 2x_2 + x_3 + x_4 - 18$ and $-y_3 = 3x_1 + 6x_2 + 2x_3 - 24$. Adding these, we obtain

$$z = -y_1 - y_2 - y_3 = 5x_1 + 10x_2 + 5x_3 + x_4 - 54.$$

The next step is to use this expression to form the new objective row:

	x_1	$*x_2$	x_3	x_4	y_1	y_2	y_3	
y_1	1	2	2	0	1	0	0	12
y_2	1	2	1	1	0	1	0	18
$**\,y_3$	3	6	2	0	0	0	1	24
	-5	-10	-5	-1	0	0	0	-54

This is the first tableau for the auxiliary problem. The entering variable is x_2 and the departing variable y_3. The second tableau is:

	x_1	x_2	$*x_3$	x_4	y_1	y_2	y_3	
$**\,y_1$	0	0	$4/3$	0	1	0	$-1/3$	4
y_2	0	0	$1/3$	1	0	1	$-1/3$	10
x_2	$1/2$	1	$1/3$	0	0	0	$1/6$	4
	0	0	$-5/3$	-1	0	0	$5/3$	-14

The entering variable is x_3 and the departing variable is y_1. The third tableau is:

	x_1	x_2	x_3	$*x_4$	y_1	y_2	y_3	
x_3	0	0	1	0	$3/4$	0	$-1/4$	3
$**\,y_2$	0	0	0	1	$-1/4$	0	$1/4$	9
x_2	$1/2$	1	0	0	$-1/4$	0	$1/4$	3
	0	0	0	-1	$5/4$	0	$5/4$	-9

The entering variable is x_4 and the departing variable is y_2. The fourth tableau is

	x_1	x_2	x_3	x_4	y_1	y_2	y_3	
x_3	0	0	1	0	$3/4$	0	$-1/4$	3
x_4	0	0	0	1	$-1/4$	0	$-1/4$	9
x_2	$1/2$	1	0	0	$-1/4$	0	$1/4$	3
	0	0	0	0	1	0	1	0

This tableau is optimal with $z = 0$. Hence we have a basic solution of the original problem, $x_1 = 0$, $x_2 = 3$, $x_3 = 3$, $x_4 = 9$.

Now Phase Two begins. To obtain an initial tableau, in the final tableau of Phase 1 delete the columns corresponding to the artificial variables y_1, y_2, y_3. The new basic variables are x_3, x_4, x_2. Replace the objective row by the entries of the original objective function, but retain 0 in the bottom right hand corner:

	x_1	x_2	x_3	x_4	
x_3	0	0	1	0	3
x_4	0	0	0	1	9
x_2	$1/2$	1	0	0	3
	-2	2	3	-2	0

Next eliminate the non-zero entries in the objective row corresponding to the basic variables x_3, x_4, x_2. This is done by adding to the objective row $(-2) \times$ row 3, $(-3) \times$ row 1 and $2 \times$ row 2. This yields the tableau:

	$*x_1$	x_2	x_3	x_4	
x_3	0	0	1	0	3
x_4	0	0	0	1	9
$**x_2$	$1/2$	1	0	0	3
	-3	0	0	0	3

The entering variable is x_1 and the departing variable is x_2. The next tableau is:

	x_1	x_2	x_3	x_4	
x_3	0	0	1	0	3
x_4	0	0	0	1	9
x_1	1	2	0	0	6
	0	6	0	0	21

This tableau is optimal with solution $x_1 = 6$, $x_2 = 0$, $x_3 = 3$, $x_4 = 9$ and $z = 21$.

In conclusion we remark that there is one possible situation that the Two Phase Method cannot handle. It could be that in the final tableau of Phase One at least one artificial variable is basic. There is a modification of the Two Phase Method to deal with this possibility. The reader is referred to a text on linear programming such as [12] or [13] for details.

Needless to say, we have merely skimmed the surface of linear programming. Recently an improvement on the simplex method known as Kamarkar's algorithm has been discovered. Again the interested reader may consult one of the above references for details.

Exercises 10.4

In the following problems use the simplex method to solve the linear programming problem or show that no optimal solution exists.

1.

maximize: $z = 3x - y$

subject to: $\begin{cases} x + 3y & \leq 6 \\ x - y & \leq 2 \\ x, y \geq 0 \end{cases}$

2.

maximize: $z = 2x + 3y$

subject to: $\begin{cases} 2x - y & \leq 6 \\ 2x + y & \leq 10 \\ x, y \geq 0 \end{cases}$

3.

minimize: $z = -2x + 3y$

subject to: $\begin{cases} x - y & \geq -2 \\ x - 2y & \leq 4 \\ x, y \geq 0 \end{cases}$

4.

minimize: $z = 3x_1 - 2x_2$

subject to: $\begin{cases} x_1 + x_2 + 2x_3 & \leq 7 \\ 2x_1 + x_2 + x_3 & \leq 4 \\ x_j \geq 0 \end{cases}$

5.

maximize: $z = x_1 + 2x_2 + x_3 - x_4$

subject to: $\begin{cases} 3x_1 + x_2 + 2x_3 - x_4 & \leq 2 \\ 2x_1 + 4x_2 - 4x_3 & \leq 4 \\ x_j \geq 0 \end{cases}$

6.

maximize: $z = x_1 + x_2 + 3x_3 - x_4$

subject to:

$\begin{cases} 2x_1 - x_2 + x_3 + x_4 & \leq 8 \\ 2x_1 + 3x_2 + 4x_4 & \leq 6 \\ 3x_1 + x_2 + 2x_3 & \leq 18 \\ x_j \geq 0 \end{cases}$

7. Use the Two Phase Method to solve the following linear programming problem, noting that only one artificial variable is needed.

maximize: $z = x_1 + 2x_2 - x_3$

subject to:

$$\begin{cases} 2x_1 + x_2 + x_3 \leq 4 \\ x_1 + x_2 + 2x_3 = 3 \\ x_j \geq 0 \end{cases}$$

Appendix

MATHEMATICAL INDUCTION

Mathematical induction is one of the most powerful methods of proof in mathematics and it is used in several places in this book. Since some readers may be unfamiliar with induction, and others may feel in need of a review, we present a brief account of it here.

The method of proof by induction rests on the following principle.

Principle of mathematical induction

Let m be an integer and let $P(n)$ be a statement or proposition defined for each integer $n \geq m$. Assume furthermore that the following hold:

(i) $P(m)$ is true;
(ii) if $P(n-1)$ is true, then $P(n)$ is true.

Then the conclusion is that $P(n)$ is true for all integers $n \geq m$.

While this may sound harmless enough, it is in fact an axiom for the integers: it cannot be deduced from the usual arithmetic properties of the integers and its validity must be assumed.

We shall give some examples to illustrate the use of this principle.

Example A.1

If n is any positive integer, prove by mathematical induction that the sum of the first n positive integers equals $\frac{1}{2}n(n+1)$.

Let $P(n)$ denote the statement:

$$1 + 2 + \cdots + n = \frac{1}{2}n(n+1).$$

415

We have to show that $P(n)$ is true for all integers $n \geq 1$. Now clearly $P(1)$ is true: it simply asserts that $1 = \frac{1}{2}(2)$. Suppose that $P(n-1)$ is true; we must show that $P(n)$ is also true. In order to prove this, we begin with $1 + 2 + \cdots + (n-1) = \frac{1}{2}(n-1)n$, which is known to be true, and then add n to both sides. This yields

$$1 + 2 + \cdots + (n-1) + n = \frac{1}{2}(n-1)n + n = \frac{1}{2}n(n+1).$$

Hence $P(n)$ is true. Therefore by the Principle of Mathematical Induction $P(n)$ is true for all $n \geq 1$.

Example A.2

Let n be any positive integer. Prove by mathematical induction that the integer $8^{n+1} + 9^{2n-1}$ is always divisible by 73.

Let $P(n)$ be the statement: 73 divides $8^{n+1} + 9^{2n-1}$. Then we easily verify that $P(1)$ is true. Assume that $P(n-1)$ is true; thus $8^n + 9^{2n-3}$ is divisible by 73. We need to show that $P(n)$ is true. The method in this example is to express $8^{n+1} + 9^{2n-1}$ in terms of $8^n + 9^{2n-3}$; thus

$$8^{n+1} + 9^{2n-1} = 8(8^n + 9^{2n-3}) + 9^{2n-1} - 8(9^{2n-3})$$
$$= 8(8^n + 9^{2n-3}) + 9^{2n-3}(9^2 - 8)$$
$$= 8(8^n + 9^{2n-3}) + 73(9^{2n-3}).$$

Since $P(n-1)$ is true, the last integer is divisible by 73. Therefore $P(n)$ is true.

Occasionally, the following alternate form of mathematical induction is useful.

Principle of mathematical induction - alternate form

Let m be an integer and let $P(n)$ be a statement or proposition defined for each integer $n \geq m$. Assume furthermore that the following hold:

(i) $P(m)$ is true;
(ii) if $P(k)$ is true for all $k < n$, then $P(n)$ is true.

Then the conclusion is that $P(n)$ is true for all integers $n \geq m$.

Example A.3

Prove that every integer $n > 1$ is a product of prime numbers.

Let $P(n)$ be the statement that n is a product of primes; here $n \geq 2$. Then $P(2)$ is certainly true since 2 is a prime. Assume that $P(k)$ is true for all $k < n$. We have to show that $P(n)$ is true. Now if n is a prime, $P(n)$ is certainly true. Assume that n is not a prime; then $n = n_1 n_2$ where n_1 and n_2 are positive integers less than n. Hence $P(n_1)$ and $P(n_2)$ are true, so both n_1 and n_2 are products of primes. Therefore $n = n_1 n_2$ is a product of primes and $P(n)$ is true. It now follows from the second form of the Principle of Mathematical Induction that $P(n)$ is true for all $n > 2$.

Exercises

1. If n is a positive integer, prove by induction that the sum of the squares of the first n positive integers equals $\frac{1}{6}n(n + 1)(2n + 1)$.

2. If n is a positive integer, prove by induction that the sum of the cubes of the first n positive integers equals $(\frac{1}{2}n(n+1))^2$.

3. Let $u_0, u_1, u_2, \ldots$ be a sequence of integers which satisfies the recurrence relation $u_{n+1} = 2u_n + 3$ and also $u_0 = 1$. Prove by induction that $u_n = 2^{n+2} - 3$.

4. Prove by induction that the number of symmetric $n \times n$ matrices over the field of two elements equals $2^{n(n+1)/2}$.

5. Use the second form of mathematical induction to prove that each integer > 1 is *uniquely* expressible as a product of primes.

ANSWERS TO THE EXERCISES

Exercises 1.1

1. $\begin{pmatrix} 2 & -3 & 4 & -5 \\ -3 & 4 & -5 & 6 \end{pmatrix}$. **2.** (a) $(-1)^{i+j-1}$; (b) $4i + j - 4$.

3. Six: $0_{12,1}, 0_{6,2}, 0_{4,3}, 0_{3,4}, 0_{2,6}, 0_{1,12}$. **4.** n should be prime.
5. Diagonal matrices.

Exercises 1.2

1. (a) $\begin{pmatrix} -3 & 6 & 1 \\ 0 & 1 & -3 \\ 2 & 5 & -6 \end{pmatrix}$;

(c) $A^2 = \begin{pmatrix} 7 & 7 & 1 \\ -2 & 0 & -1 \\ 2 & 5 & 5 \end{pmatrix}$, $A^3 = \begin{pmatrix} 9 & 22 & 14 \\ -4 & -5 & -6 \\ 12 & 14 & 1 \end{pmatrix}$.

3. A is $m \times n$ and B is $n \times m$. **4.** $A^6 = I_2$. **9.** True.
12. Numbers of books in library, lent out, lost are 7945, 1790, 265 respectively.

14. The matrix equals
$$\begin{pmatrix} 1 & 5/2 & 11/2 \\ 5/2 & 5 & -7/2 \\ 11/2 & -7/2 & 5 \end{pmatrix} + \begin{pmatrix} 0 & 1/2 & -3/2 \\ -1/2 & 0 & 5/2 \\ 3/2 & -5/2 & 0 \end{pmatrix}.$$

18. (a) The inverse is $\frac{1}{2}\begin{pmatrix} -4 & -2 \\ 3 & -1 \end{pmatrix}$; (b) not invertible.

21. $\begin{pmatrix} 0 & 1 & 0 \\ 0 & 0 & 1 \\ 0 & 0 & 0 \end{pmatrix}$.

Exercises 1.3

2. A non-zero matrix need not have an inverse.
3. $2^{n^2}, 2^{n(n+1)/2}$.

4. $A + B = \begin{pmatrix} 1 & 0 & 0 \\ 1 & 0 & 0 \\ 1 & 0 & 0 \end{pmatrix}$, $A^2 = \begin{pmatrix} 1 & 1 & 0 \\ 0 & 0 & 1 \\ 0 & 1 & 1 \end{pmatrix}$,

$AB = \begin{pmatrix} 0 & 1 & 0 \\ 0 & 0 & 1 \\ 1 & 1 & 1 \end{pmatrix}$.

7. The integer 2 does not have an inverse.

Exercises 2.1

1. $x_1 = c/3 + d/3 - 1/3$, $x_2 = 4c/3 - 2d/3 + 11/3$, $x_3 = c$, $x_4 = d$.

2. $x_1 = 2c/3 - 5/3$, $x_2 = 2c/3 + 7/3$, $x_3 = c/3 + 2/3$, $x_4 = c$.

3. Inconsistent.

4. (a) $x_1 = -c$, $x_2 = c$, $x_3 = 0$, $x_4 = 0$; (b) $x_1 = x_2 = x_3 = 0$.

5. For $t = -4$ or 3.   **6.** $t \neq -1/3$.   **7.** $n(n+1)/2$.

Exercises 2.2

1. (a) $\begin{pmatrix} 1 & -3/2 & -2 \\ 0 & 1 & 4/5 \\ 0 & 0 & 1 \end{pmatrix}$; (b) $\begin{pmatrix} 1 & 2 & -3 \\ 0 & 0 & 0 \end{pmatrix}$;

(c) $\begin{pmatrix} 1 & 2 & -3 & 0 \\ 0 & 1 & -11/5 & 1/5 \\ 0 & 0 & 0 & 1 \end{pmatrix}$.

2. (a) I_3; (b) $\begin{pmatrix} 1 & 0 & 7/5 \\ 0 & 1 & -11/5 \\ 0 & 0 & 0 \end{pmatrix}$; (c) $\begin{pmatrix} 1 & 0 & 7/5 & 0 \\ 0 & 1 & -11/5 & 0 \\ 0 & 0 & 0 & 1 \end{pmatrix}$.

5. $n(n+1)/2$.   **6.** n^2.   **7.** I_2 and $\begin{pmatrix} 1 & 1 \\ 0 & 1 \end{pmatrix}$.

8. The number of pivots equals n.

Exercises 2.3

1. (a) $\begin{pmatrix} 1 & 0 \\ 4 & 1 \end{pmatrix} \begin{pmatrix} 1 & 0 \\ 0 & -3 \end{pmatrix} \begin{pmatrix} 1 & 2 \\ 0 & 1 \end{pmatrix} \begin{pmatrix} 1 & 0 & -1 \\ 0 & 1 & 2 \end{pmatrix}$;

(b)

$$\begin{pmatrix} 1 & 0 & 0 \\ 1 & 1 & 0 \\ 0 & 0 & 1 \end{pmatrix} \begin{pmatrix} 1 & -1 & 0 \\ 0 & 1 & 0 \\ 0 & 0 & 1 \end{pmatrix} \begin{pmatrix} 1 & 0 & 0 \\ 0 & 1 & 0 \\ 0 & -1 & 1 \end{pmatrix} \begin{pmatrix} 1 & 0 & 0 \\ 0 & 1 & -2 \\ 0 & 0 & 0 \end{pmatrix}.$$

(These answers are not unique).

2.

$$\begin{pmatrix} 1 & 0 & 0 \\ 0 & 1 & 0 \\ 1 & -1 & 0 \end{pmatrix} \begin{pmatrix} 1 & 0 & 0 \\ 0 & 1 & -2 \\ 0 & 0 & 1 \end{pmatrix} \begin{pmatrix} 1 & 0 & 0 \\ 1 & 1 & 0 \\ 0 & 0 & 1 \end{pmatrix} \begin{pmatrix} 1 & 0 & 2 \\ 0 & 1 & 0 \\ 0 & 0 & 1 \end{pmatrix}$$

$$\begin{pmatrix} 1 & -1 & 0 \\ 0 & 1 & 0 \\ 0 & 0 & 1 \end{pmatrix}.$$

3. (a) $\begin{pmatrix} 1 & 0 & 0 \\ 0 & 1 & 0 \end{pmatrix}$; (b) $\begin{pmatrix} 1 & 0 & 0 \\ 0 & 1 & 0 \\ 0 & 0 & 0 \end{pmatrix}$. **5.** $n(n+1)/2$ and n^2.

6. (a) $\frac{1}{11} \begin{pmatrix} 2 & -3 \\ 1 & 4 \end{pmatrix}$; (b) $-\frac{1}{6} \begin{pmatrix} 2 & -10 & -6 \\ 3 & -6 & -3 \\ -1 & 2 & 3 \end{pmatrix}$; (c) not invertible.

7. $t = -3$ or 2. **8.** Entries on the principal diagonal must be non-zero.

Exercises 3.1

1. Odd; $-a_{11}a_{23}a_{38}a_{45}a_{52}a_{66}a_{74}a_{87}$.
2. Even; $a_{18}a_{25}a_{33}a_{42}a_{51}a_{67}a_{76}a_{89}a_{94}$.

3. 19. **4.** $n(n!) - 1$. **5.** $M_{13} = 11 = A_{13}$, $M_{23} = 7 = -A_{23}$, $M_{33} = -6 = A_{33}$. **6.** 84. **7.** (a) -40, (b) -30, (c) -36.

9. $\begin{pmatrix} 0 & 0 & 1 & 0 & 0 \\ 1 & 0 & 0 & 0 & 0 \\ 0 & 0 & 0 & 1 & 0 \\ 0 & 0 & 0 & 0 & 1 \\ 0 & 1 & 0 & 0 & 0 \end{pmatrix}$.

Exercises 3.2

1. (a) 133; (b) 132; (c) -26. **10.** $u_2 = 3$, $u_3 = 14$, $u_4 = 63$.

Exercises 3.3

2. (a) $\frac{1}{10}\begin{pmatrix} 31 \\ 24 \end{pmatrix}$; (b) $-\frac{1}{24}\begin{pmatrix} -6 & -14 & 8 \\ -15 & -11 & 8 \\ 9 & 5 & -8 \end{pmatrix}$;

(c) $\begin{pmatrix} 1 & -1 & 0 & 0 \\ 0 & 1 & -1 & 0 \\ 0 & 0 & 1 & -1 \\ 0 & 0 & 0 & 1 \end{pmatrix}$.

4. (a) $x_1 = 1$, $x_2 = 2$, $x_3 = 3$; (b) $x_1 = 1$, $x_2 = 0$, $x_3 = -2$.
7. $2x - 3y - z = 1$.

Exercises 4.1

2. (a) No; (b) no; (c) yes; (d) yes.

Exercises 4.2

1. (a) No; (b) no; (c) yes. **2.** (a) No; (b) yes; (c) yes. **3.** Yes. **4.** No.

Exercises 4.3

1. (a) Linearly independent; (b) linearly independent; (c) linearly dependent. **2.** True. **3.** True. **4.** False.
9. False. **10.** No.

Exercises 5.1

1. (a) $E_1 = 1/13(9X_1 + 3X_2 - 8X_3)$, $E_2 = 1/13(-3X_1 - X_2 + 7X_3)$, $E_3 = 1/13(-17X_1 - 10X_2 + 18X_3)$;
(b) $E_1 = -2Y_1 + 4Y_2 - Y_3$, $E_2 = 4Y_1 - 7Y_2 + 2Y_3$, $E_3 = Y_1 - 3Y_2 + Y_3$.
2. (a) $(-2 \ -1 \ 1)^T$; (b) $(1 \ -1 \ 1 \ 0)^T$ and $(-2 \ 1 \ 0 \ 1)^T$.
3. mn. **6.** $S = V$. **8.** $-4(-1 \ 1 \ 0)^T - 2(-1 \ 0 \ 1)^T$.

Exercises 5.2

1. (a) Basis of the row space is $(1 \ 0 \ 63/2), (0 \ 1 \ 18)$, basis of the column space is $(1 \ 0)^T, (0 \ 1)^T$; (b) basis of the row space is $(1 \ 0 \ 5/19 \ 25/19), (0 \ 1 \ 4/19 \ 20/19)$, basis of the column space is $(1 \ 0 \ 4)^T, (0 \ 1 \ 1)^T$.
2. (a) $1 + 5x^3/3, \ x + x^3/3, \ x^2 + x^3$;

(b) $\begin{pmatrix} 1 & 0 \\ 1/76/7 & \end{pmatrix}, \begin{pmatrix} 0 & 1 \\ 1/7 & -1/7 \end{pmatrix}$.

5. $m - r$. **6.** They are $\leq$ rank of A and $\leq$ rank of B.

Exercises 5.3

1. The subspaces generated by $(1 \ 0)^T, (1 \ 1)^T, (0 \ 1)^T$.
2. $\dim(U) = n(n+1)/2$ and $\dim(W) = n(n-1)/2$. **5.** False.
6. Let $U = < f_i \mid i = 1, \ldots, 7 >$ and $W = < f_i \mid i = 4, \ldots, 14 >$ where $f_i = x^{i-1}$.
7. $\dim(U + W) = 3$, $\dim(U \cap W) = 1$.
8. Basis for $U + W$ is $1, x, x^2, x^3$, basis for $U \cap W$ is $1 - x + x^2$.
11. $\dim(U_1) + \cdots + \dim(U_k)$. **12.** No.

14. $X_0 = \begin{pmatrix} -1/3 \\ 11/3 \\ 0 \\ 0 \end{pmatrix}$, $Y \begin{pmatrix} c/3 + d/3 \\ 4c/3 - d/3 \\ c \\ d \end{pmatrix}$.

15. $n - 1$.

Exercises 6.1

1. (a) None of these; (b) bijective; (c) surjective; (d) injective.
4. $F^{-1}(x) = \{(x+5)/2\}^{1/3}$.

Exercises 6.2

1. (a) No; (b) yes; (c) no, unless $n = 1$.

4. $\begin{pmatrix} 1 & -1 & -1 & -1 \\ 2 & 1 & -1 & 0 \\ 0 & 1 & -1 & 1 \end{pmatrix}$. **5.** $\begin{pmatrix} 1 & 0 & 0 & 0 \\ 0 & -1 & 2 & 0 \\ 0 & 0 & -3 & 6 \\ 0 & 0 & 0 & -5 \end{pmatrix}$.

6. $\begin{pmatrix} \cos 2\phi & \sin 2\phi \\ \sin 2\phi & -\cos 2\phi \end{pmatrix}$.

7. $\begin{pmatrix} 1/2 & 1/4 & -3/4 \\ 0 & 1/2 & -1/2 \\ 0 & 0 & 1 \end{pmatrix}$, $\begin{pmatrix} 2 & -1 & 1 \\ 0 & 2 & 1 \\ 0 & 0 & 1 \end{pmatrix}$.

8. $\begin{pmatrix} 6 & -7 & -2 \\ 2 & -3 & -1 \end{pmatrix}$. **9.** They have different determinants.

12. The statement is true.

Exercises 6.3

1. (a) Basis of kernel is $(-1\ 1\ 0\ 0)^T, (-1\ 0\ 1\ 0)^T, (-1\ 0\ 0\ 1)^T$, basis of image is 1; (b) basis of kernel is 1, basis of image is 1, x; (c) basis of kernel is $(-3\ 2)^T$, basis of image is $(1\ 2)^T$.
3. $\mathbf{R}^6$, $\mathbf{R}_6$ and $M(2, 3, \mathbf{R})$ are all isomorphic: $\mathbf{C}^6$ and $P_6(\mathbf{C})$ are isomorphic. **6.** True. **10.** They are not equivalent for infinitely generated vector spaces.

Exercises 7.1

1. $92.84°$. **2.** $\pm 1/\sqrt{42}(4\ 1\ -5)^T$. **3.** $1/14(1\ 2\ 3)^T$ and $1/\sqrt{14}$. **4.** $9/\sqrt{26}$. **5.** Vector product $= (-14\ -4\ 8)^T$, area $= 2\sqrt{69}$. **8.** $\det(X\ Y\ Z) = 0$. **9.** Dimension $= n - 1$ or n, according as $X \neq 0$ or $X = 0$.

11. $t\big(3-\sqrt{2}+3(\sqrt{2}+1)i \ \ (\sqrt{2}-3)(1+i) \ \ 4\big)^T$ where $i = \sqrt{-1}$ and t is arbitrary. **13.** $(X^*Y/\|Y\|^2)Y$.

Exercises 7.2

1. (a) No; (b) yes; (c) no. **4.** (a) No; (b) no; (c) yes.
8. $23 - 120x + 110x^2$. **9.** $1/105(17 \ -190 \ 331)^T$.

Exercises 7.3

2. $1/\sqrt{2}(1 \ 0 \ -1)^T$, $1/3(2 \ 1 \ 2)^T$, $1/\sqrt{18}(1 \ -4 \ 1)^T$.
3. $1/\sqrt{2}(0 \ 1 \ 1)^T$, $1/3(1 \ -2 \ 2)^T$, $1/\sqrt{18}(4 \ 1 \ -1)^T$.
4. $1/\sqrt{7}(1-6x)$, $\sqrt{5/154}(2+30x-42x^2)$. **5.** $(1/2 \ 4 \ 1/2)^T$.

6. $Q = \begin{pmatrix} 0 & 1/3 & 4/\sqrt{18} \\ 1/\sqrt{2} & -2/3 & 1/\sqrt{18} \\ 1/\sqrt{2} & 2/3 & -1/\sqrt{18} \end{pmatrix}$ and

$R = \begin{pmatrix} \sqrt{2} & 0 & 1/\sqrt{2} \\ 0 & 3 & -1/3 \\ 0 & 0 & 5/\sqrt{18} \end{pmatrix}$.

8. The product of $\begin{pmatrix} -i/\sqrt{3} & \sqrt{3/10}(2+4i)/3 \\ (1+i)/\sqrt{3} & \sqrt{3/10}(3+i)/3 \end{pmatrix}$ and
$\begin{pmatrix} \sqrt{3}(1-2i) & \sqrt{3}/3 \\ 0 & \sqrt{10/3} \end{pmatrix}$. **10.** $Q = Q'$ and $R = R'$.

Exercises 7.4

1. (a) $x_1 = 1$, $x_2 = -3/5$, $x_3 = -3/5$; (b) $x_1 = 1631/665$, $x_2 = -88/95$, $x_3 = -66/95$. **2.** $r = -70t/51 + 3610/51$.
3. $y = -4 + 7x/2 - x^2/2$. **4.** $x_1 = 13/35$, $x_2 = -17/70$, $x_3 = 1/70$. **5.** $(-8e^{-1} + 26e^{-2}) + (6e^{-1} - 18e^{-2})x$.
6. $12(\pi^2 - 10)/\pi^3 + 60(-\pi^2 + 12)x/\pi^4 + 60(\pi^2 - 12)x^2/\pi^5$.

Exercises 8.1

1. (a) Eigenvalues -2, 6; eigenvectors $t(-5 \ 3)^T$, $t(1 \ 1)^T$;
(b) eigenvalues 1, 2, 3, eigenvectors $t(-1 \ 1 \ 2)^T$, $t(-2 \ 1 \ 4)^T$,
$t(-1 \ 1 \ 4)^T$; (c) eigenvalues 1, 2, 3, 4, eigenvectors

$(2\ -4\ -1\ 1)^T, (0\ -2\ 0\ 1)^T, (0\ 0\ 1\ 1)^T, (0\ 0\ 0\ 1)^T.$ **5.** False.

7. (a) $\begin{pmatrix} -5 & 1 \\ 3 & 1 \end{pmatrix}$; (b) $\begin{pmatrix} -1 & -2 & -1 \\ 1 & 1 & 1 \\ 2 & 4 & 4 \end{pmatrix}$.

8. They should be both zero or both non-zero. **13.** Non-zero constants.

Exercises 8.2

1. (a) $y_n = 4 \cdot 3^{n+1} - 3 \cdot 4^{n+1}$, $z_n = -3^{n+1} + 4^{n+1}$;
(b) $y_n = 1/9(10 \cdot 7^n + 5(-2)^{n+1})$, $z_n = 1/9(5 \cdot 7^n - 2(-2)^{n+1})$.
2. $a_n = 1/3(a_0 + 2b_0 + 2.4^n(a_0 - b_0))$, $b_n = 1/3(a_0 + 2b_0 + 4^n(-a_0 + b_0))$: if $a_0 > b_0$, species A flourishes, and species B dies out: if $a_0 < b_0$, the reverse holds.
3. $r_n = 2/\sqrt{5}\{((1 + \sqrt{5})/2)^n - ((1 - \sqrt{5})/2)^n\}$.
4. $u_n = (2^{n+1} + (-1)^n)/3$.
5. $y_n = 1 + 2n$, $z_n = 2n$. **6.** $y_n = ((-1)^{n+1} + 1)/2$,
$z_n = (38.4^{n-1} + 3(-1)^n - 5)/30$.
7. Employed 85.7%, unemployed 14.3%. **8.** Equal numbers at each site. **9.** Conservatives 24%, liberals 45%, socialists 31%.

Exercises 8.3

1. (a) $y_1 = -c_1 e^{-5x} + 2c_2 e^x$, $y_2 = c_1 e^{-5x} + c_2 e^x$;
(b) $y_1 = c_1 e^x - c_2 e^{5x}$, $y_2 = c_1 e^x + c_2 e^{5x}$;
(c) $y_1 = c_1 e^x + c_3 e^{3x}$, $y_2 = -2c_2 e^x$, $y_3 = c_2 e^x + c_3 e^{3x}$.
2. $y_1 = e^{2x}(\cos\ x + \sin\ x)$, $y_2 = 2e^{2x} \cos\ x$.
3. $y_1 = (3c_2 x - c_1)e^{2x}$, $y_2 = (-3c_2 x + c_1 + c_2)e^{2x}$: particular solution $y_1 = 6xe^{2x}$, $y_2 = (2 - 6x)e^{2x}$.
4. $y_1 = c_1 e^x + c_2 e^{-x} + c_3 e^{3x} + c_4 e^{-3x}$, $y_2 = -c_1 e^x + 5c_2 e^{-x} + c_3 e^{3x} - 5c_4 e^{-3x}$.
5. nk. **7.** (a) $y_1 = -u_1 + u_2$, $y_2 = u_1 - 3u_2$ where $u_1 = c_1 \cosh\ \sqrt{2}x + d_1 \sinh\ \sqrt{2}x$ and $u_2 = c_2 \cosh\ 2x + d_2 \sinh\ 2x$;
(b) $y_1 = -4u_1 - u_2$, $y_2 = u_1 + u_2$ where $u_1 = c_1 \cosh\ x + d_1 \sinh\ x$ and $u_2 = c_2 \cosh\ 2x + d_2 \sinh\ 2x$.

8. $y_1 = (-1 - \sqrt{2})w_1 + (-1 + \sqrt{2})w_2$, $y_2 = w_1 + w_2$ where $w_1 = c_1 \cos ux + d_1 \sin ux$ and $w_2 = c_2 \cos vx + d_2 \sin vx$ with $u = a\sqrt{2 + \sqrt{2}}$ and $v = a\sqrt{2 - \sqrt{2}}$.

Exercises 9.1

1. (a) $1/\sqrt{2} \begin{pmatrix} 1 & 1 \\ -1 & 1 \end{pmatrix}$; (b) $\begin{pmatrix} -1/\sqrt{3} & 2/\sqrt{6} & 0 \\ 1/\sqrt{3} & 1/\sqrt{6} & -1/\sqrt{2} \\ 1/\sqrt{3} & 1/\sqrt{6} & 1/\sqrt{2} \end{pmatrix}$;

(c) $1/\sqrt{2} \begin{pmatrix} 1 & i \\ i & 1 \end{pmatrix}$, $i = \sqrt{-1}$.

8. $\begin{pmatrix} i & 0 \\ 0 & 2 \end{pmatrix}$, $i = \sqrt{-1}$.

Exercises 9.2

1. (a) Positive definite; (b) indefinite; (c) indefinite.
2. Indefinite. **6.** (a) Ellipse; (b) parabola.
7. (a) Ellipsoid; (b) hyperboloid (of one sheet).
8. (a) Local minimum at $(-4, 2)$; (b) local maximum at $(-1 - \sqrt{2}, 1 - \sqrt{2})$, local minimum at $(1 + \sqrt{2}, -1 + \sqrt{2})$, saddle points at $(\sqrt{2} - 1, \sqrt{2} + 1)$ and $(1 - \sqrt{2}, -1 - \sqrt{2})$; (c) local minimum at $(-2/5, -1/5, 3/10)$.

9. The smallest and largest values are $\frac{5-\sqrt{17}}{2}$ and $\frac{5+\sqrt{17}}{2}$ respectively.

11. The spheres have radii 0.768 and 0.434 respectively.

Exercises 9.3

1. (a) No; (b) yes; (c) yes. **2.** (a) $\begin{pmatrix} 0 & 2 \\ -3 & 0 \end{pmatrix}$; (b) $\begin{pmatrix} 0 & -3 \\ 2 & -1 \end{pmatrix}$.
3. $\dim(V') = n^2$. **5.** $2x_1' \, y_1' + 4x_2' \, y_2'$. **7.** (a) Yes; (b) no.

8. $\begin{pmatrix} 0 & 1 & 0 \\ -1 & 0 & 0 \\ 0 & 0 & 0 \end{pmatrix}$; $S = \begin{pmatrix} 1/2 & 0 & 1/2 \\ 0 & 1 & 1/2 \\ 0 & 0 & 1 \end{pmatrix}$.

Exercises 9.4

1. (a) $x - 2$; (b) $(x-2)(x-3)$; (c) $x^2 - 1$; (d) $(x-2)^2(x-3)$.

5. (a) $\begin{pmatrix} 4 & 0 \\ 0 & -1 \end{pmatrix}$; (b) $\begin{pmatrix} 2 & 1 \\ 0 & 2 \end{pmatrix}$; (c) $\begin{pmatrix} 1 & 1 & 0 \\ 0 & 1 & 1 \\ 0 & 0 & 1 \end{pmatrix}$.

6. (a) $(x-4)(x+1)$; (b) $(x-2)^2$; (c) $(x-1)^3$.

7. A must be similar to $\begin{pmatrix} I_r & 0 \\ 0 & 0_s \end{pmatrix}$ where $r + s = n$.

8. A must be similar to a block matrix with a block I_r, t blocks $\begin{pmatrix} 0 & 1 \\ 0 & 0 \end{pmatrix}$ and a block 0_s where $r + 2t + s = n$.

10.
$$\begin{pmatrix} 0 & 0 & \cdots & 0 & -a_0 \\ 1 & 0 & \cdots & 0 & -a_1 \\ 0 & 1 & \cdots & 0 & -a_2 \\ \cdot & \cdot & \cdots & \cdot & \cdot \\ 0 & 0 & \cdots & 1 & -a_{n-1} \end{pmatrix}.$$

11. $y_1 = \left(\frac{1}{2}c_2 x^2 + (c_1 + c_2)x + (c_0 + c_1)\right)e^x$, $y_2 = c_2 e^x$, $y_3 = (c_2 x + c_1)e^x$.

Exercises 10.1

1.

maximize: $p = p_1 x_1 + p_2 x_2 + p_3 x_3$

subject to : $\begin{cases} u_1 x_1 + u_2 x_2 + u_3 x_3 \leq s \\ v_1 x_1 + v_2 x_2 + v_3 x_3 \leq t \\ \qquad\qquad x_j \geq 0 \end{cases}$

2.

minimize: $e = px + qy$

subject to : $\begin{cases} a_c x + b_c y \geq m_c \\ a_f x + b_f y \geq m_f \\ a_p x + b_p y \geq m_p \\ \qquad x, y \geq 0 \end{cases}$

3.

$$\text{maximize:}\ z = -2x_1 + x_2 + x_3^+ - x_3^- - x_4$$

$$\text{subject to:}\ \begin{cases} -x_1 - 2x_2 - x_3^+ + x_3^- + x_4 \leq -5 \\ 3x_1 + x_2 - x_3^+ + x_3^- + x_4 \leq \ \ \ 4 \\ x_1, x_2, x_3^+, x_3^- \geq 0 \end{cases}$$

4.

$$\text{maximize:}\ z = -2x_1 + x_2 + x_3^+ - x_3^- - x_4$$

$$\text{subject to:}\ \begin{cases} -x_1 - 2x_2 - x_3^+ + x_3^- + x_4 + x_5 = -5 \\ 3x_1 + x_2 - x_3^+ + x_3^- + x_4 + x_6 = \ \ \ 4 \\ x_1, x_2, x_3^+, x_3^-, x_4, x_5, x_6 \geq 0 \end{cases}$$

5. (c) $z = C^T A^{-1} B$.

Exercises 10.2

4. (a) In Exercise 10.2.1 the extreme points are $(0,\ 2)$, $(1/3, 7/3), (0,\ 3)$.
(b) In Exercise 10.2.2 the extreme points are $(0,\ 0), (3,\ 0)$, $(5,\ 1), (6,\ 0)$,
5. The optimal solution is $x = 0, y = 6$.

Exercises 10.3

1. The optimal solution is $x = 3,\ y = 1$.
3. The optimal solution is $x = 0,\ y = 10$.
4. The optimal solution is $x_1 = 0,\ x_2 = 2,\ x_3 = 0$.

Exercises 10.4

1. $x = 3,\ y = 1$.
2. $x = 0,\ y = 10$.

3. No optimal solution.

4. $x_1 = 0$, $x_2 = 4$, $x_3 = 0$.

5 $x_1 = 0$, $x_2 = 4/3$, $x_3 = 1/3$, $x_4 = 0$.

6. $x_1 = 0$, $x_2 = 2/3$, $x_3 = 26/3$, $x_4 = 0$.

7. $x_1 = 0$, $x_2 = 3$, $x_3 = 0$.

BIBLIOGRAPHY

Abstract Algebra

(1) I.N. Herstein, "Topics in Algebra", 2nd ed., Wiley, New York, 1975.
(2) S. MacLane and G. Birkhoff, "Algebra", 3rd ed., Chelsea, New York, 1988.
(3) D.J.S. Robinson, "An Introduction to Abstract Algebra", De Gruyter, Berlin, 2003.
(4) Rotman, J.J. "A First Course in Abstract Algebra", 2nd ed., Prentice Hall, Upper Saddle River, NJ, 2000.

Linear Algebra

(5) C.W. Curtis, "Linear Algebra, an Introductory Approach", Springer, New York, 1984.
(6) F.R. Gantmacher, "The Theory of Matrices", 2 vols., Chelsea, New York, 1960.
(7) P.J. Halmos, "Finite-Dimensional Vector Spaces", Van Nostrand-Reinhold, Princeton, N.J., 1958.
(8) B. Kolman, "Introductory Linear Algebra with Applications", 5th ed., Macmillan, New York, 1993.
(9) S.J. Leon, "Linear Algebra with Applications", 5th ed., Prentice Hall, Upper Saddle River, NJ, 1998.
(10) G. Strang, "Linear Algebra and its Applications", 3rd ed., Harcourt Brace Jovanovich, San Diego, 1988.

Applied Linear Algebra

(11) R. Bellman, "Introduction to Matrix Analysis", 2nd ed., Society for Industrial and Applied Mathemetics, Philadelphia, 1995.
(12) H. Karloff, "Linear Programming", Birkhäuser, Boston 1991.
(13) B. Kolman and R.E. Beck, "Elementary Linear Programming with Applications", Academic Press, San Diego, 1995.

(14) B. Noble and J.W. Daniel, "Applied Linear Algebra", 3rd ed., Prentice-Hall, Englewood Cliffs, N.J., 1988.

Some Related Books of Interest

(15) W.R. Derrick and S.I. Grossman, "Elementary Differential Equations with Applications", 2nd ed., Addison-Wesley, Reading, MA, 1982.

(16) C.H. Edwards and D.E. Penney, "Elementary Differential Equations with Boundary Value Problems", 2nd ed., Prentice-Hall, Englewood Cliffs, N.J., 1989

(17) J.G. Kemeny and J.L. Snell, "Finite Markov Chains", Springer, New York, 1976.

(18) G.B. Thomas and R.L. Finney, "Calculus and Analytic Geometry", 9th ed., Addison-Wesley, Reading, MA, 1996.

Index